Study Guide

Sociology in Our Times

Diana Kendall
Baylor University

Prepared by

Shannon Carter
University of Central Florida

 WADSWORTH
CENGAGE Learning

Australia • Brazil • Japan • Korea • Mexico • Singapore • Spain • United Kingdom • United States

ISBN-13: 978-0-495-90511-0
ISBN-10: 0-495-90511-9

Wadsworth
20 Davis Drive
Belmont, CA 94002-3098
USA

Cengage Learning is a leading provider of customized learning solutions with office locations around the globe, including Singapore, the United Kingdom, Australia, Mexico, Brazil, and Japan. Locate your local office at: **www.cengage.com/global**

Cengage Learning products are represented in Canada by Nelson Education, Ltd.

To learn more about Wadsworth, visit **www.cengage.com/wadsworth**

Purchase any of our products at your local college store or at our preferred online store **www.ichapters.com**

Printed in the United States of America
1 2 3 4 5 6 7 13 12 11 10 09

Table of Contents

Preface

This Study Guide has been prepared for use with Diana Kendall's *Sociology in Our Times,* Eighth Edition. Each chapter of the Study Guide includes the following elements:

Brief Chapter Outline: The brief chapter outline provides students with a clear overview of the material that will be covered in the chapter.

Learning Objectives: These objectives will help you stay on track with the material.

Key Terms: This matching exercise will help you master the definitions of the boldface terms in each chapter.

Key People: This matching exercise will help you understand the key people in each chapter and their contributions to the core concepts.

Critical Thinking Questions: Read these questions to help guide your reading and better prepare for exams.

Fill-in-the-Blank: After reading the chapter, test your knowledge by completing this fill-in-the-blank summary exercise.

Internet Activities: Complete these online exercises to gain a broader knowledge of each chapter's subject matter as it relates to the world around you.

Student Class Activities and Projects: These activities are created to help you learn and identify the significance of sociology to your society. Some of these projects could potentially be assigned to you, and you may wish to use them on your own to practice the sociological concepts.

InfoTrac Exercises: Locate articles related to the core concepts in each chapter using keyword searches in InfoTrac College Edition.

Practice Test: Test your knowledge in this section comprised of multiple choice, true/false, and short answer questions, as well as essay suggestions.

Answer keys are provided at the end of the study guide for these sections in each chapter: Key Terms, Key People, Chapter Review, Multiple Choice Questions, and True-False Questions. Page number references are provided for the Short Answer Questions and Essay Suggestions.

1

THE SOCIOLOGICAL PERSPECTIVE

BRIEF CHAPTER OUTLINE

CHAPTER SUMMARY

Sociology is the systematic study of human society and social interaction. Sociology enables us to see how individual behavior is largely shaped by the groups we belong to and the **society** we live in. The **sociological imagination** helps us understand how seemingly personal troubles are actually related to larger social forces. Some of these larger social forces are related to *global interdependence*, a relationship in which the lives of all people are intertwined closely and any one nation's problems are part of a larger global problem. The first systematic analysis of society is found in the philosophies of early Greek thinkers, such as Plato and Aristotle. With the onset of scientific revolution, social thinkers sought to develop a scientific understanding of social life. The origins of contemporary social thinking can be traced to the scientific revolution of the late seventeenth and mid-eighteenth centuries and the Age of Enlightenment, where emphasis was placed on the individual's possession of critical reasoning and experience, coupled with widespread skepticism of the primacy of religion as a source of knowledge, and a strong opposition to traditional authority. Sociology emerged out of the social upheaval produced by **urbanization** and **industrialization** during the nineteenth and early twentieth centuries. Some early social thinkers—including **Auguste Comte**, **Harriet Martineau**, **Herbert Spencer**, and **Emile Durkheim** —emphasized social order and stability; others—including **Karl Marx**, **Max Weber**, and **Georg Simmel**—focused on conflict and social change. From its origins in Europe, sociology spread to the United States in the 1890s when departments of sociology were established at the University of Chicago and Atlanta University. Sociologists use four major contemporary sociological perspectives to examine social life: (1) **functionalist perspectives** view society as a stable, orderly system; (2) **conflict perspectives** view society as a continuous power struggle among competing groups, often based on class, race, ethnicity, and/or gender; (3) **symbolic interactionist perspectives** focus on how people make sense of their everyday social interactions; and (4) **postmodern perspectives** examine the efforts of postindustrialization, consumerism, and global communications in explaining social life in contemporary societies. While there are significant differences between the social sciences, such as sociology, anthropology, psychology, economics, and political science, their areas of research and interest overlap in that the goal of scholars, teachers, and students is to learn more about human behavior and its causes and consequences.

LEARNING OBJECTIVES

After reading Chapter 1, you should be able to:

1. Define sociology.

2. Explain how sociology helps us to better understand our social world and ourselves.

3. Distinguish between commonsense knowledge, myths, and sociological knowledge.

4. Explain and apply C. Wright Mills's concept of the sociological imagination.

5. Explain the importance of a global sociological imagination, and define and give examples of high-, middle- and low-income countries.

6. Define race, ethnicity, class, sex, and gender, and explain the importance of these terms to developing a sociological imagination.

7. Identify and discuss the key assumptions of the Age of the Enlightenment.

8. Define industrialization and urbanization, and explain the role of each in furthering sociological thought.

2

9. Explain the contributions of Auguste Comte, Harriet Martineau, Herbert Spencer, and Emile Durkheim to the emergence of sociology.

10. Explain what Durkheim meant by his use of the terms social facts and anomie.

11. Describe the main views of society presented by Karl Marx, Max Weber, and Georg Simmel.

12. Contrast Karl Marx's perspective on social change with that of Max Weber.

13. Describe the origins of sociology in the United States and discuss the role of women in early departments of sociology and social work.

14. State the major assumptions of functionalism, conflict theory, symbolic interactionism, and postmodernism, and identify the major contributors to each perspective.

15. Distinguish between microlevel and macrolevel analyses, and state which level of analysis is utilized by each of the major theoretical perspectives.

16. Relate Max Weber's concepts of rationalization and verstehen to C. Wright Mills's Sociological Imagination.

17. Explain how sociological theory helps us understand social issues like consumerism.

18. Compare sociology with other sciences, and determine areas of overlap and important differences.

KEY TERMS

(defined at page number shown and in glossary)

3

KEY PEOPLE

(identified at page number shown)

Jane Addams, p. 22
Jean Baudrillard, p. 31
Herbert Blumer, p. 27
Auguste Comte, p. 12
W.E.B Du Bois, p. 22
Emile Durkheim, p. 15
Harriet Martineau, p. 12
Karl Marx, p. 16

George Herbert Mead, p. 29
Robert K. Merton, p. 23
C. Wright Mills, p. 5, 25-26
Talcott Parsons, p. 23
Georg Simmel, pp. 20-21
Herbert Spencer, p. 14
Thorstein Veblen, p. 27
Max Weber, pp. 19, 25-26

CHAPTER OUTLINE

I. PUTTING SOCIAL LIFE INTO PERSPECTIVE
   A. Sociology is the systematic study of human society and social interaction.
   B. Why Study Sociology?
      1. **Sociology** helps us see the complex connections between our own lives and the larger, recurring patterns of our society and world.
         a. A **society** is a large social grouping that shares the same geographic territory and is subject to the same political authority and dominant cultural expectations.
         b. When we examine the world order, we become aware of *global interdependence*—a relationship in which the lives of all people are intertwined closely and any one nation's problems are part of a larger global problem.
         c. Sociological research often reveals the limitations of myths associated with *commonsense knowledge* that guides ordinary conduct in everyday life.
   C. The Sociological Imagination
      1. According to sociologist **C. Wright Mills**, the **sociological imagination** enables us to view personal troubles as public issues and recognize the relationship between the two.
      2. Personal troubles affect individuals and their networks, and require personal solutions, while public issues affect large numbers of people and may require societal solutions.
      3. Overspending, for example, can be seen as both a personal and a public issue.
II. IMPORTANCE OF A GLOBAL SOCIOLOGICAL IMAGINATION
   A. A *global sociological imagination* is necessary in the twenty-first century, and helps us understand and overcome challenges in our rapidly changing world. Sociologists distinguish between **high-income**, **middle-income** and **low-income countries**.
   B. Developing a personal **sociological imagination** requires that we take into account perspectives of other people.

4

1. People in today's world differ by *race*—a term used by many people to specify groups of people distinguished by physical characteristics such as skin color—and *ethnicity*—cultural heritage or identity of a group, based on factors such as language or country of origin.
2. They also differ by *class*—the relative location of a person or group within a larger society, based on wealth, power, prestige, or other valued resources—and by *gender*—the meanings, beliefs, and practices associated with sex differences referred to as *femininity* and *masculinity*.

III. THE ORIGINS OF SOCIOLOGICAL THINKING
   A. Sociology and the Age of Enlightenment
   1. The scientific revolution of the late seventeenth and mid-eighteenth centuries is referred to as the Age of Enlightenment.
   2. During this time period in Europe, critical reasoning and experience is emphasized; religion is viewed with skepticism.
   3. The individual's pursuit of enlightened self-interest is conducive to the welfare of society as a whole; new forms of political and economic organizations, such as democracy and capitalism, emerge.
   B. Sociology and the Age of Revolution, Industrialization, and Urbanization
   1. The Enlightenment produced an *intellectual revolution,* as well as *political* and *economic revolutions* in America and then France.
   2. **Industrialization**—the process by which societies are transformed from dependence on agriculture and handmade products to an emphasis on manufacturing and related industries—and **urbanization**—the process by which an increasing proportion of a population lives in cities rather than rural areas—contributed to the development of sociological thinking.

IV. THE DEVELOPMENT OF MODERN SOCIOLOGY
   A. Early Thinkers: Concerned with social order and stability
   1. **Auguste Comte** coined the term **sociology** and stressed the importance of **positivism**—a belief that the world can best be understood through scientific inquiry. He developed *the law of the three stages*: the theological, metaphysical, and scientific (or positive). Some consider Comte to be the founder of sociology. Both Comte and his mentor, Saint-Simon, advocated the use of the methods of the natural sciences as models for the objective study of society.
   2. **Harriet Martineau** translated and condensed Comte's works. Her most influential work was *Society in America,* in which she emphasized diversity in the United States based on race, class, and gender. She believed in racial and gender equality, enlightened reform, and creating a science of human nature and, eventually, a just society.
   3. **Herbert Spencer** used an evolutionary perspective to explain stability and change in societies. He coined the term *survival of the fittest*, equating this process of *natural selection* with progress and success. His view of society is known as **social Darwinism**, the idea that those species of animals, including humans, best adapted to their environment survive and prosper, whereas the poorly adapted die out.

5

4. According to **Emile Durkheim**, societies are built upon **social facts,** which are patterned ways of acting, thinking, and feeling that exist outside any one individual and exert social control over each person. **Anomie** is a condition in which social control becomes ineffective as a result of the loss of shared values and of a sense of purpose in society.

B. Differing Views on the Status Quo: Stability versus Change
   1. **Karl Marx** believed that conflict—especially *class conflict*—is necessary to improve society, particularly in terms of economic equality.
      a. Class conflict is the struggle between members of the capitalist class, or *bourgeoisie,* and the working class, or *proletariat.*
      b. Exploitation of workers by capitalists results in workers' *alienation*—a feeling of *powerlessness* and *estrangement* from other people and from oneself. Marx predicted that workers would overthrow capitalists, resulting in a free and classless society.
   2. **Max Weber** disagreed with Marx's idea that economics is the central force in social change, emphasizing that other factors, such as religion, influence economic systems. His concern with the growth of large-scale organizations is reflected in his work on bureaucracy. According to Weber, *rational bureaucracy*, rather than class struggle, is the most significant factor in determining the social relations between people in industrial societies. By emphasizing the goal of *value-free* inquiry and employing *verstehen*, sociologists could then understand how others see the world.
   3. **Georg Simmel** emphasized that society is best seen as a web of patterned interactions among people. By analyzing how interaction patterns vary depending on the size of the social groups and the *forms* and content of the social interaction, Simmel provided major insights into social life. He assessed the costs of industrialization and urbanization on individuals.

C. The Beginnings of Sociology in the United States
   1. The first U.S. department of sociology was at the University of Chicago. Robert E. Park studied the disintegrating influence of urbanization on social life, such as the influence of increases in the crime rate, and racial and class antagonisms.
   2. **George H. Mead** founded the symbolic interactionist perspective.
   3. **Jane Addams** wrote *Hull-House Maps and Papers,* which was used by other Chicago sociologists for the next forty years. She received a Nobel Prize for her assistance to the underprivileged.
   4. **W.E.B. Du Bois** founded the second U.S. department of sociology at Atlanta University and wrote *The Philadelphia Negro: A Social Study*, examining Philadelphia's African American community. He noted that a dual heritage creates conflict for people of color, which he referred to as *double consciousness.*

V. CONTEMPORARY THEORETICAL PERSPECTIVES
A. A **theory** is a set of logically interrelated statements that attempts to describe, explain, and (occasionally) predict social events. Theories provide a framework or perspective—an overall approach or viewpoint toward some subject—for examining various aspects of social life.

6

B. **Functionalist perspectives** are based on the idea that society is a stable, orderly system characterized by societal consensus.
   1. Societies develop social structures, or institutions that persist because they play a part in helping society survive. These institutions include: the family, education, government, religion, and economy.
   2. **Talcott Parsons** stressed that all societies must make provisions for meeting social needs in order to survive. For example, a division of labor (distinct, specialized functions) between husband and wife is essential for family stability and social order.
   3. **Robert K. Merton** distinguished between intended and unintended functions of social institutions.
      a. **Manifest functions** are intended and/or overtly recognized by the participants in a social unit.
      b. **Latent functions** are unintended functions that are hidden and remain unacknowledged by participants.
      c. *Dysfunctions* are the undesirable consequences of any element of society. For example, a dysfunction of education in the United States is the perpetuation of gender, racial, and class inequalities, while a manifest function of shopping and consuming is creating a booming economy benefiting many.
   4. Applying a **functionalist perspective** to shopping and consumption
      a. Functionalist perspectives focus on the relationships and interdependence of social institutions.
      b. Shopping and consumption create a strong economy that benefits other social institutions, such as family (employment) and education (increased funding through tax dollars).
      c. **Manifest functions** of shopping and consumption are acquiring necessary items, such as food and clothing. **Latent functions** include pleasure, relaxation, and entertainment. *Dysfunctions* include excessive spending and shoplifting.
C. According to **conflict perspectives**, groups in society are engaged in a continuous power struggle for control of scarce resources.
   1. Along with **Karl Marx**, **Max Weber** believed that economic conditions were important in producing inequality and conflict in society; however, Weber also suggested that power and prestige are other sources of inequality.
   2. **C. Wright Mills** believed that the most important decisions in the United States are made largely behind the scenes by the *power elite*, a small clique composed of the top corporate, political, and military officials. He contended that value-free sociology was impossible because social scientists must make value-related choices.
   3. **Feminist perspectives** focus on patriarchy—a system in which men dominate women. What is considered masculine is more highly valued than what is considered feminine.
   4. Applying **conflict perspectives** to shopping and consumption
      a. Conflict theorists focus on how inequalities based on race, sex, and income affect people's ability to acquire things they want and need.

7

b. Conflict perspectives stress **Veblen**'s concept of *conspicuous consumption*, or the continuous display of wealth and status through purchases.

D. Symbolic Interactionist Perspective

1. Functionalist and conflict perspectives focus primarily on **macrolevel analysis**—an examination of whole societies, large-scale social structures, and social systems. By contrast, symbolic interactionist approaches are based on a **microlevel analysis**—an examination of everyday interactions in small groups rather than large-scale social structures.

2. **Symbolic Interactionist perspectives** are based on the idea that society is the sum of the interactions of individuals and groups. Symbols help people derive meanings from social situations.

   a. **George Herbert Mead,** a founder of this perspective, emphasized that a key feature distinguishing humans from other animals is the ability to communicate in symbols—anything that meaningfully represents something else.

   b. Some interactionists focus on people's behavior while others focus on each person's interpretation or definition of a given situation.

   c. Erving Goffman created *dramaturgical analysis*, the idea that individuals go through their lives like actors performing on stage and engage in "impression management."

3. Applying **symbolic interactionist perspectives** to shopping and consumption

   a. Symbolic interactionists focus on people's face-to-face interactions and their roles in society.

   b. Individuals bring subjective meanings and past experiences with them to the social interactions that take place while shopping. Symbolic interactionists might examine the interactions that take place between a customer and cashier, and how these interactions impact future interactions.

E. Postmodern Perspectives

1. According to **postmodern perspectives**, existing theories have been unsuccessful in explaining social life in contemporary societies that are characterized by postindustrialization, consumerism, and global communications.

2. Postmodern societies are characterized by an *information explosion*, the *rise of a consumer society*, and the emergence of a *global village* in which people around the world communicate with each other by electronic technologies.

3. **Jean Baudrillard**, a key postmodern theorist, focuses on the shift away from production of goods in society to consumption of information, goods, and services.

4. Social life is not an objective reality waiting for us to discover how it works, but rather is how we think about it.

5. Applying **postmodern perspectives** to shopping and consumption

   a. Postmodern perspectives view society as a consumer society, where the economy is based on consumption instead of production.

8

> b. In postmodern society, individuals can shop in "simulated" settings, such as on TV or the Internet without having to interact with another person. This method of consumption dissolves the distinction between real and simulated interactions, where simulated interaction becomes the "real" interaction.

F. All four of these perspectives will be used in this textbook as lenses through which to view our social world.

VI. COMPARING SOCIOLOGY WITH OTHER SOCIAL SCIENCES

A. Anthropology is the study of human existence over geographic space and evolutionary time; it is divided into four main subfields: sociocultural, linguistic, archaeological, and biological anthropology.

B. Psychology is the systematic study of behavior and mental processes—what occurs in the mind.

C. Economics is the study of how the limited resources of a society are allocated among competing demands and is divided into two branches: macroeconomics and microeconomics.

D. Political Science is the study of power relations, how power is distributed, and political institutions such as the state, government, and political parties.

E. In sum, the areas of interest and research in the social sciences overlap, and the goal is to learn more about human behavior.

CRITICAL THINKING QUESTIONS

1. How could you apply the sociological imagination to a personal trouble you have experienced in your life?

2. Compare and contrast the theoretical perspectives put forth by the early sociological thinkers (Auguste Comte, Harriet Martineau, Herbert Spencer, Emile Durkheim, Karl Marx, Max Weber, and Georg Simmel). What are the strengths and weaknesses of each theorists' ideas? Which ideas are still relevant today?

3. Compare and contrast the contemporary theoretical perspectives. What are the strengths and weaknesses of each perspective? Which perspective is most relevant in your life?

4. Why do you think shopping, spending, credit card debt, and bankruptcy have become major problems for so many people? Which theoretical perspective do you think is most useful for explaining this issue? Why?

5. What do you think is the impact of Wal-Mart on people's lives in the U.S. and internationally? Which theoretical perspective do you think is most useful for analyzing the impact of Wal-Mart on the world economy?

6. In what ways does sociology differ from other social sciences?

PRACTICE TESTS

MULTIPLE CHOICE QUESTIONS

Select the response that best answers the question or completes the statement.

1. Sociology is the systematic study of
   a. intuition and commonsense knowledge.
   b. human society and social interaction.
   c. the production, distribution, and consumption of goods and services in a society.
   d. personality and human development.

2. A _____ is a large social grouping that shares the same geographic territory and is subject to the same political authority and dominant cultural expectations.
   a. territory
   b. culture
   c. city-state
   d. society

3. All of the following are reasons to study sociology, **except**
   a. sociology helps us gain a better understanding of ourselves and our social world.
   b. sociology utilizes scientific standards to study society.
   c. sociology confirms the accuracy of commonsense knowledge
   d. sociology helps us look beyond our personal experiences and gain insights into society.

4. According to C. Wright Mills, the sociological imagination refers to the ability to
   a. recognize the relationship between personal troubles and public issues.
   b. see the relationship between preliterate and literate societies.
   c. be completely objective in examining social life.
   d. seek out one specific cause for a social problem such as suicide.

5. The early sociological thinker who analyzed the consequences of industrialization and capitalism, and advocated racial and gender equality is:
   a. Harriet Martineau
   b. August Comte
   c. Herbert Spencer
   d. Karl Marx

6. The ability to provide theory and research beyond one's own country, enveloping countries all over the world is known as a _____ approach.
   a. global
   b. developed nation
   c. developing nation
   d. personal awareness

10

7. The world's _____ countries are nations with industrializing economies, particularly in urban areas, and moderate levels of national and personal income.
    a. high-income
    b. middle-income
    c. low-income
    d. subordinate-income

8. Even in high-income countries, problems of personal debt threaten economic and social stability. In the U.S.A. alone, during a one-year period (2006), _____ people filed for bankruptcy.
    a. 500,000
    b. 550,000
    c. 1 million
    d. 1.1 million

9. Many of the nations of Africa and Asia, particularly the People's Republic of China and India, where people typically work the land, are examples of _____ countries.
    a. high-income
    b. middle-income
    c. subordinate-income
    d. low-income

10. Brazil and Mexico are examples of _____ countries.
    a. high-income
    b. high middle-income
    c. middle-income
    d. low-income

11. The process by which societies are transformed from agricultural-based economies to manufacturing-based economies is referred to as:
    a. information revolution
    b. urbanization
    c. industrialization
    d. postmodernization

12. _____ viewed society as a web of patterned interactions among individuals.
    a. Karl Marx
    b. Max Weber
    c. Georg Simmel
    d. Herbert Spencer

13. The first systematic analysis of society is found in the _____.
    a. work of Auguste Comte
    b. philosophies of the Age of Reason
    c. work of the eighteenth century social philosophers
    d. philosophies of early Greek philosophers

14. _____ is the theorist who argued that our self-concept is formed through social interactions.
    a. Karl Marx
    b. George Herbert Mead
    c. Robert E. Park
    d. Jane Addams

15. In France, the Enlightenment is also referred to as the _____.
    a. Industrial Revolution
    b. French Revolution
    c. Age of Discord
    d. Age of Reason

16. Two historical factors that contributed to the development of sociological thinking were
    a. industrialization and urbanization.
    b. industrialization and immigration.
    c. urbanization and centralization.
    d. urbanization and immigration.

17. The functionalist theorist who distinguished between the manifest and latent functions of institutions is
    a. Emile Durkheim.
    b. Robert Merton.
    c. Robert E. Park.
    d. Erving Goffman.

18. French philosopher Auguste Comte's philosophy became known as _____a belief that the world can best be understood through scientific inquiry.
    a. absolutism
    b. positivism
    c. functionalism
    d. activism

19. Auguste Comte described the *law of the three stages*: he believed that knowledge began in the _____explanations were based on religion and the supernatural.
    a. theological state
    b. metaphysical stage
    c. scientific stage
    d. positive stage

20. Theoretical perspectives that focus on whole societies, large-scale social structures, and social systems are referred to as _____ analyses.
    a. macrolevel
    b. megalevel
    c. mesolevel
    d. microlevel

21. _____ is the belief that those species of animals, including human beings, best adapted to their environment survive and prosper, whereas those poorly adapted die out.
    a. Social Darwinism
    b. Social eugenics
    c. Social statics
    d. Social facts

22. _____ argued that conflict—especially class conflict—is necessary in order to produce social change and a better society.
    a. Auguste Comte
    b. Emile Durkheim
    c. Karl Marx
    d. Harriet Martineau

23. The _____ perspective examines the relationships between different social institutions.
    a. functionalist
    b. conflict
    c. symbolic interactionist
    d. postmodern

24. The first U.S. department of sociology was established at _____, where _____ was one of the best known women in the field.
    a. University of Chicago; Jane Addams
    b. Harvard University; Harriet Martineau
    c. University of California; Arlie Hochschild
    d. Princeton University; Sara McLanahan

25. Social theorist _____ observed that rapid social change can lead to a breakdown in traditional values and an increase in _____.
    a. Du Bois; subjectivity
    b. Weber; alienation
    c. Durkheim; anomie
    d. Comte; positivism

26. The word that Max Weber used to stress the need for sociologists to try to understand other people's feelings, viewpoints, and attitudes.
    a. anomie
    b. *verstehen*
    c. rationalization
    d. *gesundheit*

27. The credit card industry has contributed to the _____ process by the efficiency with which it makes loans and deals with consumers.
    a. ritualization
    b. formalization
    c. procenduralization
    d. rationalization

13

28. W. E. B. Du Bois referred to the identity conflict of being a black and an American as
    a. group consciousness.
    b. the American dilemma.
    c. false consciousness.
    d. double-consciousness.

29. A _____ is defined as a set of logically interrelated statements that attempts to describe, explain, and (occasionally) predict social events.
    a. fact
    b. theory
    c. hypothesis
    d. perspective

30. _____ perspectives are based on the assumption that society is a stable, orderly system.
    a. Functionalist
    b. Symbolic interactionist
    c. Conflict
    d. Feminist

31. _____ perspectives are based on the assumption that groups are engaged in a continuous power struggle for control of scarce resources.
    a. Functionalist
    b. Symbolic interactionist
    c. Conflict
    d. Postmodern

32. Some students meeting their friends at a mall to "hang out" would likely be viewed as a _____ of shopping and consumerism.
    a. latent function
    b. latent dysfunction
    c. manifest function
    d. manifest dysfunction

33. According to your text, all of the following are conflict theorists, **except**
    a. Max Weber.
    b. Talcott Parsons.
    c. Georg Simmel.
    d. C. Wright Mills.

34. The _____ approach directs attention to women's experiences and the importance of gender as an element of social structure.
    a. functionalist
    b. symbolic interactionist
    c. conflict
    d. feminist

14

35. Social scientist Thorstein Veblen described early wealthy U.S. industrialists as engaging in _____—the continuous public display of one's wealth and status through purchases such as expensive houses, clothing, motor vehicles, and other consumer goods.
    a. massive consumption
    b. conspicuous consumption
    c. random consumption
    d. inconspicuous consumption

36. Goffman's dramaturgical analysis represents the _____ perspective.
    a. conflict
    b. functionalist
    c. symbolic interactionist
    d. postmodern

37. Which sociological perspective would focus on the fact that our thoughts and behaviors are shaped by our social interactions with others?
    a. functionalist perspective
    b. conflict perspective
    c. symbolic interactionist perspective
    d. postmodern perspective

38. The _____ perspective focuses on individuals' interpretations of face-to-face interactions.
    a. functionalist
    b. conflict
    c. symbolic interactionist
    d. postmodern

39. The _____ perspective focuses on the shift from production to consumption.
    a. functionalist
    b. conflict
    c. symbolic interactionist
    d. postmodern

40. A _____ would more likely examine how the political process—such as the efforts of lobbyists and interest groups—affects credit card interest rates and consumer spending in the United States.
    a. sociologist
    b. economist
    c. political scientist
    d. anthropologist

15

TRUE/FALSE QUESTIONS

1. Global interdependence affects the lives of all people.
   T   F

2. All sociologists attempt to discover patterns or commonalties in human behavior.
   T   F

3. Sociology is the systematic study of behavior and mental processes.
   T   F

4. Mary Wollstonecraft is an early American sociologist, known for founding Hull House and developing a methodological technique used by sociologists.
   T   F

5. Auguste Comte is considered by some to be the "founder of sociology."
   T   F

6. Herbert Spencer coined the term "survival of the fittest."
   T   F

7. Talcott Parsons emphasized that all societies must provide for meeting social needs in order to survive.
   T   F

8. In the Marxian framework, class conflict is the struggle between the capitalist class and the working class.
   T   F

9. According to Marx, the "fetishism of commodities" describes the situation wherein the proletariat recognize that their labor gives the commodity its value.
   T   F

10. Weber's concept of rationalization refers to the increasing domination of structures devoted to efficiency, calculability, predictability, and technological control.
    T   F

11. From a functionalist perspective, leadership, decision making, and employment outside the home to support the family all constitute expressive tasks.
    T   F

12. According to Robert K. Merton, latent functions are intended and/or overtly recognized by the participants in a social unit.
    T   F

13. The conflict perspective is one unified theory using the branches of neo-Marxism and feminism.
    T   F

14. Symbolic interaction perspectives are based on a macrolevel analysis of society.
    T   F

15. The conflict perspective emphasizes individuals' interpretations of social encounters.
    T   F

FILL-IN-THE-BLANK QUESTIONS

1. A _____ is a large group that shares a geographic territory, political authority, and dominant cultural expectations.

2. _____ suggested that societies are built upon social facts.

3. _____ is considered the founder of sociology.

4. _____ studied the rationalization of society.

5. _____ created a view of society known as social Darwinism.

6. _____ focused on small social group patterns of interaction.

7. _____ coined the term "double-consciousness."

8. According to Marx _____ are exploited by the _____.

9. Weber's important book _____ evaluated the role of the Protestant Reformation in producing a social climate in which capitalism could exist and flourish.

10. The _____ analysis focuses on small groups.

11. The _____ perspective examines the emergence of a global village.

12. The _____ perspective views society as a stable, orderly system, characterized by societal consensus.

13. The _____ perspective focuses on inequalities between groups based on factors such as social class, race, ethnicity, and gender.

14. A _____ is an undesirable consequence of any element of society.

15. Thorstein Veblen described early wealthy people in the U.S. as engaging in _____, the public display of one's wealth and status through expensive goods.

SHORT ANSWER/ESSAY QUESTIONS

1. How might the sociological imagination help individuals personally and professionally? Provide an example.

2. Why were the early social thinkers concerned with social stability? Why did so many of them analyze society from the perspectives of the biological and physical sciences?

17

3. How do race, class, and gender affect, and how are they affected by, shopping and consumption?
4. How did the Age of Enlightenment and the Industrial Revolution contribute to the emergence of sociology in Europe?
5. What are the major tenets of the social sciences discussed in the text?
6. Identify the seven major early thinkers in the development of modern sociology and summarize their unique contributions to the discipline.
7. Apply each of the four contemporary theoretical perspectives to homelessness.

STUDENT CLASS PROJECTS AND ACTIVITIES

1. As discussed in your text, C. Wright Mills's sociological imagination is the ability to understand the relationship between personal troubles and public issues. Compose a list of ten personal troubles that you have recently experienced—for example, money problems, problems in a specific course, problems in interpersonal relationships, maintaining a decent grade point average, inability to find a part-time job, parental pressures, etc. Be specific in your listing. This information will be kept strictly confidential by your professor. Next, write a paragraph about an analysis of the problems. Of these, select five from the list of ten that could possibly extend beyond your personal problems and be considered a public issue. Provide specific information that indicates how these five problems have their roots in social causes. Next, choose one of the "public problems" and briefly research that specific problem, providing some statistical data that verifies that indeed this problem is a social issue. Provide a bibliographical reference for the one specific problem that you chose to research.
2. Choose ten cartoons from various sources that illustrate some sociological concepts or ideas, or any ideas of interest to sociologists as discussed in Chapter 1. Provide a copy of each of the cartoons in your paper. Write at least one paragraph for each cartoon describing what concept, idea, or area of sociology the cartoon is illustrating, as well how it illustrates that concept and why you believe it can be of interest to sociologists. Provide any additional commentary for each cartoon.
3. Select a well-known historical or contemporary sociological theorist. Write a paper that includes: (1) the background and lifestyle of the theorist, (2) other important bibliographic information on the theorist, (4) a summary of the major works and ideas of your theorist, and (5) an evaluation of the contributions your theorist makes to the field of sociology.
4. Select one of the following sociological topics: gender inequality, prostitution, international conflict and war, college football, or alcohol abuse. Explain how each of the contemporary theoretical perspectives (functionalism, conflict theory, symbolic interactionism, and postmodernism) would view the topic you selected. Write one paragraph for each theory. Then write a paragraph explaining which theory you think is most valuable for understanding the topic you selected, and explain why you chose that theory.

INTERNET ACTIVITIES

1. For an introduction to the broad field of sociology, explore the **SOCIOSITE: http://www.sociosite.net/**. From this page you can access dozens of subject areas within the discipline. The web site has hundreds of resources organized within these subject areas.
2. **A Sociological Tour Through Cyberspace** is a web site created by Professor Michael C. Kearl: **http://www.trinity.edu/~mkearl/index.html#in**. There are many resources on this site. Read the essay on credit cards. Next, read the section on Social Science Data Resources. Discover how to recognize the differences between good data and bad, misinterpreted information.
3. To conduct a more extensive analysis of one of the social thinkers introduced in this chapter, go to the web site **Women in History** to learn more about Jane Addams: **http://www.lkwdpl.org/wihohio/adda-jan.htm**. Respond to these questions: Did Jane Addams graduate from college? How did Yale University honor her? What was Hull House? Why did this program serve as a model program to assist newly arrived immigrants in the United States? Was Jane Addams the first American woman to receive the Nobel Peace Prize? Why did she receive this most prestigious international award? What was her role in the NAACP? In the Women's Suffrage Movement? Note the additional web sites on this page providing additional opportunities for you to advance your knowledge about this early founder in the discipline of sociology.

INFOTRAC COLLEGE EDITION EXERCISES

Visit the **InfoTrac College Edition** web site at: **http://www.wadsworthmedia.com/webtutor/infotrac.htm**. You will arrive at a screen that enables you to search topics.
1. Log on to the journal **<u>Social Forces</u>**. Then record both the titles and the number of research articles that focus on race, class, or gender from the past two years. How many were published in that time period? What subtopics did the articles focus on? Report your findings to the class.
2. Conduct a keyword search for **American Consumerism**. Using the resources found on InfoTrac, write an essay on the global impact of American culture.
3. Under the subject search for **Credit Cards**, there are a number of subdivisions. Go to the "usage" subdivision where you will find a number of articles. Select a few of articles to bring to class and work in groups to synthesize information with readings from the text and class presentations.
4. Put together a subject search for the term **Sociology**. There are dozens of categories and over a hundred articles. Explore the discipline and bring back to class information of interest that was not presented in class or in the assigned readings.

SOLUTIONS

MULTIPLE CHOICE QUESTIONS

1. B, p. 4	15. D, p. 11	29. B, p. 23
2. D, p. 4	16. A, p. 11	30. A, p. 23
3. C, p. 4	17. B, p. 23	31. C, p. 25
4. A, p. 5	18. B, p. 12	32. A, p. 23
5. A, p. 12	19. A, p. 12	33. B, p. 26
6. A, p. 8	20. A, p. 27	34. D, p. 26
7. B, p. 8	21. A, p. 14	35. B, p. 27
8. D, p. 9	22. C, p. 16	36. C, p. 30
9. B, p. 9	23. A, p. 23	37. C, p. 27
10. C, p. 9	24. A, p. 22	38. C, p. 27
11. C, p. 11	25. C, p. 15	39. D, p. 30
12. C, p. 20	26. B, p. 19	40. C, p. 33
13. D, p. 11	27. D, p. 19	
14. B, p. 29	28. D, p. 22	

TRUE/FALSE QUESTIONS

1. T, p. 4	6. T, p. 14	11. F, p. 23
2. T, p. 5	7. T, p. 23	12. F, p. 23
3. F, p. 4	8. T, p. 16	13. F, p. 25
4. F, p. 22	9. F, p. 16	14. F, p. 27
5. T, p. 12	10. T, p. 25	15. F, p. 27

FILL-IN-THE-BLANK QUESTIONS

1. society, p. 4
2. Emile Durkheim, p. 15
3. Auguste Comte, p. 12
4. Max Weber, p. 19
5. Herbert Spencer, p. 14
6. Georg Simmel, p. 20
7. W.E.B. Du Bois, p. 22
8. proletariats (workers), p. 22; bourgeoisie (owners), p.16
9. *The Protestant Ethic and The Spirit of Capitalism*, p. 19
10. symbolic interactionist, p. 27
11. postmodern, p.30
12. functionalist, p. 23
13. conflict, p. 25
14. dysfunction, p. 24
15. conspicuous consumption, p. 27

2
SOCIOLOGICAL RESEARCH METHODS

BRIEF CHAPTER OUTLINE

CHAPTER SUMMARY

Social research is a key part of sociology. The sociological perspective incorporates theory and research to arrive at a more accurate understanding of society and provide a factual and objective counterpoint to commonsense knowledge and ill-informed sources of information. Sociological research involves *debunking* and utilizes the *empirical approach*, which attempts to answer questions through the systematic collection and analysis of data. Theory and research form a continuous cycle that encompasses both *deductive* and *inductive approaches*. There are two main categories of sociological research: *Quantitative* research uses data that can be measured numerically, whereas *qualitative research* is based on interpretive description rather than statistics. Research models are tailored to the specific problem being investigated and the focus of the researcher, which may be quantitative or qualitative. The following are steps in the conventional quantitative approach: (1) select and define the research problem; (2) review previous research; (3) formulate the hypothesis; (4) develop the research design,;(5) collect and analyze the data; and (6) draw conclusions and report the findings. The qualitative approach is similar to the conventional research approach in that it uses the following steps: (1) formulate the problem to be studied instead of creating a hypothesis; (2) collect and analyze the data; and (3) report the results. However, it also has several unique features: (1) the research begins with a general approach rather than a highly developed plan; (2) the researcher decides when the literature review and theory application should take place; (3) the study presents a detailed view of the topic; (4) the researcher must have access to people or other resources that can provide the necessary data; and (5) the researcher uses appropriate methods for acquiring useful qualitative data. **Research methods**—systematic techniques for conducting research—include **survey research**, **secondary analysis of existing data**, **field research**, and **experiments**. Many sociologists use *triangulation*—multiple methods—in a single study in order to gain a wider scope of data and points of view. Studying human behavior raises important ethical issues for sociologists, such as the research of Zellner and Humphreys. The challenge for social research today is to find new ways to integrate knowledge, action, and all people in the research process, in order to fill in the gaps of existing knowledge about social life.

LEARNING OBJECTIVES

After reading Chapter 2, you should be able to:

1. Describe the need for systematic research in sociology.

2. Distinguish between sociology and common sense.

3. Describe the research and theory cycle from the deductive and inductive points of view.

4. Describe the six steps in the conventional research process.

5. Indicate the relationship between dependent and independent variables in a hypothesis.

6. Distinguish between a representative sample and a random sample, and explain why sampling is an integral part of quantitative research.

7. Explain why validity and reliability are important considerations in sociological research.

8. Describe the key steps in conducting qualitative research.

9. Describe the major types of surveys and indicate their major strengths and weaknesses.

10. Describe the process of secondary analysis of existing data and state the major strengths and weaknesses of this approach.

11. Describe the major methods of field research, indicate when researchers are most likely to utilize each of them, and discuss their major strengths and weaknesses.

12. Describe the structure of an experiment and distinguish between laboratory and field experiments.

13. Explain the concept of triangulation.

14. Differentiate between quantitative and qualitative research and give examples of each.

15. Discuss the major ethical concerns in sociological research and describe previous research that violated research ethics.

KEY TERMS

(Defined at page number shown and in glossary)

content analysis, p. 56
control group, p. 62
correlation, p. 62
dependent variable, p. 45
ethnography, p. 59
experiment, p. 61
experimental group, p. 62
field research, p. 58
Hawthorne effect, p. 63
hypothesis, p. 45
independent variable, p. 45
interview, p. 53

participant observation, p. 58
probability sampling, p. 47
questionnaire, p. 53
random sampling p. 46
reliability, p. 47
research methods, p. 51
respondents, p. 53
secondary analysis, p. 54
survey, p. 51
unstructured interview, p. 60
validity, p. 47

KEY PEOPLE

(identified at page number shown)

Emile Durkheim, p. 38
Barney Glaser and Anselm Strauss, p. 61

Laud Humphreys, p. 67
William F. Whyte, p. 60
William Zellner, p. 65

CHAPTER OUTLINE

I. WHY IS SOCIOLOGICAL RESEARCH NECESSARY?
   A. Common Sense and Sociological Research
      1. Sociologists obtain knowledge of human behavior through research, which results in a body of information that helps to move beyond guesswork and common sense in understanding society.
   B. Sociology and Scientific Evidence
      1. Sociology involves *debunking*—the unmasking of false ideas or opinions—in everyday and official interpretations of society.

2. Many people in society use the *normative approach* to answering questions concerning social life, which relies on religion, customs, habits, and traditions.
3. Sociologists use the *empirical approach,* which attempts to answer questions through the systematic collection and analysis of data, and is also referred to as the conventional model or scientific method. It utilizes two types of empirical studies: *descriptive studies*, which provide social facts or describe social reality, and *explanatory studies*, which attempt to explain cause and effect relationships.

C. The Theory and Research Cycle
1. A theory is a set of logically interrelated statements that attempts to describe, explain, and (occasionally) predict social events.
2. Research is the process of systematically collecting information for the purposes of testing an existing theory or generating a new one.
3. The theory and research cycle consists of *deductive* and *inductive* approaches. The *deductive approach* uses research to test a theory; the *inductive approach* generates theory through the analysis of data.
4. Theory gives meaning to research; research helps support theory.

II. THE SOCIOLOGICAL RESEARCH PROCESS
A. *Quantitative research* is based on the goal of scientific objectivity and focuses on data that can be measured in numbers, while *qualitative research* uses interpretive descriptions (words) rather than statistics (numbers) to analyze underlying meanings and patterns of social relationships.

B. The "Conventional" Research Model
1. The "conventional" research model focuses on quantitative research.
2. The steps in the conventional research method include:
   a. Selecting and defining the research problem
   b. Reviewing previous research
   c. Formulating a hypothesis (if applicable)
   d. Developing research design, selecting either *cross-sectional* or *longitudinal* time frames
   e. Collecting and analyzing data
   f. Drawing conclusions and reporting the findings
3. Important concepts in the research process:
   a. A **hypothesis** is a statement of the relationship between two or more concepts.
   b. **Variables** are concepts with measurable traits or characteristics that can change or vary from one person, time, situation, or society to another.
   c. The **independent variable** is presumed to cause or determine a dependent variable.
   d. The **dependent variable** is assumed to depend on or be caused by the independent variable(s).
   e. To use a variable, sociologists create an *operational definition*—an explanation of an abstract concept in terms of observable features that are specific enough to measure the variable.
   f. An event that occurs as a result of many factors operating in combination must be explained by *multiple causation*.

24

4. Important concepts in developing the research design:
   a. A *unit of analysis* is what or who is being studied.
   b. *Cross-sectional studies* take place at a single point in time.
   c. *Longitudinal studies* focus on processes and social change over a period of time.
5. Important concepts in collecting and analyzing data:
   a. A **random sample** is chosen by chance. Every member of an entire population being studied has the same chance of being selected.
   b. In **probability sampling**, participants are deliberately chosen by specific characteristics, such as age, race, and ethnicity.
   c. **Validity** is the extent to which a study or research instrument accurately measures what it is supposed to measure.
   d. **Reliability** is the extent to which a study or research instrument yields consistent results.

C. A Qualitative Research Model
   1. Qualitative research differs from quantitative research in several ways:
      a. Qualitative research is often used when the research question does not lend itself to numbers and statistical methods.
      b. Researchers may engage in *problem formulation* instead of creating a hypothesis.
      c. This type of research is often built on a collaborative approach in which the "subjects" are active participants in the design process, not just passive objects to be studied.
      d. Researchers tend to gather data in natural settings, such as where the person lives or works, rather than in a laboratory or other research settings.
      e. Data collection and analysis frequently occur concurrently, and the analysis draws heavily on the language of the persons studied, not the researcher.
   2. Qualitative research follows the conventional research approach, but has several unique features:
      a. The researcher begins with a general approach rather than a highly detailed plan.
      b. The researcher has to decide when the literature review and theory application should take place.
      c. The study presents a detailed view of the topic.
      d. Access to people or other resources that can provide the necessary data is crucial.
      e. Appropriate research methods are important for acquiring useful qualitative data.

III. RESEARCH METHODS
   A. **Research methods** are strategies or techniques for systematically conducting research.
   B. A **survey** is a poll in which researchers gather facts or attempt to determine the relationship between facts. Surveys are often used to gather information about

25

people's attitudes and behaviors. Survey data are collected by using self-administered questionnaires, personal interviews, and/or telephone surveys.

1. **Respondents** are persons who provide data for analysis through interviews or questionnaires.

2. A **questionnaire** is a printed research instrument containing a series of items for the participant's response. Questionnaires may be self-administered by respondents or administered by interviewers in face-to-face encounters or by telephone.

3. An **interview** is a data-collection encounter in which an interviewer asks the respondent questions and records the answers. Survey research often uses *structured interviews*, in which the interviewer asks questions from a standardized questionnaire.

4. Some strengths of survey research include the ability to describe the characteristics of the large population and an enhanced ability to do *multivariate analysis*—research involving more than two independent variables.

5. Several weaknesses of survey research include the concern of validity, that respondents may not be truthful, and the belief that the data do not always constitute "hard facts" that other analysts might use.

C. In **secondary analysis of data**, researchers use existing material and analyze data that were originally collected by others.

1. Existing data sources include public records, official reports of organizations or government agencies, surveys taken by researchers in universities and private corporations, books, magazines, newspapers, radio, television, and personal documents.

2. **Content analysis** is the systematic examination of *cultural artifacts* or various forms of communication to extract thematic data and draw conclusions about social life.

3. Several strengths of secondary analysis include that the data are readily available and inexpensive, the chance of bias may be reduced, and there is the possibility of using longitudinal data.

4. Some weaknesses include that the data may be incomplete, unauthentic, or inaccurate, and *coding* the data may be difficult.

D. **Field research** is the study of social life in its natural setting: observing and interviewing people where they live, work, or play.

1. In **participant observation**, researchers collect systematic observations while being part of the activities of the groups they are studying.

2. A *case study* is an in-depth, multifaceted investigation of a single event, person, or social grouping; a *collective case study* involves multiple cases.

3. An **ethnography** is a detailed study of the life and activities of a group of people by researchers who may live with that group over a period of years.

4. An **unstructured interview** is an extended, open-ended interaction between an interviewer and an interviewee.

5. *Grounded theory* is an *inductive* method of theory construction whereby data collection and analysis occur at the same time.

26

6. Strengths of field research include providing a wealth of information to generate theories, enabling the study of social processes over time, and affording the opportunity to study race, ethnicity, and gender over time.
7. Some weaknesses of field research include that findings may not be generalized to a larger population, the data collected are descriptive, rather than precise measurements, and cause and effect relationships cannot be depicted.
E. **Experiments**—carefully designed situations in which the researcher studies the impact of certain variables on subjects' attitudes or behavior—typically require that subjects be divided into two groups:
1. The **experimental group** contains the subjects who are exposed to the independent variable (the experimental condition) to study its effect on them.
2. The **control group** contains the subjects who are not exposed to the independent variable.
3. Experiments may be conducted in a *laboratory* or *natural setting*.
4. At the conclusion of the research, the experiment and control groups are compared to see if they differ in relation to the dependent variable, and the hypothesis about the relationship of the two variables is confirmed or rejected. **Correlation** exists when two variables in an experiment are associated more frequently than could be expected by chance.
5. Some strengths of the controlled experiment include the researcher's ability to isolate the experiment and the possibility of continued replication.
6. Some weaknesses of experiments include the rigid control and manipulation of variables, the artificial setting, and the problem of generalization to other groups. Another weakness is known as the **Hawthorne effect**—changes in the subject's behavior caused by the researcher's presence or by the subject's awareness of being studied.
F. Multiple Methods: Triangulation
1. Many sociologists utilize *triangulation*—the use of multiple approaches in a single study.
2. Triangulation refers to both multiple research methods and multiple data sources.
IV. ETHICAL ISSUES IN SOCIOLOGICAL RESEARCH
A. The ASA Code of Ethics
1. The American Sociological Society (ASA) has a Code of Ethics that sets forth certain basic standards sociologists must follow in conducting research.
2. Sociologists are committed to adhering to this code and to protecting research participants; however, many ethical issues arise that cannot be resolved easily.
B. The Zellner Research
1. The research sought to determine if some automobile accidents were actually suicides, which he called "autocides."
2. To recruit respondents, he misrepresented the reasons for his study.
C. The Humphreys Research
1. This research studied homosexual acts in public restrooms in parks.

2. Humphreys did not ask the permission of his subjects, nor did he inform them that they were being studied. His subsequent personal interviews with them were held under false pretenses.
D. Research Summary
   1. These examples of research demonstrate the difficulties of resolving ethical issues.
   2. The challenge today is to find new ways to integrate knowledge and action, and to include all people in the research process in order to fill the gaps in our existing knowledge about social life.

CRITICAL THINKING QUESTIONS

1. What are the strengths and weaknesses of each research method described in this chapter?
2. Both quantitative and qualitative research methods are important to sociology. What kinds of research questions are best answered by quantitative research? What kinds of research questions are best answered by qualitative research? How do the two types of research compliment each other?
3. If you were conducting a research project on shopping patterns among college students, which research method would you use? Why? What kind of sampling technique would you use? Why?
4. What are the advantages and disadvantages of each sampling technique described in this chapter?
5. Regarding research and ethics, what do you think researchers should and should not be able to do in order to generate knowledge about a topic? For example, is deception acceptable as long as no one gets hurt? Why or why not?
6. Which research method do you think provides the most useful information on suicide? Why?

PRACTICE TESTS

MULTIPLE CHOICE QUESTIONS

Select the response that best answers the question or completes the statement:

1. The _____ approach involves gathering evidence to answer questions through systematic collection and analysis of data.
   a. empirical
   b. descriptive
   c. normative
   d. evaluative

28

2. A type of research that describes social reality or provides facts about the social world is:
   a. explanatory
   b. descriptive
   c. evaluative
   d. observational

3. In the _____ approach, the researcher begins with a theory and uses research to test the theory.
   a. inductive
   b. deductive
   c. model
   d. command

4. The first step in the conventional research process is
   a. reviewing the literature.
   b. selecting a research model.
   c. formulating the hypothesis.
   d. selecting and defining the research problem.

5. When concepts have two or more degrees or values, they are referred to as
   a. hypotheses.
   b. triangulation.
   c. variables.
   d. theories.

6. Which of the following can serve as a variable in a study?
   a. age
   b. sex
   c. ethnic background
   d. all of these choices

7. In Emile Durkheim's study of suicide, the degree of social integration was the
   a. operational definition.
   b. dependent variable.
   c. independent variable.
   d. spurious correlation.

8. In a medical study, lung cancer could be the _____ variable, while smoking could be the _____ variable.
   a. dependent; independent
   b. independent; dependent
   c. valid; reliable
   d. reliable; valid

9. In creating an operational definition, a sociologist
   a. duplicates research in a precise manner.
   b. analyzes the findings of the research.
   c. makes a concept or variable measurable.
   d. conducts field research.

10. Suppose we are investigating the primary causes of suicide in the late 1990s. Upon looking into recent cases of suicide, we find out that a number of the people had lost their jobs; that they had been unemployed off and on for the past ten years; that they had no religious affiliation; and that a number of them had been divorced within the past five years. The analysis reflects what the text terms:
    a. singular determination
    b. multivariate involvement
    c. plural association
    d. multiple causation

11. A _____ sample refers to a situation in which every member of an entire population has the same chance of being selected for a study.
    a. selective
    b. random
    c. probability
    d. snowball

12. _____ is the extent to which a study or research instrument accurately measures what it is supposed to measure; _____ is the extent to which a study or research instrument yields consistent results.
    a. Validity, replication
    b. Replication, validity
    c. Validity, reliability
    d. Reliability, validity

13. The national census is an example of _____ .
    a. survey research
    b. a field experiment
    c. secondary analysis
    d. participant observation research

14. Research involving more than two independent variables is known as:
    a. multivariate analysis
    b. secondary analysis
    c. random sampling
    d. field analysis

15. Researchers who use existing material and analyze data that were originally collected by others are engaged in:
    a. unethical conduct
    b. primary analysis
    c. secondary analysis
    d. survey analysis

16. Observation, ethnography, and case studies are examples of:
    a. survey research
    b. experiments
    c. secondary analysis of existing data
    d. field research

30

17. In an experiment, the subjects in the control group
    a. are exposed to the independent variable.
    b. are not exposed to the independent variable.
    c. are exposed to the dependent variable.
    d. are not exposed to the dependent variable.

18. A sports sociologist finds that body temperature increases as the "pumping iron" increases. In this example, body temperature would be the:
    a. dummy variable.
    b. null hypothesis.
    c. dependent variable.
    d. independent variable.

19. _____ exists when two variables change together in a predictable direction.
    a. Correlation
    b. Triangulation
    c. Multiple causation
    d. Quantification

20. The American Sociological Association _____ sets forth certain ethical standards sociologists must follow when conducting research.
    a. Policies and Procedures
    b. Research Guidebook
    c. Procedural Report
    d. Code of Ethics

21. Sociologists use _____ to obtain knowledge of human behavior, which moves beyond guesswork and common sense.
    a. research
    b. myths
    c. common sense beliefs
    d. science

22. Laud Humphreys's research on homosexual conduct in "tearooms" would be considered unethical according to today's standards because
    a. Humphreys did not ask permission to study his subjects.
    b. homosexuality is a taboo topic that is inappropriate for research.
    c. Humphreys's study participants were under the age of 18.
    d. Humphreys inflicted physical harm on his study participants.

23. The _____ approach, referred to as the conventional model or the scientific method, attempts to answer questions through the systematic collection and analysis of data.
    a. explanatory
    b. descriptive
    c. empirical
    d. normative

31

24. Sociologists would use a(n) _____ study of suicide in answering the question of why African American men over age sixty-five have a significantly lower rate of suicide than white males in the same age bracket.
    a. objective
    b. evaluative
    c. descriptive
    d. explanatory

25. _____ attempt to describe social reality or provide facts about some group, practice, or event.
    a. Explanatory studies
    b. Observational studies
    c. Descriptive studies
    d. Evaluative studies

26. With _____ research, the goal is scientific objectivity, and the focus is on data that can be measured numerically.
    a. explanatory
    b. qualitative
    c. quantitative
    d. normative

27. _____ is a term used to describe a research study that uses multiple sources of data collected through different research methods.
    a. Multimethod
    b. Triangulation
    c. Longitudinal
    d. Cross-sectional

28. In a hypothesis, the researcher assumes the _____ to be caused by another variable.
    a. independent variable
    b. dependent variable
    c. multiple variable
    d. null variable

29. In this sociology course, your professor may have specified the points needed, or the average test grades needed, to earn an "A" in the course, thus providing you a(n) _____ of a "grade" in the course.
    a. operational definition
    b. explanatory definition
    c. qualitative explanation
    d. correlated definition

30. A(n) _____ is an extended, open-ended conversation between an interviewer and interviewee in which the interviewer has a general plan of inquiry but no specific set of questions that must be answered.
    a. survey
    b. questionnaire
    c. experiment
    d. unstructured interview

31. When events are too complex to be caused by one variable, sociologists explain those causes in terms of _____.
    a. pluralistic association
    b. multiple causation
    c. multiple associations
    d. multiplicity causation

32. _____ studies are based on observations that take place at a single point in time; these studies focus on behavior or responses at a specific moment.
    a. Participant
    b. Cross-sectional
    c. Longitudinal
    d. Sectional

33. In _____, participants are deliberately chosen because they have specific characteristics.
    a. probability sampling
    b. validity sampling
    c. random sampling
    d. reliable sampling

34. Which of the following methods would most likely be used to study the everyday lives and experiences of homeless adults in Washington, D.C.?
    a. experiment
    b. survey
    c. field research
    d. secondary analysis

35. Students in this sociology class develop a research instrument to measure happiness among the students enrolled in the university. You are concerned with the issue of _____ in that you want your research instrument to accurately measure what it is suppose to measure.
    a. representation
    b. predictability
    c. reliability
    d. validity

36. Which of the following methods would most likely be used to study the attitudes of the American public on global warming?
    a. an experiment
    b. participant observation research
    c. a survey
    d. a case study

37. A method of theory construction created by Barney Glaser and Anselm Strauss in which data collection and analysis occur at the same time is called
    a. grounded theory.
    b. deductive.
    c. validity.
    d. probability construction.

38. Secondary analysis is referred as _____ research because it has no impact on the people being studied.
    a. obtrusive
    b. unobtrusive
    c. non-opinionated
    d. reliable

39. A researcher would probably use _____ in studying how gender is depicted in award-winning children's books.
    a. survey research
    b. content analysis
    c. field research
    d. an experiment

40. _____ is the study of social life in its natural setting.
    a. Field research
    b. A natural experiment
    c. Secondary analysis
    d. A case study

TRUE/FALSE QUESTIONS

1. Descriptive studies attempt to explain cause and effect relationships.
   T   F

2. The normative approach answers questions through systematic collection and analysis of data.
   T   F

3. Theory gives meaning to research; research helps support theory.
   T   F

4. In the deductive approach, the researcher collects data and then generates theories from the analysis of that data.
   T   F

34

5. With qualitative research, the focus is on data that can be measured numerically.
   T   F

6. In a hypothesis, the independent variable is presumed to cause or determine a dependent variable.
   T   F

7. Cross-sectional studies examine what has happened over a period of time.
   T   F

8. In random sampling, participants are chosen because they have certain characteristics.
   T   F

9. Validity is when a study gives consistent results to different research over time.
   T   F

10. In quantitative research, the researcher begins with a general approach rather than a detailed plan.
    T   F

11. In probability sampling, researchers select a few participants with certain characteristics and then ask those participants to provide names of others with similar characteristics.
    T   F

12. In survey research, every person of that population must be interviewed.
    T   F

13. Participant observation is a quantitative research method.
    T   F

14. Sociologist William F. Whyte's classic study of Boston's low-income Italian neighborhood is an example of ethnography.
    T   F

15. The Hawthorne effect refers to changes in the subject's behavior caused by the researcher's presence or by the subject's awareness of being studied.
    T   F

FILL-IN-THE-BLANK QUESTIONS

1. Sociology involves _____, which challenges false or mistaken ideas or opinions.

2. A(n) _____ is a statement of the relationship between two or more concepts.

3. _____ studies compare data over a period of time.

4. In _____ sampling, participants are deliberately chosen because they have specific characteristics.

5. According to Durkheim, _____ suicide results from a lack of shared values or purpose.

6. Specific strategies for systematically conducting research are known as _____.

7. Persons who provide data for analysis through interviews or questions are called _____.

8. Researchers use _____ research to study social life in its natural setting.

9. _____ research uses descriptions and interpretations to analyze meanings and patterns of social relationships.

10. The controversial research of _____ examined "tearoom trade."

11. In conducting research, a(n) _____ makes it possible to measure a concept or a variable.

12. A _____ is a printed research instrument containing a set of questions.

13. A _____ is a data collection procedure in which an interviewer asks the respondent questions and records the answers.

14. In _____, researchers analyze data that were collected by others.

15. A _____ is a detailed study of a group of people by researchers who may live with that group over a period of years.

SHORT ANSWER/ESSAY QUESTIONS

1. What is the difference between qualitative and quantitative research? Give an example of each.
2. What is the relationship between theory and research?
3. What are the steps in the conventional research process?
4. List and describe the five basic types of research methods.
5. Identify four ethical concerns that must be considered when conducting research, and give three examples of controversial research discussed in your textbook.

STUDENT CLASS PROJECTS AND ACTIVITIES

1. Over a period of two or three weeks, collect five articles from newspapers or magazines reporting the results of a public opinion poll. (Examples: New York Times, Gallup, CNN, etc.). Evaluate the polls and respond in writing to these questions: (1) Were the questions fully reported? (2) What is the size of the sample, its composition (i.e., gender, age, race, nationality, region of country), and the method of selection (i.e., was it random, stratified, etc.)? (3) How were the data collected (telephone, mail, etc.)? (4) What was the wording of the questions asked in the poll? (5) Summarize the findings of the poll. (6) Did you find any problems of the poll? Was it scientifically designed? (7) Include any other information in your paper appropriate to this assignment. (8) Provide bibliographic reference for the polls you have selected. (9) Evaluate this project.

2. Conduct a survey of any 20 people and ask these questions: (1) What is sociology? (2) What do sociologists do? (3) What is your opinion of sociologists? (4) Have you had a course in sociology? If so, where? (5) Have you had a course in a social science? (6) Do you know a sociologist? If so, who? (7) Include any other kind of information you may want to obtain from your subjects. (8) Include the names, gender, ages, and, if students, college majors of your subjects. (9) Provide a summary and an evaluation of this project in the writing of your paper.

3. Conduct an informal poll, collecting data from 10 to 20 students enrolled in your college or university. (1) Have your subjects list and briefly explain their top 10 frustrations. (2) Summarize your findings and list the top ten frustrations according to your subjects. (3) Provide an analysis of this comparison. (4) Provide the ages and gender of your subjects in your written paper. (5) Provide a conclusion and an evaluation of this project.

INTERNET ACTIVITIES

1. Learn more about the United States by accessing the **U.S. Census web site**: **http://www.census.gov/index.html** . The best beginning site is the American FactFinder. This is a more user-friendly entry point to see reports that are being generated from data collected during the 2000 Census, with an update through 2006. Note that some statistics have been revised as recently as 2008.

2. More students are using Internet sources for their class research projects. Before you join the herd, access the **Guide for Citing Electronic Information**, **http://www.wpunj.edu/wpcpages/library/citing.htm**, provided by William Paterson University of New Jersey. The site contains guidelines for citing everything from e-mails to online journal articles. Also included are links to other related web pages.

3. The **National Opinion Research Center**, **http://www.norc.uchicago.edu/projects/gensoc.asp**, at the University of Chicago is one of the largest survey research organizations. On this site you can access information about the NORC, past and present. Next, gather information on the General Social Survey. Then, go to the FAQ link on this site and bring to class a basic description of the GSS.

4. Some research organizations openly admit that their research is guided by a particular set of political values. One such example is the **Cato Institute**, **http://www.cato.org/**, a politically conservative research organization guided by libertarian principles. Go to their home page, click on daily commentaries, and then select a topic that interests you. Can you find examples of research that is guided by conservative principles? How does this compare with the standards of objectivity that are discussed in the text?

5. When you start thinking about the difficulties of researching highly subjective social phenomena, consider accessing the **World Database of Happiness**: **http://www.eur.nl/fsw/research/happiness/**. This site is an ongoing register of scientific research related to subjective appreciation of life. Can you find out which nation is cited as the most happy? Can you discover the degree of unhappiness of nations, as measured by the rate of suicide?

INFOTRAC COLLEGE EDITION EXERCISES

Visit the **InfoTrac College Edition** web site at:
http://www.wadsworthmedia.com/webtutor/infotrac.htm. You will arrive at a screen that enables you to search topics.

1. Look up the keywords **suicide research** on InfoTrac. Among the articles should be those that address social factors. Bring articles to class and be prepared to discuss them in small groups. Determine what kinds of research methods were used in each study.

2. The **Hawthorne effect** is a phenomenon that you may hear referenced in other courses and reading assignments. A recent re-examination of the phenomenon and the original research contains some new surprises. Use InfoTrac to look up the keywords **Hawthorne effect** (be sure of the spelling) and read the article. Construct a list of new facts contained in the report and bring the report to class.

3. Why are surveys chosen over other methods of collecting data? Research the keyword **survey** and select a survey study to examine. Read the research report and then try to determine why a survey was used.

SOLUTIONS

MULTIPLE CHOICE QUESTIONS

1. A, p. 40
2. B, p. 41
3. B. p. 42
4. D, p. 45
5. C, p. 45
6. D, p. 45
7. C, p. 45
8. A, p. 45
9. C, p. 45
10. D, p. 45
11. B, p. 46-47
12. C, p. 47
13. A, p. 51-52
14. A, p. 54

15. C, p. 54-55
16. D, p. 58
17. B, p. 62
18. C, p. 62
19. A, p. 62
20. D, p. 65
21. A, p. 41
22. A, p. 67
23. B, p. 41
24. D, p. 41
25. C, p. 41
26. C, p. 43
27. B, p. 64
28. B, p. 45

29. A, p. 45
30. D, p. 60
31. B, p. 45
32. B, p. 46
33. A, p. 47
34. C, p. 58
35. D, p. 47
36. C, p. 51
37. A, p. 61
38. B, p. 55
39. B, p. 56
40. A, p. 58

TRUE/FALSE QUESTIONS

1. F, p. 41
2. F, p. 40
3. T, p. 41-42
4. F, p. 42
5. F, p. 43

6. T, p. 45
7. F, p. 46
8. F, p. 46
9. F, p. 47
10. F, p. 45-46

11. F, p. 47
12. F, p. 46-47
13. F, p. 58
14. T, p. 60
15. T, p. 63

FILL-IN-THE-BLANK-QUESTIONS

1. debunking, p. 40
2. hypothesis, p. 45
3. Longitudinal, p. 46
4. probability, p. 47
5. anomic, p. 46
6. research methods, p. 51
7. respondents, p. 52
8. field, p. 58
9. Qualitative, p. 43
10. Laud Humphreys, p. 67
11. operational definition, p. 45
12. questionnaire, p. 52
13. interview, p. 53
14. secondary analysis, p. 54-55
15. ethnography, p. 59

3
CULTURE

BRIEF CHAPTER OUTLINE

CHAPTER SUMMARY

Culture is the knowledge, language, values, customs, and material objects that are passed from person to person and from one generation to the next. At the macrolevel, culture can be a stabilizing force or a source of discord, conflict, and even violence. At the microlevel, culture is essential for individual survival. Sociologists distinguish between **material culture**—the physical creations of society—and **nonmaterial culture**—the abstract or intangible human creations of society (such as **symbols**, **language**, **values**, and **norms**). According to the **Sapir-Whorf hypothesis**, language shapes our perception of reality. For example, language may create and reinforce inaccurate perceptions based on gender, race, ethnicity, or other human attributes. Cultural change and diversity are intertwined. In the United States, diversity is reflected through race, ethnicity, age, sexual orientation, religion, occupation, and so forth.

40

Culture shock refers to the anxiety people experience when they encounter cultures radically different from their own. **Ethnocentrism**—or the practice of judging all other cultures by one's own culture—is based on the assumption that one's own culture is superior to others. An alternative to ethnocentrism is **cultural relativism**—the belief that the behaviors and customs of a society must be examined within the context of its own culture. *High culture* appeals primarily to elite audiences, whereas **popular culture** is created to appeal to members of the middle and working classes. Political and religious leaders in some cultures fear and oppose **cultural imperialism**, the extensive infusion of one nation's culture into other cultures, while others point to numerous cross-cultural influences. Sociologists regard culture as a central ingredient in human behavior; however, they use different theoretical perspectives in their research of culture. A *functional analysis* of culture assumes that a common language and shared values help produce consensus and harmony. A *conflict perspective* views culture as a mechanism that may be used by certain groups to maintain their privilege and exclude others from society's benefits. *Symbolic interactionists* suggest that people create, maintain, and modify their culture as they go about their everyday activities. *Postmodern theorists* analyze the impact of media and advanced technology on culture, concluding that culture is now produced on a screen, and suggest that the complexity and diversity in the world have resulted in multiple cultures. As we look toward even more diverse and global cultural patterns in the future, it is important to apply the sociological imagination not only to our society, but to the entire world as well.

LEARNING OBJECTIVES

After reading Chapter 3, you should be able to:

1. Define culture.

2. Describe how culture can be both a stabilizing force and a source of conflict in societies.

3. Describe the importance of culture in determining how people think and act on a daily basis.

4. Explain the various components of culture.

5. Describe the importance of language and relate to the Sapir-Whorf hypothesis.

6. List and briefly explain ten core values in U.S. society.

7. Contrast ideal and real culture, and give examples of each.

8. State the definition of norms, and distinguish between folkways, mores, and laws.

9. Describe subcultures and countercultures; give examples of each.

10. State the definitions for culture shock, ethnocentrism, and cultural relativism, and explain the relationship between these three concepts.

11. Distinguish between discovery, invention, and diffusion as means of cultural change.

12. Explain why the rate of cultural change is uneven.

13. Distinguish between high culture and popular culture, and between fads and fashion.

14. Describe the functionalist, conflict, symbolic interactionist, and postmodernist perspectives on culture.

41

KEY TERMS

(defined at page number shown and in glossary)

counterculture, p. 89
cultural imperialism, p. 91
cultural lag, p. 84
cultural relativism, p. 90
cultural universals, p. 76
culture, p. 72
culture shock, p. 89
diffusion, p. 85
discovery, p. 85
ethnocentrism, p. 90
folkways, p. 83
invention, p. 85
language, p. 77

laws, p. 84
material culture, p. 75
mores, p. 83
nonmaterial culture, p. 75
norms, p. 83
popular culture, p. 90
sanctions, p. 83
Sapir-Whorf hypothesis, p. 78
subculture, p. 87
symbol, p. 77
taboos, p. 83
technology, p. 75
values, p. 80

KEY PEOPLE

(identified at page number shown)

Jean Baudrillard, p. 96
Pierre Bourdieu, p. 90
Bronislaw Malinowski, p. 93

George Murdock, p. 76
Edward Sapir and Benjamin Whorf, p. 78

CHAPTER OUTLINE

I. CULTURE AND SOCIETY IN A CHANGING WORLD
   A. **Culture** is the knowledge, language, values, customs, and material objects that are passed from person to person and from one generation to the next in a human group or society.
   B. Culture is essential for our individual survival and our communication with other people.
   C. As our society becomes more diverse and communication among members of international cultures becomes more frequent, the need to appreciate diversity and understand how people in other cultures view their world increases.
   D. **Material culture** consists of the physical or tangible creations that members of a society make, use, and share, while **nonmaterial culture** consists of the abstract or intangible human creations of society that influence people's behavior.
   E. According to anthropologist **George Murdock**, **cultural universals** are customs and practices that occur across all societies. Examples include appearance, activities, social institutions, and customary practices.

II. COMPONENTS OF CULTURE
  A. A **symbol** is anything that meaningfully represents something else.
  B. **Language** is defined as a set of symbols that express ideas and enable people to think and communicate with one another.
     1. Language and Social Reality
        a. According to the **Sapir-Whorf hypothesis**, language shapes our view of reality.
        b. Most sociologists contend that language may *influence* but does not *determine* social reality.
     2. Language and Gender
        a. Examples of situations in which the English language ignores women include using the masculine gender to refer to human beings in general, and nouns that show the gender of the person we expect in a particular occupation.
        b. Words have positive connotations when relating to male power, prestige, and leadership; when related to women, they convey negative overtones of weakness, inferiority, and immaturity.
     3. Language, Race, and Ethnicity
        a. Language may create and reinforce our perceptions about race and ethnicity by transmitting preconceived ideas about the superiority of one category of people over another.
        b. The "voice" of verbs may devalue contributions of members of some racial-ethnic groups.
        c. Adjectives can have different meanings when used in certain contexts.
  C. **Values** are collective ideas about what is right or wrong, good or bad, and desirable or undesirable in a particular culture.
     1. Ten core U.S. values identified by sociologist Robin M. Williams are:
        a. Individualism
        b. Achievement and success
        c. Activity and work
        d. Science and technology
        e. Progress and material comfort
        f. Efficiency and practicality
        g. Equality
        h. Morality and humanitarianism
        i. Freedom and liberty
        j. Racism and group superiority
     2. *Value contradictions* are values that conflict with one another or are mutually exclusive (achieving one makes it difficult to achieve another). All societies have value contradictions.
     3. *Ideal culture* refers to the values and standards of behavior that people in a society profess to hold; *real culture* refers to the values and standards of behavior that people actually follow.
  D. **Norms** are established rules of behavior or standards of conduct.
     1. *Formal norms* are written and involve specific punishments for violations whereas *informal norms* are unwritten standards of behavior. **Sanctions** are rewards for appropriate behaviors and punishments for inappropriate behaviors.

43

2. **Folkways** are everyday customs that may be violated without serious consequences within a particular culture.
3. **Mores** are strongly held norms that may not be violated without serious consequences within a particular culture. **Taboos** are mores so strong that their violation is considered to be extremely offensive.
4. **Laws** are formal, standardized norms that have been enacted by legislatures and are enforced by formal sanctions.

III. TECNOLOGY, CULTURAL CHANGE, AND DIVERSITY
   A. Cultural change is continual in societies, and changes are often set in motion by these processes:
      1. **Technology** refers to the knowledge, techniques, and tools that allow people to transform resources into a usable form, and the knowledge and skills required to use what is developed.
      2. **Cultural lag** is a gap between the technical development (**material culture**) of a society and its moral and legal institutions (**nonmaterial culture**).
      3. **Discovery** is the process of learning about something previously unknown or unrecognized.
      4. **Invention** is the process of combining existing cultural items into a new form.
      5. **Diffusion** is the transmission of cultural items or social practices from one group or society to another.
   B. Cultural Diversity
      1. *Cultural diversity* refers to the wide range of cultural differences found between and within nations.
         a. Homogeneous societies are made up of people who share a common culture.
         b. Heterogeneous societies are made up of people who are dissimilar.
      2. A **subculture** is a group of people who share a distinctive set of cultural beliefs and behaviors that differ in some significant way from that of the larger society. Examples include Old Order Amish and Chinatowns. Some people who share a common racial, language, or national background identify themselves as members of a specific ethnic subculture, whereas others do not.
      3. A **counterculture** is a group that strongly rejects dominant societal values and norms and seeks alternative lifestyles. Examples include skinheads and members of some paramilitary militias.
   C. **Culture shock** is the disruption people feel when they encounter cultures radically different from their own and believe they cannot depend on their own taken-for-granted assumptions about life.
   D. **Ethnocentrism** is the assumption that one's own culture and way of life are superior to all others. **Cultural relativism** is the belief that the behaviors and customs of a society must be viewed and analyzed within the context of its own culture.

IV. A GLOBAL POPULAR CULTURE?
    A. *High culture* consists of activities usually patronized by elite audiences, while **popular culture** consists of activities, products, and services that are assumed to appeal primarily to members of the middle and working class.
    B. **Pierre Bourdieu**'s *culture capital theory* argues that high culture is a device used by the dominant class to exclude subordinate classes.
    C. Forms of popular culture:
        1. A *fad* is a temporary but widely copied activity followed enthusiastically by large numbers of people. Fads include *object fads*, *activity fads*, *idea fads*, and *personality fads*.
        2. A *fashion* is a currently valued style of behavior, thinking, or appearance that is longer lasting and more widespread than a fad.
        3. **Cultural imperialism** is the extensive intrusion of one nation's culture into other nations, and is viewed as a threat with many countries becoming westernized.
        4. Others argue that if a global culture comes into existence, it will include components from many societies and cultures.
V. SOCIOLOGICAL ANALYSIS OF CULTURE
    A. Functionalist theorists view culture as one element that helps societies function.
        1. **Bronislaw Malinowski** suggested that culture helps people meet *biological needs* (food), *instrumental needs* (education), and *integrative needs* (religion and art).
        2. Functionalists believe societies where people share a common language and core values are more likely to have consensus and harmony.
    B. Conflict theorists suggest that values and norms help create and sustain the privileged position of the powerful in society.
        1. According to Karl Marx, ideas are *cultural relations* of a society's most powerful members, who use *ideology* to maintain their positions of dominance.
    C. Symbolic interactionists believe people create, maintain, and modify culture as they go about their everyday activities; symbols make communication with others possible by providing shared meanings.
    D. According to postmodern theorists, no single perspective can grasp the complexity and diversity of the social world; we should speak of *cultures* rather than *culture*.
        1. **Jean Baudrillard** believes that culture today is based on *simulation* rather than reality, resulting in a state of *hyperreality*, where the simulation of reality becomes more real than the thing itself.
VI. CULTURE IN THE FUTURE
    A. In future decades, the issue of cultural diversity will increase in importance.
    B. Some of the most important changes in cultural patterns may include television and radio, films and videos, and electronic communications including computers and cyberspace, which will continue to accelerate the flow of information. However, most of the world's population will not participate in this technological revolution.

45

C. The study of culture not only helps us understand our own "tool kit" of symbols, stories, rituals, and worldviews, but also expands our insights to include those of other people of the world who seek strategies for enhancing their own quantity and quality of life.

CRITICAL THINKING QUESTIONS

1. In what ways is culture necessary for individual survival? In what ways is culture necessary for the continuation of society?
2. Which theoretical perspective do you think is most useful for understanding culture? Why? What are the strengths and weaknesses of each theory's analysis of culture?
3. Pierre Bourdieu's cultural capital theory argues that individuals have unequal access to high culture based on their social class. What kind of culture were you exposed to when you were growing up? What kind of culture do you think you would have been exposed to if you grew up in a different social class?
4. Do you think the malling of China is an example of cultural imperialism or cultural diffusion? What is the boundary between these two concepts?
5. In what ways do you think increasing cultural diversity, globalization, and advances in technology will change culture in the U.S. in the future? How will it change culture worldwide?

PRACTICE TESTS

MULTIPLE CHOICE QUESTIONS

Select the response that best answers the question or completes the statement.

1. _____ consists of knowledge, language, values, customs, and material objects.
   a. Social structure
   b. Society
   c. Culture
   d. Social organization

2. All of the following statements regarding culture are true, **except**:
   a. culture is essential for our survival
   b. culture is essential for our communications with other people
   c. culture is fundamental for the survival of societies
   d. culture reflects human instincts

3. In terms of their functions, cultural universals are
   a. useful because they ensure the smooth and continued operation of society.
   b. the result of attempts by a dominant group to impose its will on a subordinate group.
   c. independent from functional necessities.
   d. very similar in form from one group to another and from one time to another within the same group.

4. Most contemporary sociologists believe that _____ account for virtually all of our behavior patterns.
   a. drives
   b. culture and social learning
   c. personality characteristics
   d. instincts

5. Regarding the relationship between language and gender, the text points out that
   a. the pronouns *he* and *she* are seldom used in everyday conversation by most people.
   b. the English language largely has been purged of sexist connotations.
   c. the English language ignores women by using the masculine form to refer to human beings in general.
   d. words in the English language typically have positive connotations when relating to female power, prestige, and leadership.

6. From a(n) _____ perspective, a shared language is essential to a common culture.
   a. functionalist
   b. conflict
   c. feminist
   d. interactionist

7. Books, art, music, and buildings are examples of
   a. nonmaterial culture
   b. material culture
   c. folkways
   d. values

8. _____ is a set of symbols that expresses ideas and enables people to think and communicate with each other.
   a. Material culture
   b. Technology
   c. A folkway
   d. Language

9. All of the following are examples of U.S. folkways, **except**:
   a. using underarm deodorant
   b. brushing our teeth
   c. avoiding sexual relationships with siblings
   d. wearing appropriate clothing for specific occasions

10. The most common type of formal norms are:
    a. folkways
    b. mores
    c. sanctions
    d. laws

47

11. Sociological research suggests that racial and ethnic stereotypes and beliefs about the superiority of one category of people over another are perpetuated through
    a. language.
    b. instincts.
    c. cultural relativism.
    d. taboos.

12. When one part of a culture changes at a faster pace than another, _____ occurs.
    a. cultural diffusion
    b. cultural lag
    c. cultural shock
    d. cultural diversity

13. _____ is the process of reshaping existing cultural items into a new form.
    a. Discovery
    b. Diffusion
    c. Invention
    d. Restoration

14. The growing popularity of eating sushi in the U.S. is an example of:
    a. cultural diffusion
    b. cultural lag
    c. cultural relativism
    d. cultural imperialism

15. The disorientation that people feel when they encounter cultures radically different from their own is referred to as:
    a. cultural diffusion
    b. cultural relativism
    c. cultural disorientation
    d. culture shock

16. _____ is based on the assumption that one's own culture is superior to others.
    a. Ethnocentrism
    b. Cultural relativism
    c. Cultural diffusion
    d. Cultural indifference

17. American anthropologist Marvin Harris has pointed out that the Hindu taboo against killing cattle is very important to the economic system in India. This view exemplifies:
    a. ethnocentrism
    b. cultural relativism
    c. cultural diffusion
    d. cultural indifference

48

18. Some view the widespread infusion of the English language into non-English speaking countries as a form of
    a. cultural relativism.
    b. cultural imperialism.
    c. cultural diversity.
    d. cultural lag.

19. Which theoretical perspective suggests that ideas are used by agents of the ruling class to maintain their positions of dominance in a society?
    a. symbolic interactionist
    b. postmodernist
    c. conflict
    d. structural functionalist

20. Classical music, opera, and ballet are examples of
    a. popular culture
    b. high culture
    c. global culture
    d. subculture

21. An example of a(n) _____ is a spider building a web because of basic biological needs.
    a. instinct
    b. impulse
    c. drive
    d. reflex

22. A(n) _____ is an unlearned, biologically determined, involuntary response to some physical stimuli.
    a. instinct
    b. reflex
    c. impulse
    d. drive

23. Biologically determined impulses common to all members of a species that satisfy needs such as sleep, food, water, and sexual gratification are known as:
    a. instincts
    b. drives
    c. reflexes
    d. reactions

24. According to Pierre Bourdieu, individuals possess _____ when they appreciate and understand high culture.
    a. ideology
    b. cultural relativism
    c. cultural imperialism
    d. cultural capital

25. _____consists of the physical or tangible creations that members of a society make, use, and share.
    a. Technology
    b. Nonmaterial culture
    c. A cultural symbol
    d. Material culture

26. Sociologists define _____ as the knowledge, techniques, and tools that make it possible for people to transform resources into usable forms, and the knowledge and skills required to use them after they are developed.
    a. nonmaterial culture
    b. technology
    c. material culture
    d. normative culture

27. _____ consists of the abstract or intangible human creations of society that influence people's behavior.
    a. Nonmaterial culture
    b. Material culture
    c. A symbol
    d. Technology

28. A central component of _____ culture is _____, which are the mental acceptance or conviction that certain things are true or real.
    a. material; values
    b. nonmaterial; beliefs
    c. nonmaterial; values
    d. material; beliefs

29. Anthropologist George Murdock compiled a list of over seventy _____—customs and practices that occur across all societies.
    a. morals
    b. beliefs
    c. cultural universals
    d. symbols

30. _____ are shared ideas about what is right or wrong, good or bad, and desirable or undesirable in a particular culture.
    a. Norms
    b. Symbols
    c. Beliefs
    d. Values

31. A _____ is anything that meaningfully represents something else.
    a. symbol
    b. belief
    c. value
    d. norm

32. The _____ suggests that language not only expresses our thoughts and perceptions but also influences our perception of reality.
    a. Murdock-Williams hypothesis
    b. Sapir-Whorf hypothesis
    c. Ogburn-Mead hypothesis
    d. Sumner-Spencer hypothesis

33. Recent data gathered by the U.S. Census Bureau indicate that approximately _____ percent of the people in this country speak a language other than English at home.
    a. 5
    b. 10
    c. 15
    d. 20

34. _____ theorists view language as a source of power and social control; it perpetuates inequalities between people and groups because words are used to "keep people in their place."
    a. Symbolic interactionist
    b. Postmodern
    c. Structural functionalist
    d. Conflict

35. Postmodern theorist Jean Baudrillard argues that culture today is based on media simulations of reality that result in _____, or the perception that the simulation of reality is more real than the reality itself.
    a. hyperreality
    b. diffusion
    c. ideal culture
    d. real culture

36. According to sociologist Robin Williams, each of the following is a core value in U.S. culture **except:**
    a. individualism
    b. progress and material comfort
    c. racism and group superiority
    d. belief in God and Christianity

37. "The United States stands for equal opportunity for all." This statement illustrates _____ culture
    a. ideal
    b. real
    c. diverse
    d. universal

38. Jon, a college student, believes in the idea of success, but does not study or attend classes as regularly as he could in order to achieve a high grade point average. His behavior illustrates _____ culture.
   a. ideal
   b. real
   c. material
   d. universal

39. Chinese Americans in San Francisco, Korean Americans and Puerto Ricans in New York, and Mexican Americans in San Antonio are examples of _____.
   a. countercultures
   b. majority subcultures
   c. ethnic subcultures
   d. popular cultures

40. According to _____ theorists, much of what has been written around the world is Eurocentric.
   a. postmodern
   b. conflict
   c. symbolic interaction
   d. structural functional

TRUE/FALSE QUESTIONS

1. Culture is composed of people, whereas a society is composed of ideas.
   T   F

2. Culture is essential for individuals, but not as fundamental for the survival of societies.
   T   F

3. Humans have a number of basic instincts.
   T   F

4. Popular culture refers to activities, products, and services that are assumed to appeal to middle and working class people.
   T   F

5. The subject matter of jokes is a cultural universal.
   T   F

6. As a result of their significance, symbols can simultaneously produce loyalty and animosity.
   T   F

7. According to the Sapir-Whorf hypothesis, language shapes the view of reality of its speakers.
   T   F

8. Functionalist theorists view language as a source of power and social control that perpetuates inequalities in society.
   T   F

9. Some sociologists argue that the development of shopping malls with U.S. retailers in China is an example of cultural imperialism because it promotes the purchase and popularity of American cultural symbols.
   T   F

10. Unlike folkways, mores have strong moral and ethical connotations that may not be violated without serious consequences.
    T   F

11. Popular culture is the U.S.A.'s second largest export to other countries.
    T   F

12. Fads tend to be longer lasting and more widespread than fashions.
    T   F

13. An example of cultural imperialism would be if most world cultures became westernized.
    T   F

14. Chinatowns and other ethnic subcultures help first generation immigrants adapt to abrupt cultural change.
    T   F

15. Symbolic interactionist theorists view popular culture as an entity that holds society together.
    T   F

16. Terms such as "fireman," "airline stewardess," and "maid" are examples of gender-neutral terms.
    T   F

17. Sanctions are informal norms that may be violated without serious consequences within a particular culture.
    T   F

18. According to Box 3.1, round-shaped foods are believed to symbolize family unity in some cultures.
    T   F

19. Conflict theorists view language as a source of power and a means of social control.
    T   F

20. A society's core values must be cohesive and not contradict each other.
    T   F

FILL-IN-THE-BLANK QUESTIONS

1. The physical or tangible creations that members of a society make, use, and share are examples of _____.
2. A(n) _____ is anything that meaningfully represents something else.
3. The _____ hypothesis argues that language shapes the view of reality of its speakers.
4. Collective ideas about what is right or wrong, good or bad are known as _____.
5. _____ culture refers to the values that people profess to hold, while _____ culture refers to values and standards that people actually follow.
6. _____ sanctions are rewards for appropriate behavior.
7. Mores so strong that their violation is considered to be extremely offensive are known as _____.
8. Societies wherein people are dissimilar in "social characteristics" such as religion, income, or race/ethnicity are referred to as _____ societies.
9. A group that strongly rejects dominant societal values within a culture is known as a(n) _____.
10. A(n) _____ includes clothing, behavior, thinking, or appearance that is longer lasting and more widespread than a fad.
11. _____ culture consists of classical music, opera, ballet, and live theater usually patronized by the elite.
12. _____ culture in the United States is thought to be "home grown."
13. According to Pierre Bourdieu's _____ theory, high culture is a device used by the dominant class to exclude the subordinate class.
14. A(n) _____ is a category of people who share some distinguishing attribute, belief, value and/or norm that set them apart from the dominant culture.
15. According to Jean Baudrillard, the world of culture is based on _____, not reality itself.

SHORT ANSWER/ESSAY QUESTIONS

1. What are the major types of norms in our society? Give an example of each.
2. Define the following concepts and explain how each could reflect diversity within our society: (1) counterculture; (2) subculture; (3) nonmaterial culture; (4) fads; (5) fashions; and (6) symbols.
3. Explain the relationship between: (a) language and social reality; (b) language and gender; and (c) language, race, and ethnicity.
4. What is culture? How do sociologists distinguish between culture and society?
5. Which values of those discussed in this chapter seem to be the most important and enduring in the United States today? Which would seem to be the most important in the future?
6. What does the "lived experience" introduction at the beginning of this chapter say about the importance of food in becoming an American?

STUDENT CLASS PROJECTS AND ACTIVITIES

1. Examine the lyrics to ten popular songs that have to do with identity and the problems of teenagers in America. Write your responses to the following: (1) What aspects of culture do the lyrics represent or focus on? (2) What norms do the lyrics promote? (3) What are the themes of the lyrics? (4) What type of music accompanies the lyrics, and what are some of the impacts of the musical sounds? (5) Is the music geared toward or meant to represent a particular U.S. subculture? If so, which subculture? What does the music say about this subculture? (6) Overall, what does this music tell you about culture in the U.S.?

2. Write a paper on this issue: "Are nonindustrial cultures an 'endangered species' in the world today?" In this paper, discuss the localizing and globalizing forces of all cultures today. Examine why, in previously obscure cultures, the native cultures appear to be eroding and becoming acculturated into the norms of the industrialized world. Is this globalizing force welcomed by the few remaining aboriginal cultures? Why or why not? Is it functional for them? Why or why not? Select a specific culture where this process may or may not be occurring. For example, you could research the Australian Aborigine; the Yanomamo; some of the native groups in Venezuela and other South American cultures; the Mayan Indians of Mexico; the Bushmen in Africa; the Zulu in South Africa; the IK in Uganda, Kenya, and the Sudan; the Wajos, Bugis, and Makassarese of Sulawesi; the Navaho or Hopi in the United States; etc. Provide bibliographic references for your data and any other information that you may find relevant to your topic.

3. Write a paper on an event that is considered to be one of the top ten most important cultural events worldwide in the history of humankind. You can choose from one of the top ten of the following list, or you may choose another event, if you prefer. The top ten, according to the staff of Time Magazine: (1) the Crusades; (2) the signing of the Magna Carta; (3) the Black Death; (4) the Renaissance; (5) the discovery of the Americas; (6) the Reformation; (7) the Industrial Revolution; (8) the American and French Revolutions; (9) World War II; and (10) the rise and fall of Communism. Explain why you chose a specific event, and fully describe, discuss, and evaluate that event. Provide a historical background to that specific event and include in your discussion the reason this event was most influential in worldwide history. In your writing of this project, you must include bibliographic references.

4. Since the 9/11 terrorist attacks in the United States, the word *jihad* has sparked debates all over the world. Research the origin of the word, the definition of the word, and various uses of the word. Explain how some acts of world terrorism are related to the term and specifically how/why terrorism is directed against the United States and the Western world. Does the term evoke acts and thoughts of ethnocentrism? Explain. Next, conduct a survey of 20 to 25 people, either on your campus or in your community. Ask these questions: (1) What does *jihad* mean to you? (2) What do you think this word means to most people in the United States? (3) What do you think this word means, or implies, to Muslims in the United States? To Muslims in the Middle East and other parts of the world? (4) Provide a summary and analysis of your results.

INTERNET ACTIVITIES

1. To read about current issues in postmodern culture, go to the leading electronic journal of interdisciplinary thought on contemporary cultures, published by The Johns Hopkins University, **Postmodern Culture** (PMC), **http://www3.iath.virginia.edu/pmc/contents.all.html**, and click on the current issue. Explore the recently published articles. Select one to read, critique it, and share your report in class.
2. *The Voice of the Shuttle*, **http://vos.ucsb.edu/index-netscape.asp**, is one of the most extensive collections of cultural studies resources. This site designates the intersection between cultural criticism/theory and selective resources in sociology, media studies, postcolonial studies, economics, literature, and other fields chosen to represent the alignments that now signify "culture" for the contemporary humanities. This is a great research site for further information on cultural studies and will link you to a number of sociology sites that might be useful in future research projects.
3. Find out who lives in the *Global Village*: **http://www.empowermentresources.com/info2/theglobalvillage.html**. It answers the question, "What would the world look like if it contained just 1000 people?" The responses are indeed provocative.
4. **Cultural Studies Central** is the first site that you should explore if you are interested in cultural studies: **http://www.culturalstudies.net/**. It is an academic site where you can find journals, articles, papers, bibliographies, reading lists, theorists, and critics. This is a good site for resource material for papers and reports.

INFOTRAC COLLEGE EDITION EXERCISES

Visit the **InfoTrac College Edition** web site at**: http://www.wadsworthmedia.com/webtutor/infotrac.htm**. You will arrive at a screen that allows you to search topics.

1. Go to InfoTrac and look up the concept **social norms**. You should first view the encyclopedia excerpt. From the text of the definition, jump to the definition of other related concepts (blue colored links). Follow the related concepts and develop a string of at least five branching off of the social norms definition where you began. Cut and paste these encyclopedia definitions to create a single document that provides a broader perspective on social norms.
2. Discover the meaning of the concept **ethnocentrism** from reading its use in the periodical literature found on InfoTrac. Could you explain why this phenomenon develops in a group of sixth graders?
3. Look up the subject **technology culture** on InfoTrac. Scan several of the articles. Create a list of problems that appear to be related to technological aspects of culture.

SOLUTIONS

MULTIPLE CHOICE QUESTIONS

1. C, p. 72
2. D, p. 73
3. A, p. 76
4. B, p. 73-74
5. C, p. 78
6. A, p. 93
7. B, p. 75
8. D, p. 77
9. C, p. 83
10. D, p. 84
11. A, p. 79
12. B, p. 84
13. C, p. 85
14. A, p. 85

15. D, p. 89
16. A, p. 90
17. B, p. 90
18. B, p. 91
19. C, p. 93-95
20. B, p. 90
21. A, p. 73
22. B, p. 73
23. B, p. 74
24. D, p. 90-91
25. D, p. 75
26. B, p. 75
27. A, p. 75
28. B, p. 76

29. C, p. 76
30. D, p. 80
31. A, p. 77
32. B, p. 78
33. D, p. 80
34. D, p. 93-94
35. A, p. 96
36. D, p. 80-81
37. A, p. 82-83
38. B, p. 82-83
39. C, p. 88
40. A, p. 90

TRUE/FALSE QUESTIONS

1. F, p. 72
2. F, p. 73
3. F, p. 73
4. T, p. 90
5. F, p. 76
6. T, p. 77
7. T, p. 78

8. F, p. 77
9. T, p. 92
10. T, p. 83
11. T, p. 90
12. F, p. 91
13. T, p. 91
14. T, p. 88

15. F, p. 95-96
16. F, p. 78
17. F, p. 83
18. T, p. 73
19. T, p. 93
20. F, p. 82

FILL-IN-THE-BLANK QUESTIONS

1. material culture, p. 75
2. symbol, p. 77
3. Sapir-Whorf, p. 78
4. values, p. 80
5. Ideal, p. 82; real, p. 82
6. Positive, p. 83
7. taboos, p. 83
8. heterogeneous, p. 85
9. counterculture, p. 89
10. fashion, p. 91
11. High, p. 90
12. popular, p. 90
13. cultural capital, p. 90-91
14. subculture, p. 87
15. simulation, p. 96

57

4

SOCIALIZATION

BRIEF CHAPTER OUTLINE

CHAPTER SUMMARY

Socialization is the lifelong process through which individuals acquire a self-identity and the physical, mental, and social skills needed for survival in society. Socialization is essential for the individual's survival and for human development; it also is essential for the survival and stability of society. People are a product of two forces: heredity and social environment. Most sociologists agree that while biology dictates our physical makeup, the social environment largely determines how we develop and behave. Humans need social contact to develop properly. Cases of isolated children have shown that people who are isolated during their formative years fail to develop their full emotional and intellectual capacities, and that social contact is essential in developing a self. A variety of psychological and sociological theories have been developed to explain child abuse but also to describe how a positive process of socialization occurs. Psychological theories focus primarily on how the individual's personality develops. **Sigmund Freud** developed the psychoanalytic perspective, while **Erik Erikson**, drawing from Freud's theory, identified eight psychosocial stages of development. **Jean Piaget** pioneered the field of cognitive development, which emphasizes the intellectual development of children. The stages of moral development were developed by **Lawrence Kohlberg,** and later criticized and reformulated by **Carol Gilligan**. Sociological perspectives examine how people develop an awareness of self and learn about their culture. This **self-concept** is not present at birth; it arises in the process of social experience. **Charles Horton Cooley** developed the image of the **looking-glass self** to explain how people see themselves through their imagination of the perceptions of others. **George Herbert Mead** linked the idea of **self-concept** to **role-taking** and to learning the rules of social interaction. Ecological perspectives emphasize cultural or environmental influences on human development. When children do not have a positive environment in which to develop a self-concept, it becomes difficult to form a healthy social self. According to sociologists, **agents of socialization**—including families, schools, peer groups, and mass media—teach us what we need to know in order to participate in society. Gender, race-ethnicity, and social class are all determining factors in the lifelong socialization process. **Anticipatory socialization**—the process by which knowledge and skills are learned for future roles—often occurs before achieving a new status. **Resocialization**—the process of learning new attitudes, values, and behaviors, either voluntarily or involuntarily—sometimes takes place in **total institutions**. In today's global society, because of the rapid pace of technological change, we must learn how to anticipate and consider the consequences of the future.

LEARNING OBJECTIVES

After reading Chapter 4, you should be able to:

1. Define socialization and explain why this process is essential for the individual and society.

2. Explain why cases of isolated children are important to understanding of the socialization process.

3. Distinguish between sociological and sociobiological perspectives on the development of human behavior.

4. Explain Freud's views on the conflict between individual desires and the demands of society.

5. Describe the stages of psychosocial development proposed by Erikson.

6. Outline the stages of cognitive development as set forth by Piaget.

7. Describe Mead's concept of the generalized other and explain socialization as an interactive process.

8. Describe and be able to apply Cooley's concept of the looking-glass self.

9. Explain the stages of development according to Erikson's developmental theory.

10. Compare and contrast the moral development theories of Kohlberg and Gilligan.

11. State the major agents of socialization and describe their effects on children's development.

12. Explain what is meant by gender socialization and racial socialization.

13. Outline the stages of the life course and explain how each stage varies based on gender, race/ethnicity, class, and positive or negative treatment.

14. Discuss the issue of ageism and its impact on older individuals.

15. Describe the process of resocialization and explain why it often takes place in a total institution.

KEY TERMS

(defined at page number shown and in glossary)

ageism, p. 130
agents of socialization, p. 117
anticipatory socialization, p.128
ego, p. 109
gender socialization, p. 125
generalized other, p. 114
id, p. 109
looking-glass self, p. 113
peer group, p. 119
racial socialization, p. 127

resocialization, p. 131
role-taking, p. 113
self-concept, p. 113
significant other, p. 114
social devaluation, p. 130
socialization, p. 104
sociobiology, p. 105
superego, p. 109
total institution, p. 131

KEY PEOPLE

(identified at page number shown)

Urie Bronfenbrenner, p. 117
Charles Horton Cooley, p. 113
Erik H. Erikson, p. 109-110
Sigmund Freud, p. 109

Carol Gilligan, p. 112
Lawrence Kohlberg, p. 112
George Herbert Mead, p. 113
Jean Piaget, p. 110-111

CHAPTER OUTLINE

I. WHY IS SOCIALIZATION IMPORTANT AROUND THE GLOBE?
   A. **Socialization** is the lifelong process of social interaction through which individuals acquire a self-identity and the physical, mental, and social skills needed for survival in society.
   B. Human Development: Biology and Society
      1. Every human being is a product of biology, society, and personal experiences, or heredity and environment.
      2. **Sociobiology** is the systematic study of how biology affects social behavior.
   C. Problems Associated with Social Isolation and Maltreatment
      1. Social environment is a crucial part of an individual's socialization; people need social contact with others in order to develop properly.
      2. Research on nonhuman primates demonstrates the detrimental effects of social isolation.
      3. Isolated children illustrate the importance of socialization.
      4. The most frequent form of child maltreatment is child neglect.
II. SOCIAL PSYCHOLOGICAL THEORIES OF HUMAN DEVELOPMENT
   A. Freud's psychoanalytic perspective suggested that human behavior and personality originate from unconscious forces within individuals.
      1. The **id** is the component of personality that includes all of the individual's basic biological drives and needs that demand immediate gratification.
      2. The **ego** is the rational, reality-oriented component of personality that imposes restrictions on the innate pleasure seeking drives of the id.
      3. The **superego** consists of the moral and ethical aspects of personality. When a person is well-adjusted, the ego successfully manages the opposing forces of the id and the superego.
   B. **Erik H. Erikson** identified eight psychosocial stages of development, each of which is accompanied by a crisis or potential crisis that involves transitions in social relationships:
      1. Trust versus Mistrust (birth to age 1)
      2. Autonomy versus Shame and Doubt (1-3 years)
      3. Initiative versus Guilt (3-5 years)
      4. Industry versus Inferiority (6-11 years)
      5. Identity versus Role Confusion (12-18 years)
      6. Intimacy versus Isolation (18-35 years)
      7. Generativity versus Self-absorption (35-55 years)
      8. Integrity versus Despair (maturity and old age)
   C. **Jean Piaget**'s theory of cognitive development relates to changes over time in how people think. Piaget was interested in how children obtain, process, and use information. He believed that children experience four stages of cognitive development:
      1. *Sensorimotor Stage* (birth to age 2)—children understand the world only though sensory contact and immediate action because they cannot engage in symbolic thought or use language.

61

2. *Preoperational Stage* (ages 2-7)—children begin to use words as mental symbols and to develop the ability to use mental images.
3. *Concrete Operational Stage* (ages 7-11)—children think in terms of tangible objects and actual events; they also can draw conclusions about the likely physical consequences of an action without always having to try it out.
4. *Formal Operational Stage* (age 12 through adolescence)—adolescents are able to engage in highly abstract thought and understand places, things, and events they have never seen. Beyond this point, changes in thinking are a matter of degree rather than in the nature of their thinking.

D. **Lawrence Kohlberg** classified moral reasoning into three sequential levels:
1. *Preconventional Level* (ages 7-10)—children give little consideration to the views of others.
2. *Conventional Level* (age 10 through adulthood)—children initially believe that behavior is right if it receives wide approval from significant others, including peers, and then through a law and order orientation, based on how one conforms to rules and laws.
3. *Postconventional Level* (few adults reach this stage)—people view morality in terms of individual rights. At the final stage of moral development, "moral conduct" is judged by principles based on human rights that transcend government and laws.

E. Gilligan's View on Gender and Moral Development
1. One of the major critics of Kohlberg's work was psychologist **Carol Gilligan**, who noted that Kohlberg's model was based solely on male responses.
2. She noted that, because of differences in socialization and life experiences, men and women approach moral issues differently; men become more concerned with law and order, and women analyze social relationships and social consequences to behavior.
3. She argues that men are socialized to make moral decisions based on a justice perspective, whereas women are socialized to make moral decisions based on a care and responsibility perspective.
4. Her research identified three states in female moral development, which move from (1) selfish concerns, (2) concerns over responsibilities toward others, and (3) concern over doing the greatest good for both herself and others.

III. SOCIOLOGICAL THEORIES OF HUMAN DEVELOPMENT
A. Cooley, Mead, and Symbolic Interactionist Perspectives
1. Without social contact, we cannot form a **self-concept**—the totality of our beliefs and feelings about ourselves.
2. According to **Charles Horton Cooley's looking-glass self**, a person's sense of self is derived from the perceptions of others through a three-step process:
   a. We imagine how our personality and appearance will look to other people.
   b. We imagine how other people judge the appearance and personality that we think we present.
   c. We develop a self-concept.

62

3. **George Herbert Mead** linked the idea of self-concept to **role-taking**—the process by which a person mentally assumes the role of another person in order to understand the world from that person's point of view.
   a. **Significant others** are those persons whose care, affection, and approval are especially desired and who are most important in the development of the self; these individuals are extremely important in the socialization process.
   b. Mead divided the self into the "I"—the subjective element of the self that represents the spontaneous and unique traits of each person—and the "me"—the objective element of the self, which is composed of the internalized attitudes and demands of other members of society and the individual's awareness of those demands.
   c. Mead outlined three stages of self-development:
      i. *Preparatory Stage*—children largely imitate the people around them.
      ii. *Play Stage (*from about age 3 to 5)—children learn to use language and other symbols, thus making it possible for them to pretend to take the roles of specific people.
      iii. *Game Stage*—children understand not only their own social position but also the positions of others around them. At this time, the child develops a **generalized other**—an awareness of the demands and expectations of the society as a whole or of the child's subculture.
4. Some recent symbolic interactionist research emphasizes that socialization is a collective process in which children are active and creative agents, not passive recipients in the socialization process.
   a. Children are capable of actively constructing their own shared meanings as they acquire language skills and accumulate interactive experiences.
   b. Some researchers believe that the peer group is the most significant public realm for children.
5. Ecological perspectives emphasize cultural or environmental influences on human development.
   a. *Ecological systems theory*, set forth by Urie Bronfenbrenner, focuses on the interactions a child has with other people and situations.
   b. The ecological systems are as follows: (1) the *microsystem*, (2) the *mesosystem*, (3) the *exosystem*, and (4) the *macrosystem*.

IV. AGENTS OF SOCIALIZATION
   A. **Agents of socialization** are the persons, groups, or institutions that teach us what we need to know in order to participate in society. These are the most pervasive agents of socialization in childhood.
   B. The family is the most important agent of socialization in all societies.
      1. Functionalists emphasize that families are the primary agents for the procreation and socialization of children, as well as the primary source of emotional support.
      2. Conflict theorists stress that socialization reproduces the class structure in the next generation.

3. The social construction/symbolic interactionist perspective suggests that children affect their parents' lives and change the overall household environment.

C. The school has played an increasingly important role in the socialization process as the amount of specialized technical and scientific knowledge has expanded rapidly.
   1. Schools teach specific knowledge and skills; they also have a profound effect on a child's self-image, beliefs, and values.
   2. From a functionalist perspective, schools are responsible for: (1) **socialization**—teaching students to be productive members of society; (2) transmission of culture; (3) social control and personal development; and (4) the selection, training, and placement of individuals on different rungs in the society.
   3. According to conflict theorists such as Samuel Bowles and Herbert Gintis, much of what happens in school amounts to a hidden curriculum—the process by which children from working-class and lower-income families learn to be neat, to be on time, to be quiet, to wait their turn, and to remain attentive to their work—attributes that are important for later roles in the work force.
   4. Symbolic interactionists might focus on how daily interactions and practices in school affect the construction of students' beliefs, practices, and values.

D. A **peer group** is a group of people who are linked by common interests, equal social position, and (usually) similar age.
   1. Peer groups function as agents of socialization by contributing to our sense of "belonging" and our feelings of self-worth.
   2. Individuals must earn their acceptance with their peers by meeting the group's demands for a high level of conformity to its norms, attitudes, speech, and dress codes.

E. The *mass media* is an agent of socialization that has a profound impact on both children and adults.
   1. The media function as socializing agents in several ways: (1) they inform us about events; (2) they introduce us to a wide variety of people; (3) they provide an array of viewpoints on current issues; (4) they make us aware of products and services that, if we purchase them, supposedly will help us to be accepted by others; and (5) they entertain us by providing the opportunity to live vicariously through other people's experiences.
   2. Recent studies have shown that, on average, U.S. children spend more time per year in front of television sets, computers, and video games than they spend per year in school.
      a. Television has been praised for offering children numerous positive, educational, and prosocial experiences; but blamed for children having lower grades in school, reading fewer books, exercising less, and becoming overweight.
      b. The influence of the Internet on young people continues to be a concern; young people make up the fastest-growing segment of Internet users.

64

     c. Cultural studies scholars and some postmodern theorists believe that "media culture" has in recent years dramatically changed the sociological process for very young children.

V. GENDER AND RACIAL-ETHNIC SOCIALIZATION

  A. **Gender socialization** is the aspect of socialization that contains specific messages and practices concerning the nature of being female or male in a specific group or society.

    1. Families, schools, and sports tend to reinforce traditional roles through gender socialization.

    2. From an early age, media, including children's books, television programs, movies, and music provide subtle and not-so-subtle messages about masculine and feminine behavior.

  B. **Racial socialization** is the aspect of socialization that contains specific messages and practices concerning the nature of one's racial or ethnic status as it relates to: (1) personal and group identity, (2) intergroup and interindividual relationships, and (3) position in the social hierarchy.

VI. SOCIALIZATION THROUGH THE LIFE COURSE

  A. Socialization is a lifelong process: each time we experience a change in status, we learn a new set of rules, roles, and relationships.

    1. Even before we enter a new status, we often participate in **anticipatory socialization**—the process by which knowledge and skills are learned for future roles.

    2. The most common categories of age are infancy, childhood, adolescence, and adulthood (often subdivided into young adulthood, middle adulthood, and older adulthood).

  B. During infancy and early childhood, family support and guidance are crucial to a child's developing self-concept; however, some families reflect the discrepancy between cultural ideals and reality—children grow up in a setting characterized by fear, danger, and risks that are created by parental neglect, emotional maltreatment, or premature economic and sexual demands.

  C. Anticipatory socialization for adult roles is often associated with adolescence; however, some young people may plunge into adult responsibilities at this time.

  D. In early adulthood (usually until about age 40), people work toward their own goals of creating meaningful relationships with others, finding employment, and seeking personal fulfillment. Wilbert Moore divided workplace, or occupational, socialization into four phases:

    1. career choice

    2. anticipatory socialization

    3. conditioning and commitment

    4. continuous commitment

  E. Between the ages of 40 and 65, people enter middle adulthood, and many begin to compare their accomplishments with their earlier expectations.

  F. Late adulthood can be divided into three categories: the "young old" (65-74), the "old-old" (75-85), and the "oldest-old" (86 and older).

    1. In late adulthood, some people are quite happy while others are not.

2. Difficult changes in adult attitudes and behavior may occur in the last years of life when people experience decreased physical ability and **social devaluation**—when a person or group is considered to have less social value than other groups.
3. Negative images of older people reinforce **ageism**, defined as prejudice and discrimination against people on the basis of age.

G. It is important to note that not everyone goes through certain passages or stages of a life course at the same age and that life course patterns are strongly influenced by race, ethnicity, gender, and social class.

VII. RESOCIALIZATION

A. **Resocialization** is the process of learning a new and different set of attitudes, values, and behaviors from those in one's previous background and experience.
1. *Voluntary resocialization* occurs when we enter a new status of our own free will (e.g., medical or psychological treatment or religious conversion).
2. *Involuntary resocialization* occurs against a person's wishes and generally takes place within a **total institution**—a place where people are isolated from the rest of society for a set period of time and come under the control of the officials who run the institution. Examples include military boot camps, prisons, concentration camps, and some mental hospitals.

VIII. SOCIALIZATION IN THE FUTURE

A. Families are likely to remain the institution that most fundamentally shapes and nurtures personal values and self-identity.
B. However, parents increasingly may feel overburdened by this responsibility, especially without societal support—such as high-quality, affordable child care—and more education in parenting skills.
C. A central issue facing parents and teachers as they socialize children is the growing dominance of the media and other forms of technology.
D. With the rapid pace of technological change in the socialization process, we must anticipate—and consider the consequences of—the future.

CRITICAL THINKING QUESTIONS

1. Why is socialization important for individual survival and development? Why is socialization important for the survival and stability of society?
2. Which theory of human development discussed in the text do you think is most accurate? Why? What are the strengths and weaknesses of each theory?
3. Carol Gilligan critiqued Lawrence Kohlberg's theory of moral development for being gender biased. Describe Kohlberg's theory and Gilligan's critique. Which theory do you think is more accurate? Why? Relate your conclusion to gender socialization.
4. How do you think your life would be different if your home environment was so unpleasant that you decided to run away from home? How do you think the experiences of homeless teens differ in different countries? What kinds of policies and programs do you think could reduce the incidence of teen runaways and homelessness in the U.S. and other countries?

5. Your textbook analyzes the family, school, peer groups, and mass media as the most pervasive agents of socialization in childhood. What other agents of socialization can you think of that are important during childhood? What did you learn as a child from these agents of socialization?

6. Think about the gender and racial socialization you received as a child. In what ways do you think your gender impacted your racial socialization? In what ways did your race impact your gender socialization? In what ways did your social class impact your gender and racial socialization?

PRACTICE TESTS

MULTIPLE CHOICE QUESTIONS

Select the response that best answers the question or completes the statement.

1. _____ is the lifelong process of social interaction through which individuals acquire a self-identity.
   a. Human development
   b. Socialization
   c. Behavior modification
   d. Imitation

2. The systematic study of how biology affects social behavior is known as:
   a. sociophysiology
   b. sociobiology
   c. sociology
   d. social psychology

3. Harry and Margaret Harlow's experiment with rhesus monkeys demonstrated that
   a. food was more important to the monkeys than warmth, affection, and physical comfort.
   b. monkeys cannot distinguish between a nonliving "mother substitute" and their own mother.
   c. socialization is not important to rhesus monkeys; their behavior is purely instinctive.
   d. without socialization, young monkeys do not learn normal social or emotional behavior.

4. In discussing child maltreatment, the text points out that
   a. many types of neglect constitute child maltreatment.
   b. the extent of this problem has been exaggerated by the media.
   c. most sexual abuse perpetrators are punished by imprisonment.
   d. most child maltreatment occurs in families living below the poverty line.

5. According to Sigmund Freud, the _____ consists of the moral and ethical aspects of personality.
   a. id
   b. ego
   c. superego
   d. libido

6. Which of the following is **not** one of Jean Piaget's stages of cognitive development?
   a. preoperational
   b. concrete operational
   c. formal operational
   d. post operational

7. The stages of moral development were initially set forth by _____ and then criticized by _____.
   a. Charles Horton Cooley, George Herbert Mead
   b. Lawrence Kohlberg, Carol Gilligan
   c. Jean Piaget, Lawrence Kohlberg
   d. Erik Erikson, Jean Piaget

8. According to Erik Erikson's psychosocial stage of development, old age is characterized by:
   a. trust versus mistrust
   b. intimacy versus isolation
   c. integrity versus despair
   d. identity versus role confusion

9. The theories of Charles Horton Cooley and George Herbert Mead can best be classified as _____ perspectives.
   a. symbolic interactionist
   b. functionalist
   c. conflict
   d. feminist

10. According to Charles Horton Cooley, we base our perception of who we are on how we think other people see and evaluate us. He referred to this perspective as the:
    a. self-fulfilling prophecy
    b. generalized other
    c. looking-glass self
    d. significant other

11. All of the following are stages in Mead's theory of self-development, **except** the:
    a. anticipatory stage
    b. game stage
    c. play stage
    d. preparatory stage

12. The _____ refers to the child's awareness of the demands and expectations of society as a whole or of the child's subculture.
    a. looking-glass self
    b. id
    c. ego
    d. generalized other

13. According to Urie Bronfenbrenner's ecological systems theory, all of the following are ecological systems **except**:
    a. ecosystem
    b. microsystem
    c. mesosystem
    d. exosystem

14. Sociologist Melvin Kohn has suggested that _____ is one of the strongest influences on what and how parents teach their children.
    a. race/ethnicity
    b. religion
    c. social class
    d. age

15. Currently, about _____ percent of all U.S. preschool children are in day care of one kind or another.
    a. 15
    b. 25
    c. 60
    d. 75

16. As agents of socialization, peer groups are thought to "pressure" children and adolescents because
    a. individualism is encouraged and rewarded in these groups.
    b. individuals must earn their acceptance with their peers by conforming to the group's norms.
    c. individuals are encouraged to put friendship above material possessions.
    d. individuals are discouraged from making long-term friends in the peer group.

17. According to Patricia Hill Collins, _____ play an important role in the gender socialization and motivation of African American children, especially girls.
    a. "other mothers"
    b. substitute mothers
    c. aunts
    d. siblings

18. The process by which knowledge and skills are learned for future roles is known as:
    a. resocialization
    b. anticipatory socialization
    c. cybersocialization
    d. expectant socialization

19. Social devaluation is most likely to be experienced during this stage of the life course:
    a. infancy and childhood
    b. adolescence
    c. early adulthood
    d. older adulthood

20. All of the following are examples of voluntary resocialization, **except**:
    a. becoming a student
    b. going to prison
    c. becoming a Buddhist
    d. joining Alcoholics Anonymous

21. Our _____ is our perception about what kind of person we are.
    a. psychological self
    b. looking-glass self
    c. me
    d. self-identity

22. All of the following statements regarding socialization are TRUE, **except:**
    a. Socialization is essential for the individual's survival and for human development.
    b. Socialization is essential for the survival and stability of society.
    c. Socialization enables a society to "reproduce" itself by passing on cultural content from one generation to the next.
    d. Socialization is a learning process that takes place primarily during childhood.

23. _____ is considered to be the pioneer of sociobiology.
    a. Sigmund Freud
    b. Edward O. Wilson
    c. Kingsley David
    d. Urie Bronfenbrenner

24. Sociologist Kingsley David was interested in the case of Anna, a child who was kept in an attic-like room, because
    a. he was studying parental attitudes of young, unmarried mothers.
    b. he wanted to know more about what happens when a child is raised in isolation.
    c. he was attempting to determine how children develop a generalized other.
    d. he was examining the relationship between sexual motives and human behavior.

25. The case of Genie, an isolated child, illustrates that
    a. with proper therapy, children who have been isolated can become a part of the mainstream.
    b. children who have experienced extreme isolation do not live long enough to reach adulthood.
    c. children who experience social isolation and neglect may be unable to function independently in society.
    d. isolated children can recover from any physical damages.

26. Drawing from Freud's theory, _____ identified eight psychosocial stages of development, reasoning that each stage is accompanied by a crisis or potential crisis that involves transitions in social relationships.
    a. Erik H. Erikson
    b. Charles Horton Cooley
    c. Jean Piaget
    d. George Herbert Mead

27. In Erik Erikson's theory, during the _____ stage of psychosocial development, children gain a feeling of control over their behavior, develop a variety of physical and mental abilities, and begin to assert their independence.
    a. trust versus mistrust
    b. autonomy versus shame and doubt
    c. initiative versus guilt
    d. industry versus inferiority

28. According to Jean Piaget's cognitive development theory, children think in terms of tangible objects and actual events during the _____ stage.
    a. formal operational
    b. preoperational
    c. concrete operational
    d. sensorimotor

29. According to Carol Gilligan, the key weakness of Lawrence Kohlberg's work is that it
    a. underestimates human potential for immorality.
    b. was based on only male subjects.
    c. ignores key social psychological insights.
    d. overemphasizes the subconscious mind.

30. The sum total of perceptions and feelings that an individual has of being a distinct, unique person represents the:
    a. personality
    b. psyche
    c. self
    d. individual orientation

71

31. According to the concept of the looking-glass self, our looking-glass self is
    a. who we actually are.
    b. based on our own perception of self.
    c. based on how we think people see us.
    d. based on our correct perceptions.

32. The "I" is the _____ element of self.
    a. subjective
    b. reflective
    c. objective
    d. conjective

33. _____ refers to the process by which a person mentally assumes the role of another person in order to understand the world from that person's point of view.
    a. Role exploration
    b. Role assumption
    c. Role-searching
    d. Role-taking

34. Urie Bronfenbrenner's _____ theory focuses on the overall context in which child development occurs.
    a. conflict
    b. postmodern
    c. ecological systems
    d. symbolic interaction

35. The theorist who created the psychoanalytic perspective of human development is:
    a. Sigmund Freud
    b. Erik Erikson
    c. Jean Piaget
    d. Lawrence Kohlberg

36. According to George Herbert Mead, when children between the ages of 3 and 5 play "house" or "firefighter," they are in the _____ stage of self-development.
    a. game
    b. play
    c. preparatory
    d. me

37. According to Mead, "selves"
    a. are present at birth.
    b. reflect only significant others.
    c. can never change.
    d. can only exist in relation to other selves.

72

38. Research on the influence of television on children has concluded that children who spend considerable time watching television often
    a. have lower grades in school.
    b. read fewer books.
    c. exercise less.
    d. all of the above

39. Sociologists use the concept of _____ to refer to the aspect of socialization that contains specific messages and practices concerning the nature of being female or male in a specific group or society.
    a. sex role socialization
    b. gender socialization
    c. feminism and masculine
    d. peer group socialization

40. The most important aspects of one's racial identity are
    a. influenced by the media.
    b. influenced by one's peers.
    c. influenced by one's neighbors.
    d. influenced by one's family.

TRUE/FALSE QUESTIONS

1. Around the globe, the process of socialization is essential for the survival of both the society and the individual.
   T   F

2. According to Lawrence Kohlberg's theory of moral development, the conventional level is characterized by concern over the perceptions of peers and conforming to the rules.
   T   F

3. Unlike humans, nonhuman primates such as monkeys and chimpanzees do not need social contact with others of their species in order to develop properly.
   T   F

4. Agents of socialization are the people, groups, and institutions that teach us what we need to know to participate in society.
   T   F

5. The cases of "Anna" and "Genie" make us aware of the importance of the socialization process because they show the detrimental effects of social isolation and neglect.
   T   F

6. Sigmund Freud theorized that our personalities are largely unconscious—hidden away outside our normal awareness.
   T   F

7. Erik Erikson suggested that if infants receive good care and nurturing from their parents, they will develop a sense of trust that is important to their future psychosocial development.
   T   F

8. According to George H. Mead, our looking-glass self is based on our perception of how other people think of us.
   T   F

9. According to Cooley, the self develops through social interactions with others.
   T   F

10. The family is the most important agent of socialization in all societies.
    T   F

11. Conflict theorists assert that students have different experiences in the school system depending on their class, racial/ethnic background, gender, and the neighborhood in which they live.
    T   F

12. Peer pressure does not affect preschool children.
    T   F

13. Gender socialization contributes to people's beliefs about what the "preferred" sex of a child should be and influences our beliefs about acceptable behaviors for males and females.
    T   F

14. Scholars have found that ethnic values begin to crystallize among children at about the age of adolescence.
    T   F

15. Occupational socialization tends to be most intense immediately after a person makes the transition from school to the workplace.
    T   F

16. Military boot camps, jails, and prisons are all examples of total institutions.
    T   F

17. According to Box 4.1, all states have reporting requirements for child maltreatment.
    T   F

18. According to Box 4.2, paying for child care is a large concern in most other high-income countries, as in the United States.
    T   F

19. According to Urie Bronfenbrenner's ecological systems theory, the exosystem involves the impact of public policy, such as child care and health care legislation, on the child.
    T   F

74

20. Negative images of older people reinforce prejudice and discrimination against them, known as ageism.
    T   F

FILL-IN-THE-BLANK QUESTIONS

1. The lifelong process of social interaction through which individuals acquire a self-identity is defined as _____.

2. According to Piaget's theory of cognitive development, children begin to use words as symbols and form mental images during the _____ stage.

3. According to Freud, the _____ is made up of the individual's basic biological drives and needs.

4. The _____ is the totality of our beliefs and feelings about ourselves.

5. According to Cooley, the _____ refers to the way in which a person's sense of self is derived from the perceptions of others.

6. According to Mead, the _____ refers to the child's awareness of the demands and expressions of the society as a whole or of the child's subculture.

7. A _____ group is a group of people who are linked by common interest, equal social position, and similar age.

8. Gender _____ is important in determining what we think the "preferred" sex of a child should be and our ideas about acceptable behaviors for females and males.

9. The process by which knowledge and skills are learned for future roles is termed _____.

10. The process wherein a person or group is considered to have less social value than other persons or groups is known as _____.

11. _____ are the persons, groups, or institutions that teach us what we need to know in order to participate in society.

12. A(n) _____ is a place where people are isolated from the rest of society for a set period of time and are under the control of the officials who run the institution.

13. _____ are those persons whose care, affection and approval are especially desired and who are most important in the development of the self.

14. _____ is the aspect of socialization that contains specific messages and practices concerning the nature of one's racial or ethnic status.

15. _____ is the process of learning a new and different set of attitudes, values, and behaviors from those in one's background and previous experiences.

SHORT ANSWER/ESSAY QUESTIONS

1. Explain the significance of socialization to human development. In your answer, explain the relevance of cases of isolated children and research with nonhuman primates.
2. Describe the differences between the sociological theories of human development and the psychological theories of human development. Provide examples of each in your answer.
3. How do we develop a self-concept according to sociologists?
4. What is Lawrence Kohlberg's perspective on moral development, and what is Carol Gilligan's criticism of his perspective?
5. Describe each of the stages of socialization over the life course and the main developmental tasks associated with each stage.

STUDENT CLASS PROJECTS AND ACTIVITIES

1. Think back to your childhood and list the four or five televisions programs that you watched most often. Next, list the four or five programs you watch most often today. Respond in writing to these questions: (1) What values were demonstrated in each of the programs? (2) Were there any values presented that contradict your parents' values? If so, what were/are they? (3) What kinds of images do the programs portray of people of different races, genders, and social classes? (4) What similarities do you see between the programs you watched as a child and the programs you watch now? What differences do you see? (5) How do you personally evaluate each of these television programs? Include any other information pertinent to this project.
2. Write a five to ten page biography of your life, with special emphasis on your own socialization processes. Include responses to the following questions: (1) What were the major agents of socialization in your life; (2) Who, specifically (by name and status) were the "significant others" in your life? (3) What type of values did your significant others advocate? (4) In what ways did your significant others shape your self-image and goals? Provide any additional information in the writing of your biography.

INTERNET ACTIVITIES

1. To gain more insight on two of the most influential sociologists and their concepts of "the self"—Charles Horton Cooley's concept of the "looking-glass self" and George Herbert Mead's concepts of the "genesis of self," the "I," and the "Me"—use the **Google** search engine: **http://www.google.com**. Type in the name of Charles Horton Cooley first; select the section on the collection of his articles and books. Note in Cooley's own words, "the looking-glass self" theory. Then select the biography of Charles Horton Cooley to more fully understand the major societal forces that shaped his life. Next, using the same search engine, type in the name George Herbert Mead. Surf the many sites available on Mead, ranging from the various biographies to his major writings. Note the impact of the works of Cooley and

76

John Dewey on Mead's ideas. Click on the section listing Mead's collection of articles and read about his concept of the "genesis of self." Also focus on Mead's theory of role-taking and the process by which the self is constructed and refined.

2. Michael Kearl's fabulous **sociology web gateway**, **http://www.trinity.edu/mkearl/fam-kids.html**, addresses issues related to parents and children. There are connections to pages that directly address socialization. This is an excellent resource when you are looking for more information for a paper/presentation or want to bring additional facts to a class discussion.

3. An enormous sociological gateway site, the **SOCIOSITE**, **www.sociosite.net** , has a topical index that will introduce you to dozens of subjects. Once you access this address, use the topical index and search for sites in the social psychology category. Most of the web sites in this category are resource or gateway sites. You can construct a *Social Psychology Web Site Fact Sheet*. Related web sites can be listed, categorized, and summarized. Pay special attention to those sites related to aspects of socialization.

4. The **Mass Media & Society** web site, **http://www.public.asu.edu/~zeyno217/365/notes1.html**, is a concise presentation of some of the most important concepts of mass media and society. Read the section on "Mass Media and Socialization." How does the media influence the "Sociological Imagination"? Why is "television one of the primary socializing agents of today's society"?

INFOTRAC COLLEGE EDITION EXERCISES

Visit the **InfoTrac College Edition** web site at: **http://www.wadsworthmedia.com/webtutor/infotrac.htm**. You will arrive at a screen that enables you to search topics.

1. Bring to class an article that addresses a subcategory of socialization.

Gender	Racial/Ethnic	Cross-cultural	Family	School
Media	Life Course	Peers	Self concept	Role taking

2. You can use InfoTrac to conduct a search about **Carol Gilligan**, a pioneer in the field of feminist-oriented developmental psychology. Articles include several reviews of her book, *Between Voice and Silence: Women and Girls, Race and Relationship*. Compare the reviews from a number of divergent sources.

3. Use InfoTrac to locate biographical sketches of the major psychologists and sociologists introduced in this chapter. InfoTrac provides access to encyclopedia excerpts containing biographical information. Put together a one-page sketch of each theorist. Portraits can be exported from the Internet and inserted in the document. If you are taking psychology and sociology courses in the future, resources such as these might prove useful.

SOLUTIONS

MULTIPLE CHOICE QUESTIONS

1. B, p. 104	15. C, p. 119	29. B, p. 112
2. B, p. 105	16. B, p. 119	30. C, p. 113
3. D, p. 106	17. A, p. 126	31. C, p. 113
4. A, p. 108	18. B, p. 128	32. A, p. 113
5. C, p. 109	19. D, p. 130	33. D, p. 113
6. D, pp. 110-111	20. B, p. 131	34. C, p. 117
7. B, p. 112	21. D, p. 113	35. A, p. 109
8. C, p. 109-110	22. D, p. 104	36. B, p. 113
9. A, p. 113	23. B, p. 105	37. D, p. 113
10. C, p. 113	24. B, p. 107	38. D, p. 125
11. A, p. 113	25. C, p. 107	39. B, p. 125
12. D, p. 113	26. A, p. 109	40. D, p. 127
13. A, p. 117	27. B, pp. 109-110	
14. C, p. 118	28. C, pp. 110-111	

TRUE/FALSE QUESTIONS

1. T, p. 104	8. F, p. 113	15. T, p. 130
2. T, p. 112	9. T, p. 113	16. T, p. 131
3. F, pp. 105-106	10. T, p. 117	17. T, p. 105
4. T, p. 117	11. T, p. 119	18. F, p. 120
5. T, p. 107	12. F, p. 119	19. F, p. 117
6. T, p. 109	13. T, p. 125	20. T, p. 130
7. T, pp. 109-110	14. F, p. 127	

FILL-IN-THE-BLANK QUESTIONS

1. socialization, p. 104
2. preoperational, p. 110-111
3. id, p. 109
4. self-concept, p. 113
5. looking-glass self, p. 113
6. generalized other, p. 114
7. peer, p. 119
8. socialization, p. 104
9. anticipatory socialization, p. 128
10. social devaluation, p. 130
11. Agents of socialization, p. 117
12. total institution, p. 131
13. Significant others, p. 114
14. Racial socialization, p. 127
15. Resocialization, p. 131

5
SOCIETY, SOCIAL STRUCTURE, AND INTERACTION

BRIEF CHAPTER OUTLINE

SOCIAL STRUCTURE: THE MACROLEVEL PERSPECTIVE
COMPONENTS OF SOCIAL STRUCTURE
    Status
    Role
    Groups
    Social Institutions
SOCIETIES, TECHNOLOGY, AND SOCIOCULTURAL CHANGE
    Hunting and Gathering Societies
    Horticultural and Pastoral Societies
    Agrarian Societies
    Industrial Societies
    Postindustrial Societies
STABILITY AND CHANGE IN SOCIETIES
    Durkheim: Mechanical and Organic Solidarity
    Tönnies: *Gemeinschaft* and *Gesellschaft*
    Social Structure and Homelessness
SOCIAL INTERACTION: THE MICROLEVEL PERSPECTIVE
    Social Interaction and Meaning
    The Social Construction of Reality
    Ethnomethodology
    Dramaturgical Analysis
    The Sociology of Emotions
    Nonverbal Communications
FUTURE CHANGES IN SOCIETY, SOCIAL STRUCTURE, AND INTERACTION

CHAPTER SUMMARY

Social structure and interaction are critical components of everyday life. At the microlevel, **social interaction**—the process by which people act toward or respond to other people—is the foundation of meaningful relationships in society. At the macrolevel, **social structure** is the stable pattern of social relationships that exist within a particular group or society. This structure includes **status**, **roles**, **groups**, and **social institutions**. Social researchers have identified five types of societies based on various levels of subsistence technology: **hunting and gathering**, **horticultural**, and **pastoral**, **agrarian**, **industrial**, and **postindustrial**. Sociologists have theorized the impact of changes in social structure on social solidarity, such as Durkheim's concepts of **mechanical** and **organic solidarity** and Tönnies' *Gemeinschaft* and *Gesellschaft*. In examining the relationship between social structure and homelessness, several suggestions for homelessness are offered; however the answers derived are often based upon perceptions of reality in a society. Social interaction within a society is guided by certain shared meanings across situations. The microlevel perspective focuses on the social interactions among individuals, especially face-to-face encounters. Not everyone interprets social interaction situations in the same way. Race/ethnicity, gender, and social class often influence perceptions of meaning. The **social construction of reality** refers to the process by which our perception of reality is shaped by the subjective meaning we give to an experience. **Ethnomethodology** is the study of the commonsense knowledge people use to understand the situations in which they find themselves. **Dramaturgical analysis** is the study of social interaction that compares everyday life to a theatrical presentation. **Presentation of self** refers to efforts to present our own self to others in ways that are most favorable to our interests or image. *Feeling rules* shape the appropriate emotions for a given role or specific situation. Social interaction also is marked by **nonverbal communication**, which is the transfer of information between people without the use of speech. Macrolevel and microlevel analyses are essential in determining how our social structures should be shaped in the future so that they can respond to pressing social needs.

LEARNING OBJECTIVES

After reading Chapter 5, you should be able to:

1. State the definition of social structure and explain why it is important for individuals and society.

2. List and explain each of the components of social structure.

3. State the definition of status, distinguish between ascribed and achieved statuses, and explain the relationship between the two.

4. Explain what is meant by master status and provide examples.

5. Define role expectation, role performance, role conflict, role strain, and role exiting, and give an example of each.

6. Explain the significance of groups and distinguish between primary and secondary groups.

7. State the definition for social institution and name the major institutions found in contemporary society.

8. Evaluate functionalist and conflict perspectives on the nature and purpose of social institutions.

9. Define formal organization and explain why many contemporary organizations are known as "people-processing" organizations.

10. Identify and describe the five major types of societies.

11. Compare Emile Durkheim's typology of mechanical and organic solidarity with Ferdinand Tönnies' G*emeinschaft* and *Gesellschaft*.

12. Describe and provide examples of the social construction of reality.

13. Describe ethnomethodology and note its strengths and weaknesses.

14. Describe Goffman's dramaturgical analysis and explain what he meant by presentation of self.

15. Explain what is meant by the sociology of emotions and describe sociologist Arlie Hochschild's contribution to this area of study.

16. Define nonverbal communication and personal space and explain how these concepts relate to our interactions with others.

17. Summarize the most significant ways that homelessness is experienced.

KEY TERMS

(defined at page number shown and in glossary)

achieved status, p. 140
agrarian societies, p.150
ascribed status, p. 140
dramaturgical analysis, p. 160
ethnomethodology, p. 159
formal organization, p. 147
Gemeinschaft, p. 155
Gesellschaft, p. 155
horticultural societies, p. 150
hunting and gathering
   societies, p. 149
impression management
   (presentation of self) , p. 160
industrial societies, p.151
master status, p. 141
mechanical solidarity, p. 154
nonverbal communication, p. 165
organic solidarity, p. 154
pastoral societies, p. 150

personal space, p. 166
postindustrial societies, p. 152
primary group, p. 146
role, p. 144
role conflict, p. 144
role exit, p. 146
role expectation, p. 144
role performance, p. 144
role strain, p. 145
secondary group, p. 147
self-fulfilling prophecy, p. 158
social construction of reality, p. 158
social group, p. 147
social institution, p. 147
social interaction, p. 137
social structure, p. 138
status, p. 139
status symbol, p. 142

KEY PEOPLE

(identified at page number shown)

Emile Durkheim, p. 154
Harold Garfinkel, p. 159
Erving Goffman, p. 160

Arlie Hochschild, p. 161-163
Gerhard Lenski and Jean Lenski, p. 149
Ferdinand Tönnies, p. 155

CHAPTER OUTLINE

I. SOCIAL STRUCTURE: THE MACROLEVEL PERSPECTIVE
   A. **Social interaction** is the process of how people act toward or respond to other people.
   B. **Social structure** is the complex framework of social institutions (such as the economy) and social practices (such as rules and social roles) that make up a society and organize and establish limits on people's behavior.
      1. This structure is essential for the survival of society and for the well-being of individuals.
      2. This structure provides for a social web of familial support and social relationships that connects each person to the larger society.
      3. Functionalists believe social structure is important because it creates order and stability.
      4. Conflict theorists believe social structure reflects a system of domination in society and that deeper, underlying structures must be analyzed.
   C. Social structure creates boundaries that define which persons or groups will be the "insiders" and which will be the "outsiders."
      1. *Social marginality* is the state of being part insider and part outsider in the social structure. Social marginality results in stigmatization.
      2. A *stigma* is any physical or social attribute or sign that so devalues a person's social identity that it disqualifies that person from full social acceptance.
II. COMPONENTS OF SOCIAL STRUCTURE
   A. A **status** is a socially defined position in a group or society characterized by certain expectations, rights, and duties.
      1. Ascribed and achieved statuses:
         a. An **ascribed status** is a social position conferred at birth or received involuntarily later in life. Examples include race/ethnicity, age, and gender.
         b. An **achieved status** is a social position a person assumes voluntarily as a result of personal choice, merit, or direct effort. Examples include occupation, education and income.
         c. Ascribed statuses have a significant influence on the achieved statuses we occupy.
      2. A **master status** is the most important status a person occupies; it dominates all of the individual's other statuses and is the overriding ingredient in determining a person's general social position (e.g., being poor or rich is a master status).

82

3. **Status symbols** are material signs that inform others of a person's general social position. Examples include a wedding ring or a Rolls-Royce automobile.
B. A **role** is a set of behavioral expectations associated with a given status.
   1. **Role expectation**—a group's or society's definition of the way a specific role ought to be played—may sharply contrast with **role performance**—how a person actually plays the role.
   2. **Role conflict** occurs when incompatible role demands are placed on a person by two or more statuses held at the same time (e.g., a woman whose role of full-time employee may conflict with her role as mother).
   3. **Role strain** occurs when incompatible demands are built into a single status that a person occupies (e.g., married women in the workforce having many responsibilities at home). Sexual orientation, age, and occupation frequently are associated with role strain.
   4. **Role exit** occurs when people disengage from social roles that have been central to their self identity (e.g., ex-convicts, ex-nuns, retirees, and divorced women and men).
C. A **social group** consists of two or more people who interact frequently and share a common identity and feeling of interdependence.
   1. A **primary group** is a small, less specialized group in which members engage in face-to-face, emotion-based interactions over an extended period of time (e.g., one's family, close friends, and school or work related peer groups).
   2. A **secondary group** is a larger, more specialized group in which the members engage in more impersonal, goal-oriented relationships for a limited period of time (e.g., schools, churches, the military, and corporations).
   3. A *social network* is a series of social relationships that link an individual to others.
   4. A **formal organization** is a highly structured group formed for the purpose of completing certain tasks or achieving specific goals (e.g., colleges, corporations, and the government).
D. A **social institution** is a set of organized beliefs and rules that establishes how a society will attempt to meet its basic social needs. Examples include the family, religion, education, the economy, the government, mass media, sports, science and medicine, and the military.
III. SOCIETIES, TECHNOLOGY, AND SOCIOCULTURAL CHANGE
A. **Gerhard and Jean Lenski** theorized that societies change over time through a process of *sociocultural evolution* as they gain new technologies.
B. **Hunting and gathering societies**, one type of preindustrial society, use simple technology for hunting animals and gathering vegetation; their technology is limited to tools and weapons used for basic subsistence.
C. **Horticultural societies** and **pastoral societies** are also types of preindustrial societies.
   1. **Pastoral societies** are based on technology that supports the domestication of large animals to provide food, and emerged in mountainous regions and

83

areas with low amounts of annual rainfall. They typically remain nomadic, as they regularly seek new grazing lands and water sources for their animals.

    2. **Horticultural societies** are based on technology that supports the cultivation of plants to provide food. These societies emerged in more fertile areas that were better suited for growing plants through the use of hand tools.

D. **Agrarian societies** use the technology of large-scale farming, including animal-drawn or energy-powered plows and equipment, to produce their food supply; most of the world's population lives in agrarian societies that are in various stages of industrialization.

E. **Industrial societies** are based on technology that mechanizes production; this mode of production transforms predominantly rural and agrarian societies into urban and industrial societies.

F. **Postindustrial societies** are societies in which technology supports a service-and-information-based economy; knowledge is viewed as the basic source of innovation and policy formulation; education becomes an important social institution; and scientific research becomes institutionalized.

IV. STABILITY AND CHANGE IN SOCIETIES

A. Sociologists **Emile Durkheim** and **Ferdinand Tönnies** developed typologies to explain how stability and change occur in the social structure of societies.

B. Durkheim's Typology

    1. **Mechanical solidarity** refers to the social cohesion in preindustrial societies where there is minimal division of labor and people feel united by shared values and common social bonds.

    2. **Organic solidarity** refers to the social cohesion found in industrial (and perhaps postindustrial) societies in which people perform very specialized tasks and feel united by their mutual dependence.

C. Tönnies' Typology

    1. The *Gemeinschaft* is a traditional society in which social relationships are based on personal bonds of friendship and kinship, and on intergenerational stability. Relationships are based on ascribed statuses.

    2. The *Gesellschaft* is a large, urban society in which social bonds are based on impersonal and specialized relationships, with little long-term commitment to the group or consensus on values. Relationships are based on achieved statuses.

D. Social Structure and Homelessness—different explanations are provided for homelessness in *Gesellschaft* societies.

    1. Many people believe the homeless have made bad decisions which eventually led to homelessness; they are responsible for the consequences of their actions.

    2. Some sociologists note that homelessness is rooted in poverty; it is a result of steady, across-the-board lowering of the standard of living of the working class and lower class—a social class phenomenon.

V. SOCIAL INTERACTION: THE MICROLEVEL PERSPECTIVE

A. Symbolic interactionists use a microsociological approach, analyzing the impact of social institutions on our daily lives.

B. Social interaction within a given society has certain shared meanings across situations; however, everyone does not interpret social interaction rituals in the same way. Interpretations are based on race/ethnicity, gender, and social class, as well as individual experiences.

C. The **social construction of reality** is the process by which our perception of reality is shaped, largely by the subjective meaning that we give to an experience. Our *definition of the situation* can result in a **self-fulfilling prophecy**—a false belief or prediction that produces behavior that makes the original false belief come true.

D. **Ethnomethodology** is the study of the commonsense knowledge that people use to understand the situations in which they find themselves.
   1. **Harold Garfinkel** created ethnomethodology.
   2. This approach examines existing patterns of conventional behavior in order to uncover people's *background expectancies*, that is, their shared interpretation of objects and events, as well as the actions they take as a result.
   3. To uncover people's background expectancies, ethnomethodologists frequently conduct *breaching experiments* in which they break "rules" or act as though they do not understand some basic rule of social life so that they can observe other people's responses.

E. **Dramaturgical analysis** is the study of social interaction that compares everyday life to a theatrical presentation.
   1. This perspective was initiated by **Erving Goffman**, who suggested that day to day interactions have much in common with being on stage or in a dramatic production.
   2. Most of us engage in **impression management**, or **presentation of self**—people's efforts to present themselves to others in ways that are most favorable to their own interests or image.
   3. Social interaction, like a theater, has a *front stage*—the area where a player performs a specific role before an audience—and a *backstage*—the area where a player is not required to perform a specific role because it is out of view of a given audience.

F. The Sociology of Emotions
   1. **Arlie Hochschild** suggests that we acquire a set of *feeling rules*, which shape the appropriate emotions for a given role or specific situation.
   2. *Emotional labor* occurs when employees are required by their employers to feel and display only certain carefully selected emotions.
   3. Gender, class, and race are related to the expression of emotions necessary to manage one's feelings.

G. **Nonverbal communication** is the transfer of information between persons without the use of words (e.g. facial expressions, head movements, body positions, and other gestures).
   1. Nonverbal communication often supplants and regulates verbal communication.
   2. *Deference*—the symbolic means by which subordinates give a required permissive response to those in power—is often conveyed through facial expressions, eye contact, and touching.

3. **Personal space** is the immediate area surrounding a person that the person claims as private. Age, gender, kind of relationship, and social class are important factors in allocation of personal space. Power differentials between people are reflected in personal space and privacy.

VI. FUTURE CHANGES IN SOCIETY, SOCIAL STRUCTURE, AND INTERACTION

A. The social structure in the U.S. has been changing rapidly in recent decades (e.g., more possible statuses for persons to occupy and roles to play than at any other time in history).

B. Ironically, at a time when we have more technological capability, more leisure activities and types of entertainment, and vast quantities of material goods available for consumption, many people experience high levels of stress, fear for their lives because of crime, and face problems such as homelessness.

C. While some individuals and groups continue to show initiative in trying to solve our pressing problems, the future of this country rests on our collective ability to deal with major social problems at both the macrolevel (structural) and the microlevel of society.

CRITICAL THINKING QUESTIONS

1. Compare and contrast functionalist, conflict, and symbolic interactionist perspectives of social structure. Which theory do you think is most useful for understanding social structure? Why?

2. What are your ascribed and achieved statuses? In what ways have your ascribed statuses impacted your achieved statuses?

3. In what ways do your race, class, and gender impact your perception of reality? How do you think U.S. society would look different to you if you were a different race, class, and gender?

4. Think of a social situation you have been in recently (e.g., an argument with a friend or partner, a meeting with a professor, etc.). Apply each of the following to that social situation: social construction of reality, ethnomethodology, dramaturgical analysis. Which idea provides the most insight into the social situation you analyzed? Why?

5. What do you think are some of the main causes of homelessness? What can be done to eliminate homelessness in the U.S.? What are some things you can do to help?

6. In what ways do you show domination or subordination through your nonverbal communication with different people?

PRACTICE TESTS

MULTIPLE CHOICE QUESTIONS

Select the response that best answers the question or completes the statement.

1. The stable patterns of social relationships that exist within a particular group or society are referred to as:
   a. social structure
   b. social interaction
   c. social dynamics
   d. social constructions of reality

2. A recently arrived immigrant often experiences:
   a. social depletion
   b. social marginality
   a. social disintegration
   b. social misappropriation

3. A _____ is a socially defined position in a group or society.
   a. location
   b. set
   c. status
   d. role

4. Maxine is an attorney, a mother, a resident of California, and a Jewish American. All of these socially defined positions constitute her:
   a. role pattern
   b. role set
   c. ascribed statuses
   d. status set

5. A master status
   a. historically has been held only by men.
   b. is comprised of all of the statuses that a person occupies at a given time.
   c. is the most important status a person occupies.
   d. is a social position a person always assumes voluntarily.

6. Wedding rings, designer jeans, and expensive cars are examples of:
   a. conspicuous consumption
   b. status symbols
   c. status markers
   d. master status indicators

7. _____ is/are the dynamic aspect of a status.
   a. Role
   b. Norms
   c. Groups
   d. People

8. Sarah is a full-time attorney and mother of two children. When one of her children needs to stay home due to illness, she must figure out how to negotiate her identities of attorney and mother. This is an example of role:
    a. strain
    b. distancing
    c. conflict
    d. confusion

9. According to the text, lesbians and gay men often experience role _____ because of the pressures associated with having an identity heavily stigmatized by the dominant cultural group.
    a. strain
    b. distancing
    c. conflict
    d. confusion

10. When a homeless person is able to become a domiciled person, sociologists refer to the process as:
    a. role disengagement
    b. role exit
    c. role engulfment
    d. role relinquishment

11. All of the following are examples of a primary group, **except**:
    a. family
    b. close friends
    c. peer groups
    d. students in a lecture hall

12. Which of the following statements **best** describes the characteristics of a secondary group?
    a. a small, less specialized group in which members engage in impersonal, goal-oriented relationships over an extended period of time
    b. a small, less specialized group in which members engage in face-to-face, emotion-based interactions over an extended period of time
    c. a larger, more specialized group in which members engage in face-to-face, emotion-based interactions over an extended period of time
    d. a larger, more specialized group in which members engage in impersonal, goal-oriented relationships over an extended period of time

13. The National Law Center on Homelessness and Poverty and other caregiver groups that provide services for the homeless and those in need are examples of:
    a. primary groups
    b. formal organizations
    c. informal organizations
    d. social networks

14. Sociologists use the term _____ to refer to a set of organized beliefs and rules that establish how a society will attempt to meet its basic social needs.
    a. social structure
    b. social expectations
    c. social networking
    d. social institutions

15. Which one of the following types of societies is more egalitarian?
    a. hunting and gathering
    b. horticultural and pastoral
    c. agrarian
    d. industrial

16. Most of the world's population lives in _____ societies.
    a. horticultural and pastoral
    b. postindustrial
    c. agrarian
    d. industrial

17. Emile Durkheim referred to the social cohesion found in industrial societies as:
    a. organic solidarity
    b. mechanical solidarity
    c. *Gemeinschaft*
    d. *Gesellschaft*

18. A young person being told repeatedly that she or he is not a good student and, as a result, stops studying and receives failing grades is an example of a(n):
    a. selective perception
    b. objective deduction
    c. subjective reality
    d. self-fulfilling prophecy

19. Ethnomethodologist Harold Garfinkel assigned different activities to his students to see how breaking the unspoken rules of behavior created confusion. His research involved a series of:
    a. shared expectancies
    b. breaching experiments
    c. dramaturgical analyses
    d. impression managements

20. According to sociologist Arlie Hochschild, people acquire a set of _____, which shape the appropriate emotions for a given role or specific situation.
    a. role expectations
    b. emotional experiences
    c. feeling rules
    d. nonverbal cues

21. _____ studies the ongoing methods people use to create reality and produce their own world.
    a. Self-fulfilling prophecy
    b. Sociology of emotions
    c. Dramaturgical analysis
    d. Ethnomethodology

22. In recent years, _____ have accounted for 40 percent of the homeless population. This is the fastest growing category of homeless persons in the United States.
    a. families with children
    b. divorced females
    c. families without children
    d. single African American males

23. John is a mediocre basketball player, but his friends and coach keep telling him he is the best player they have ever seen. John internalizes the idea that he is a great player, begins to try harder, and becomes one of the best players on the team. This is an example of:
    a. impression management
    b. social construction of reality
    c. ethnomethodology
    d. self-fulfilling prophecy

24. Which of the following statements regarding homelessness is true?
    a. All homeless people are unemployed.
    b. Some homeless people have attended college and graduate school.
    c. Most homeless people are heavy drug users.
    d. Most homeless people are mentally ill.

25. _____ theorists maintain that, in capitalistic societies where a few people control the labor of many, the social structure reflects a system of relationships of domination among categories of people, such as owner-worker and employer-employee.
    a. Postmodern
    b. Symbolic interactionist
    c. Functionalist
    d. Conflict

26. A convicted criminal, wearing a prison uniform, is an example of a person who has been _____; the uniform says that the person has done something wrong and should not be allowed unsupervised outside the prison walls.
    a. stigmatized
    b. marginalized
    c. dislabeled
    d. alienated

27. Sociologist Robert Park coined the term _____, to refer to a state of being partially an insider and an outsider in a society.
    a. alienation
    b. stigmatization
    c. social marginality
    d. anomie

28. _____ is a group's or society's definition of the way a specific role *ought* to be played.
    a. Status expectation
    b. Collective judgment
    c. Role expectation
    d. Role performance

29. _____ is how a person actually plays the role.
    a. Role expectation
    b. Role performance
    c. Role behavior
    d. Status set

30. When individuals choose to include and exclude certain information from their profiles on networking websites such as Facebook and Twitter in an effort to manipulate others' opinions of them, they are engaging in:
    a. impression management
    b. social construction of reality
    c. ethnomethodology
    d. self-fulfilling prophecy

31. Many of us build _____ that involve our personal friends in primary groups and our acquaintances in secondary groups. They are a social web that links an individual to others.
    a. cultural bonds
    b. cohesive structures
    c. social networks
    d. cliques

32. In _____ societies, the division of labor between men and women in the middle and upper classes became more distinct: men were responsible for being "breadwinners"; women were seen as "homemakers."
    a. hunting and gathering
    b. industrial
    c. agrarian
    d. horticultural and pastoral

33. _____ are characterized by an information explosion and an economy in which large numbers of people provide or apply information or are employed in service jobs (such as fast-food server or health care worker).
   a. Industrial societies
   b. Agrarian societies
   c. Horticultural and pastoral societies
   d. Postindustrial societies

34. Sociologist Emile Durkheim believed that people in preindustrial societies feel a more or less _____.
   a. organic solidarity
   b. cohesive bonding
   c. direct bonding
   d. mechanical solidarity

35. _____ theorists emphasize that social institutions exist because they perform five essentials tasks: replacing members; teaching new members; producing, distributing, and consuming goods and services; preserving order; and providing and maintaining a sense of purpose.
   a. Premodern
   b. Conflict
   c. Functionalist
   d. Symbolic interactionist

36. _____ were the first societies to exhibit gender inequality.
   a. Hunting and gathering societies
   b. Industrial societies
   c. Horticultural and pastoral societies
   d. Agrarian societies

37. Some symbolic interactionists believe that there is very little shared reality beyond that which is socially created. They refer to this as the _____—the process by which our perception of reality is largely shaped by the subjective meaning that we give to an experience.
   a. social construction of reality
   b. objectification of social reality
   c. subjective assessment of reality
   d. self-fulfilling prophecy

38. Sociologist _____ was concerned about social solidarity in large, urban societies where social bonds are based on impersonal relationships.
   a. Ferdinand Tönnies
   b. Emile Durkheim
   c. Harold Garfinkel
   d. Erving Goffman

39. A/an _____ is a false belief or prediction that produces behavior that makes the originally false belief come true.
    a. definition of the situation
    b. interactionist dialogue
    c. reality impression
    d. self-fulfilling prophecy

40. Physical space is an important component of nonverbal communication. Anthropologist Edward Hall asserted that _____ is the immediate area surrounding a person that the person claims as private.
    a. intimate space
    b. social space
    c. public space
    d. personal space

TRUE/FALSE QUESTIONS

1. Most homeless persons are on the streets by choice or because they were deinstitutionalized by mental hospitals.
    T   F

2. Sociologists use the term "status" to refer only to highlevel positions in society.
    T   F

3. Historically, the most common master statuses for women have related to positions in the family, such as daughter, wife, and mother.
    T   F

4. Role performance does not always match role expectation.
    T   F

5. Role conflict occurs when incompatible role demands are placed on a person by two or more statuses held at the same time.
    T   F

6. Role distancing occurs when social identities are inconsistent with one's own identity.
    T   F

7. Role exit usually does not require a creation of a new identity.
    T   F

8. Close friends would be an example of a primary group.
    T   F

9. Gerhard and Jean Lenski argued that societies change over time through a process of sociocultural evolution.
    T   F

10. Social networks work the same way for men and women and for people from different racial/ethnic groups.
T   F

11. Agrarian societies are based on technology that supports the cultivation of plants to provide food.
T   F

12. In an industrialized society, the family decreases in significance as the economic, educational, and political institutions grow in size.
T   F

13. Relationships in *Gemeinschaft* societies are based on achieved rather than ascribed status.
T   F

14. Race/ethnicity, gender, and social class play a part in the way that social interaction is interpreted.
T   F

15. Age, gender, and social class are strong influences in the allocation of personal space.
T   F

16. An achieved status is a social condition conferred at birth.
T   F

17. According to Box 5.1, most homeless people choose to be homeless.
T   F

18. According to Box 5.2, "Framing Homelessness in the Media," episodic framing refers to news stories that focus primarily on statistics about the homeless population and recent trends in homelessness.
T   F

19. According to Box 5.3, "public space protection" has become an issue not only in the United States, but also in other countries.
T   F

20. According to Box 5.4, college students are leading the way in helping the homeless in some communities.
T   F

FILL-IN-THE-BLANK QUESTIONS

1. The process by which people act toward or respond to other people is defined as _____.

2. A status gained as a result of personal ability is defined as a/an _____ status.

3. _____ societies use simple technology for acquiring food and live in small groups of about 24 to 40 people.

94

4. Societies based on technology that supports the domestication of large animals to provide food are known as _____ societies.

5. _____ is the sociologist who created ethnomethodology.

6. The _____ refers to the idea that people analyze a social context in which they find themselves, determine what is in their best interest, and adjust their attitudes and actions accordingly.

7. The study of social interaction that compares everyday life to a theatrical presentation is known as _____.

8. _____ refers to people's efforts to present themselves to others in ways that are most favorable to their own interests or image.

9. The transfer of information between persons without the use of words is called _____.

10. _____ occurs when an individual disengages from a social role that has been central to their identity, as in the case of divorce.

11. _____ is the study of the commonsense knowledge that people use to understand the situations in which they find themselves.

12. Sociologist _____ used the term _____ in referring to the area where a player performs a specific role before an audience.

13. Sociologist _____ referred to the social cohesion in preindustrial societies where there is minimal division of labor and people feel united by shared values as a _____.

14. Individuals engage in _____ when they have to put forth effort to feel and display the socially appropriate emotion while at work.

15. According to Goffman, the _____ is the area where a player is not required to perform a specific role because it is out of the view of a given audience.

SHORT ANSWER/ESSAY QUESTIONS

1. What are the major components of social structure? Give examples of each.
2. What are the five major components of societies? List the major characteristics of each type. In which preindustrial society is social inequality the highest? The lowest? Why?
3. How is homelessness related to the social structure of a society? In what type of society do you think there might be more homelessness? Why?
4. What do sociologists mean by the social construction of reality? Give some examples.
5. What are some of the projected future changes in the social structure in the United States? How should our society react to these changes?
6. According to the "Things We Take for Granted" photo essay section of this chapter, how were social structures and institutions affected by Hurricane Katrina, and how were the people in this area directly affected?
7. Why are social institutions and social interaction so important in defining who we are and what we want to become?

STUDENT CLASS PROJECTS AND ACTIVITIES

1. Utilizing the concept of social networks as discussed in the text, trace your own social networks. Construct diagrams linking yourself to your family and other relatives, and to your friends and other acquaintances. After constructing the diagrams, complete the following: (1) Describe the members in your network (who are they, and what is their gender, approximate age, and current employment, if applicable?). (2) How and why are they linked to you? (3) Summarize the impact of each person in your network; are the ties weak or strong? Why? (4) What persons in your network are most important now? Why? (5) What persons in your network will be especially important to you in future years? Why? (6) Is there any potential for anyone in your network to harm you in any manner? What can/should you do about this? (7) Is there any potential for anyone in your network to provide you with special advantages for a potential career?

2. As described in the text, changes in social structure have a dramatic impact on individuals, groups, and societies. Many nations in the world today are experiencing radical changes as they attempt to modernize, while many contemporary societies are experiencing great change as they become more complex with the introduction of new technology. Select any country that is experiencing such change, utilizing at least one of the typologies mentioned in the text (that of Durkheim or Tönnies). Examine the changes taking place in the country you selected from both a microlevel and a macrolevel perspective. Trace the changes in (1) statuses, (2) roles, (3) groups, (4) social institutions, and (5) the society for both the individual and the country as a whole. You should also explain the effects these changes will have on the future options and opportunities of people living in that particular country.

3. The text uses homelessness as a concrete example to help you understand the concept of social structure. Providing effective solutions to this problem is even more difficult. Many politicians suggest that one reason for the increase in homelessness is the shortage of available low-income housing. Research and write an issue paper on this topic: "Should more low-income housing be created to help the homeless?" In your paper, provide the following: (1) a background of this problem; (2) factual information germane to homelessness; (3) the positions of opposing viewpoints; (4) arguments for and against the issue; and (5) your own position and resolution of homelessness. Suggested note: You may attempt a synthesis of opposing positions; leave the issue essentially unresolved; provide a new position; or provide a completely different, innovative resolution.

INTERNET ACTIVITIES

1. **Emotions: Observing the Fusion Between Body, Mind, and Society:** **http://www.trinity.edu/mkearl/spsy-emo.html**. This is a fantastic web site that will help you understand the connections between sociology, psychology, and physiology as they address the topic of emotions. The site contains links to research articles, survey data, and an online bibliography.
2. **Civil Practices Network** (CPN) is a collaborative and nonpartisan project dedicated to bringing together diverse social organizations in communities and institutional settings across the United States in order to revitalize and strengthen major infrastructures throughout the country. From the web site, **www.cpn.org**, discover some of these rebuilding strategies.
3. Learn more about homelessness from the **National Coalition for the Homeless** web site, **http://www.nationalhomeless.org/factsheets/**. Read the factsheets to gather more information about who is homeless and what causes homelessness. Click on "projects" to find out about current projects designed to help the homeless.

INFOTRAC COLLEGE EDITION EXERCISES

Visit the **InfoTrac College Edition** web site at: **http://www.wadsworthmedia.com/webtutor/infotrac.htm**. You will arrive at a screen that enables you to search topics.

1. Search for articles using the subject guide **social structure**. You will get 254 periodical references. Choose the article entitled: "Pathways To Adulthood In Changing Societies: Variability and Mechanisms in Life Course Perspective." The article can help you connect the curriculum content to your own life. The author, Michael J. Shanahan, suggests new lines of inquiry that focus on the interplay between agency and social structures in the shaping of lives.
2. Look for research related to the study of **emotions**. Bring copies of articles of interest to class. This is an effective way to learn about empirical research, research methods, science, and the sociology of emotions.
3. These three articles were accessed using InfoTrac. Each is a good beginning point for a discussion on **social structure and interaction**. Each can inspire the sociological imagination. Find other research articles that help you understand the powerful influence of society on you individual life.
   - "Restoring American cultural institutions." Jerry L. Martin. *Society*. Jan-Feb 1999. v36 i2. p35(6).
   - "Responses to humiliation." (The Decent Society) Arthur Ripstein. *Social Research*. Spring 1997. v64 n1. p90(22).
   - "On moral issues Americans prefer persuasion." (The Communication Network survey). *Society*. Nov-Dec 1997. v35 n1. p2(1).

SOLUTIONS

MULTIPLE CHOICE QUESTIONS

1. A, p. 138	15. A, p. 149	29. B, p. 144
2. B, p. 139	16. C, p. 150	30. A, p. 160
3. C, p. 139	17. A, p. 154	31. C, p. 147
4. D, p. 140	18. D, p. 158	32. B, p. 151
5. C, p. 141	19. B, p. 159	33. D, p. 152
6. B, p. 142	20. C, p. 162	34. D, p. 154
7. A, p. 144	21. D, p. 159	35. C, p. 148
8. C, p. 144	22. A, p. 138	36. D, p. 150
9. A, p. 145	23. D, p. 158	37. A, p. 158
10. B, p. 146	24. B, p. 140	38. A, p. 155
11. D, p. 146	25. D, p. 148	39. D, p. 158
12. D, p. 147	26. A, p. 139	40. D, p. 166
13. B, p. 147	27. C, p. 139	
14. D, p. 147	28. C, p. 144	

TRUE/FALSE QUESTIONS

1. F, p. 140	8. T, p. 146	15. T, p, 166
2. F, p. 139	9. T, p. 149	16. F, p. 140
3. T, p. 141	10. F, p. 147	17. F, p. 159
4. T, p. 144	11. F, p. 150	18. F, p. 143
5. T, p. 144	12. T, p. 151	19. T, p. 156
6. T, p. 144	13. F, p. 155	20. T, p. 166
7. F, p. 146	14. T, p. 165	

FILL-IN-THE-BLANK QUESTIONS

1. social interaction, p. 137
2. achieved, p. 140
3. Hunting and gathering, p. 149
4. pastoral, p. 150
5. Harold Garfinkel, p. 159
6. definition of the situation, p. 163
7. dramaturgical analysis, p. 160
8. impression management, p. 160
9. nonverbal communication, p. 165
10. Role exit, p. 146
11. Ethnomethodology, p. 159
12. Erving Goffman; front stage, p. 160
13. Ferdinand Tönnies; *Gemeinschaft*, p. 155
14. emotional labor, p. 162
15. backstage, p. 160-161

6
GROUPS AND ORGANIZATIONS

BRIEF CHAPTER OUTLINE

CHAPTER SUMMARY

Groups are a key element of our social structure and much of our social interaction takes place within them. A *social group* is a collection of two or more people who interact frequently, share a sense of belonging, and have a feeling of interdependence. Social groups may be either *primary groups*—small, personal groups in which members engage in emotion-based interactions over an extended period—or *secondary groups*—larger, more specialized groups in which members have less personal and more formal, goal-oriented relationships. All groups set boundaries to indicate who does and who does not belong: an **ingroup** is a group to which we belong and with which we identify; an **outgroup** is a group we do not belong to or perhaps feel hostile toward. The size of a group is one of its most important features. The smallest groups are **dyads**—groups composed of two members—and **triads**—groups of three. In order to maintain ties with a group, many members are willing to conform to norms established and reinforced by group members. Three types of *formal organizations*—highly structured secondary groups formed to achieve specific goals in an efficient manner—are normative, coercive, and utilitarian organizations. A **bureaucracy** is a formal organization characterized by hierarchical authority, division of labor, explicit procedures, and impersonality in personnel concerns. The **iron law of oligarchy** refers to the tendency of organizations to become a bureaucracy ruled by the few. A recent movement to humanize bureaucracy has focused on developing human resources. The best organizational structure for the future is one that operates humanely and includes opportunities for all, regardless of race, gender, or class.

LEARNING OBJECTIVES

After reading Chapter 6 you should be able to:

1. Distinguish between aggregates, categories, and groups from a sociological perspective.

2. Distinguish between primary and secondary groups and explain how people's relationships differ in each.

3. State definitions for ingroup, outgroup, and reference group and describe the significance of these concepts in everyday life.

4. Explain functionalist, conflict, symbolic interactionist, and postmodern perspectives on the purposes of groups.

5. Describe the significance of group size and distinguish between a dyad and a triad.

6. Distinguish between the two functions of leadership and the three major styles of group leadership.

7. Explain the concepts of group conformity and groupthink, summarize early research on the two concepts, and elaborate on reasons why they can be dangerous for organizations.

8. Explain the contributions of Solomon Asch and Stanley Milgram to our understanding about group conformity and obedience to authority.

9. Explain and apply social exchange and rational choice theories.

10. Compare normative, coercive, and utilitarian organizations and describe the nature of membership in each.

11. Summarize Max Weber's perspective on rationality and outline his ideal characteristics of bureaucracy.

12. Explain the relationship between bureaucratic hierarchies and oligarchies and define the iron law of oligarchy.

13. Describe the informal structure in bureaucracies and list its positive and negative aspects.

14. Discuss the major shortcomings of bureaucracies and their effects on workers, clients or customers, and levels of productivity.

15. Compare and evaluate U.S. and Japanese models of organization.

KEY TERMS

(defined at page number shown and in glossary)

aggregate, p. 174

authoritarian leaders, p.182

bureaucracy, p. 189

bureaucratic personality, p. 194

category, p. 174

conformity, p. 182

democratic leaders, p. 182

dyad, p. 179

expressive leadership, p. 182

goal displacement, p. 194

groupthink, p. 186

ideal type, p. 190

informal side of a bureaucracy, p. 191

ingroup, p. 175

instrumental leadership, p. 181

iron law of oligarchy, p. 195

laissez-faire leaders, p. 182

network, p. 177

outgroup, p. 175

rationality, p. 186

reference group, p. 176

small group, p. 179

triad, p. 179

KEY PEOPLE

(identified at page number shown)

Solomon Asch, p.183

Charles H. Cooley, p. 174

Irving Janis, p. 186

Stanley Milgram, p. 184

Georg Simmel, p. 179

William Graham Sumner, p. 175

Max Weber, p. 189

CHAPTER OUTLINE

I. SOCIAL GROUPS
   A. Groups, Aggregates, and Categories
      1. A *social group* is a collection of two or more people who interact frequently with one another, share a sense of belonging, and have a feeling of interdependence.
      2. An **aggregate** is a collection of people who happen to be in the same place at the same time but share little else in common.
      3. A **category** is a number of people who may never have met one another but share a similar characteristic (e.g., men, Native Americans, college students).
   B. Types of Groups
      1. Primary and Secondary Groups
         a. According to Charles H. Cooley, a *primary group* is a small group whose members engage in face-to-face, emotion-based interactions over an extended period of time (e.g., *significant others*).
         b. A *secondary group* is a larger, more specialized group in which the members engage in more impersonal, goal-oriented relationships for a limited period of time (e.g., students in a class).

2. **Ingroups** and **Outgroups**
   a. According to **William Graham Sumner**, an **ingroup** is a group to which a person belongs and feels a sense of identity with.
   b. An **outgroup** is a group to which a person does not belong and toward which the person may feel a sense of competitiveness or hostility.
   c. Ingroup and outgroup distinctions may promote classism, racism, and sexism because ingroup members typically view themselves positively and members of outgroups negatively.
3. **Reference Groups**
   a. A **reference group** is a group that strongly influences a person's behavior and social attitudes, regardless of whether that individual is an actual member.
   b. Reference groups help us explain why our behavior and attitudes sometimes differ from those of our membership groups; we may accept the values and norms of a group with which we identify rather than one to which we belong.
4. **Networks**
   a. A **network** is a web of social relationships that link one person with other people and, through them, other people they know.
   b. Networks extend far, not only to people you know, but also to people that know of you.

II. GROUP CHARACTERISTICS AND DYNAMICS
   A. Theoretical Analyses of Groups
      1. Functionalists view groups as important in meeting people's *instrumental needs* (e.g., task-oriented) and *expressive needs* (e.g., emotional).
      2. Conflict theorists point out that groups involve a series of power relationships, whereby the needs of individual members may not be served equally.
      3. Symbolic interactionists focus on how the size of groups influences the types of interactions that take place among members.
      4. Postmodern theorists analyze the superficiality and lack of depth in social relationships formed in groups and organizations (e.g., scripted interactions between fast-food employees and customers).
   B. Group Size
      1. A **small group** is collectivity small enough for all members to be acquainted with one another and to interact simultaneously.
      2. According to **Georg Simmel**, small groups have distinctive interaction patterns that do not exist in larger groups.
         a. In a **dyad**—a group composed of two members—the active participation of both members is crucial for the group's survival, and members have an intense bond and sense of unity not found in most larger groups.
         b. When a third person is added to a dyad, a **triad**—a group composed of three members—is formed, and the nature of the relationship and interaction patterns change; for example two members may unite to create a *coalition*—an alliance created in an attempt to reach a shared objective or goal, which sometimes can subject the third member to group pressure to conform.

102

3. As group size increases, members tend to specialize in different tasks, and communication patterns change.
C. Group Leadership
   1. Leaders are responsible for directing plans and activities so that the group completes its task or fulfills its goals.
   2. Leadership functions:
      a. **Instrumental leadership** is goal- or task-oriented; if the underlying purpose of a group is to complete a task or reach a particular goal, this type of leadership is most appropriate.
      b. **Expressive leadership** provides emotional support for members; this type of leadership is most appropriate when harmony, solidarity, and high morale are needed.
   3. Leadership styles:
      a. **Authoritarian leaders** make all major group decisions and assign tasks to members.
      b. **Democratic leaders** encourage group discussion and decision making through consensus building.
      c. **Laissez-faire leaders** are only minimally involved in decision making and encourage group members to make their own decisions.
D. Group Conformity
   1. **Conformity** is the process of maintaining or changing behavior to comply with the norms established by a society, subculture, or other group.
   2. In a series of experiments, **Solomon Asch** found that the pressure toward group conformity was so great that participants were willing to contradict their own best judgment if the rest of the group disagreed with them, thus demonstrating the power that groups have in producing *compliance*.
   3. **Stanley Milgram** conducted a series of controversial experiments on people's obedience to authority and found that most participants continued to follow orders from a person in a position of authority, even when they believed they were inflicting pain on someone else. Milgram's study showed that obedience to authority may be more common than most people would like to believe.
   4. The research of John Pryor provides a contemporary example of group conformity, which demonstrates that sexual harassment is more likely to occur when it is encouraged (or not actively discouraged) by others.
E. **Groupthink**
   1. **Irving Janis** coined the term **groupthink** to describe the process by which members of a cohesive group arrive at a decision that many individual members privately believe is unwise.
   2. The tragic launch of the space shuttle *Columbia* is an example of this process; although some individuals raised concerns about the safety of the shuttle, the group as a whole chose not to act on these concerns.
F. Social Exchange/Rational Choice Theories
   1. Social exchange/rational choice theories focus on the process by which actors settle on one optimal outcome out of a range of possible choices.

103

2. Social exchange theories are based on the assumption of *self-interest* as the basic motivating factor, and that people seek to maximize rewards and minimize punishments.
3. Rational choice theories posit that individual *actors* make choices based on the *resources* available to them.

III. FORMAL ORGANIZATIONS IN GLOBAL PERSPECTIVE
   A. A *formal organization* is a highly structured secondary group formed for the purpose of achieving specific goals in the most efficient manner (e.g., corporations, schools, and government agencies).
   B. Types of Formal Organizations
      1. Amitai Etzioni classified formal organizations into three categories based on the nature of membership.
      2. We voluntarily join *normative organizations* when we want to pursue some common interest or to gain personal satisfaction or prestige from being a member. Examples include political parties, religious organizations, and college social clubs.
      3. People do not voluntarily become members of *coercive organizations*—associations people are forced to join. Examples include total institutions, such as boot camps and prisons.
      4. We voluntarily join *utilitarian organizations* when they can provide us with a material reward we seek. Examples include colleges and universities, and the workplace.
   C. Bureaucracies
      1. **Bureaucracy** is an organizational model characterized by a hierarchy of authority, a clear division of labor, explicit rules and procedures, and impersonality in personnel matters.
      2. According to **Max Weber**, bureaucracy is the most "rational" and efficient means of attaining organizational goals because it contributes to coordination and control. **Rationality** is the process by which traditional methods of social organization, characterized by informality and spontaneity, gradually are replaced by efficiently administered formal rules and procedures.
      3. An **ideal type** is an abstract model which describes the recurring characteristics of some phenomenon; the ideal characteristics of bureaucracy include:
         a. Division of Labor—Each member has a specific status with certain assigned tasks to fulfill.
         b. Hierarchy of Authority—A chain of command that is based on each lower office being under the control and supervision of a higher one.
         c. Rules and Regulations—Standardized rules and regulations establish authority within an organization and usually are provided to members in a written format.
         d. Qualifications-Based Employment—Hiring of staff members and professional employees is based on specific qualifications; individual performance is evaluated against specific standards; and promotions are based on merit as spelled out in personnel policies.

104

e. Impersonality—Interaction is based on status and standardized criteria rather than personal feelings or subjective factors.
4. A contemporary example of Weber's theory of bureaucracy and rationality is George Ritzer's theory of "McDonaldization," which identifies four dimensions of rationality—efficiency, predictability, emphasis on quantity over quality, and control through nonhuman technologies—that are found globally in fast-food restaurants and other businesses.
5. The **informal side of bureaucracy** is composed of those aspects of participants' day-to-day activities and interactions that ignore, bypass, or do not correspond with the official rules and procedures of the bureaucracy.
   a. The informal structure also has been referred to as *work culture* because it includes the ideology and practices of workers on the job; workers create this work culture in order to conform, resist, or adapt to the constraints of their jobs, as well as to guide and interpret social relations on the job.
   b. There are positive and negative aspects of informal structure in organizations; traditional management theories emphasize control (or eradication) of informal groups; the *human relations approach* views informal networks as adaptive behaviors.

D. Problems of Bureaucracies
   1. Inefficiency and Rigidity
      a. **Goal displacement** occurs when the rules become an end in themselves (rather than a means to an end), and organizational survival becomes more important than achievement of goals
      b. The term **bureaucratic personality** is used to describe those workers who are more concerned with following correct procedures than they are with getting the job done correctly.
   2. Resistance to Change
      a. Once created, bureaucratic organizations tend to resist change, making them difficult to eliminate and contributing to bureaucratic enlargement.
      b. Often bureaucracies promote people from within the bureaucracy who may not be qualified for the positions, which can produce incompetence.
   3. Perpetuation of Race, Class, and Gender Inequalities
      a. The bureaucratic structure was created specifically for middle- and upper-middle-class white men, which makes it difficult for others to be fully integrated.

E. Bureaucracy and Oligarchy
   1. Bureaucracy generates an enormous degree of unregulated and often unperceived social power in the hands of a very few leaders.
   2. According to Robert Michels, this results in the **iron law of oligarchy**—a bureaucracy ruled by a few people.

IV. ALTERNATIVE FORMS OF ORGANIZATION
   A. "Humanizing" bureaucracy includes: (1) less rigid, hierarchical structures and greater sharing of power and responsibility; (2) encouragement of participants to share their ideas and try new approaches; and (3) efforts to reduce the number of people in dead-end jobs and help people meet outside family responsibilities while still receiving equal treatment inside the organization.

B. The Japanese model of organization has been widely praised for its innovative structure, which (until recently) has included:
   1. Long-term employment and company loyalty—Workers were guaranteed permanent employment after an initial probationary period.
   2. Quality Circles—Small workgroups that meet regularly with managers to discuss the group's performance and working conditions.

V. ORGANIZATIONS IN THE FUTURE
   A. There is a lack of consensus among organizational theorists about the "best" model of organization; however, some have suggested a *horizontal* model in which both hierarchy and functional or departmental boundaries largely would be eliminated.
   B. In the horizontal structure, a limited number of senior executives would still exist in support roles (such as finance and human resources); everyone else would work in multidisciplinary teams that would perform core processes (e.g., product development or sales).
   C. It is difficult to determine what the best organizational structure for the future might be; however, everyone can benefit from humane organizational environments that provide opportunities for all people regardless of race, gender, or class.

CRITICAL THINKING QUESTIONS

1. In what ways does the creation of ingroups and outgroups contribute to social inequality?
2. Explain the functionalist, conflict, symbolic interactionist, and postmodern views of groups. Which theoretical perspective do you think is most useful for understanding groups? Why?
3. Think about groups you have participated in. What kinds of leaders did these groups have? What kind of leadership style(s) did the leaders use? How effective were they? Could they have been more effective with a different style of leadership? Why or why not?
4. What examples can you think of other than those described in the textbook of conformity and groupthink?
5. Evaluate social exchange and rational choice theories. What are the strengths and weaknesses of each?
6. What organization can you think of that best meets the "ideal type" of bureaucracy? How does this organization meet the characteristics outlined by Weber?
7. What kind of organization do you think will be most effective in the future? How can organizations equalize opportunities based on race, class, and gender?

PRACTICE TESTS

MULTIPLE CHOICE QUESTIONS

Select the response that best answers the question or completes the statement.

1. According to the "lived experience" introduction in this chapter, Facebook is
   a. a "fun" way to get to know other people.
   b. a way to avoid actual human contact.
   c. a way to join online groups with similar interests.
   d. all of the above

2. A(n) _____ is a collection of people who happen to be in the same place at the same time, while a(n) _____ is a number of people who may never have met one another but share a similar characteristic.
   a. aggregate; category
   b. category; aggregate
   c. social group; aggregate
   d. category; social group

3. John had thought about trying out for the football team, but because he is not very athletic, he decides to hang out with the "stoners" at his school. He harbors some feelings of hostility for the football team. For John, the football team represents a(n):
   a. reference group
   b. secondary group
   c. ingroup
   d. outgroup

4. A _____ is an alliance created in an attempt to reach a shared objective or goal.
   a. triad
   b. coalition
   c. dyad
   d. reference group

5. _____leadership is goal- or task-oriented; _____leadership provides emotional support for members.
   a. Authoritarian; democratic
   b. Authoritarian; laissez-faire
   c. Expressive; instrumental
   d. Instrumental; expressive

6. A group leader at a business-related seminar is only minimally involved with the decisions made by the group and encourages members to make their own choices. This illustrates a(n) _____ leader.
   a. expressive
   b. democratic
   c. laissez-faire
   d. authoritarian

7. In one of Solomon Asch's experiments, subjects were asked to compare the length of lines printed on a series of cards without knowing that all other research "subjects" actually were assistants to the researcher. The results of Asch's experiments revealed that
    a. 85 percent of the subjects routinely chose the correct response regardless of the assistants' opinions.
    b. 33 percent of the subjects routinely chose to conform to the opinion of the assistants by giving an incorrect answer.
    c. 10 percent of the subjects routinely chose to conform to the opinion of the assistants by giving an incorrect answer.
    d. the opinion of the assistants had no influence on the subjects' opinions.

8. The experiments by Stanley Milgram found that
    a. obedience to authority may be less common than most people would like to believe.
    b. many of the subjects questioned the ethics of the experiment.
    c. most participants obeyed an authority figure even when they believed their actions were causing another person physical pain.
    d. most of the subjects were afraid to conform because the use of electrical current was involved.

9. _____ occurs when members of a group arrive at a decision that many individuals privately believe is unwise.
    a. Groupthink
    b. Obedience to authority
    c. Compliance
    d. The iron law of oligarchy

10. All of the following are types of formal organizations, **except** _____ organizations.
    a. normative
    b. anomic
    c. coercive
    d. utilitarian

11. Membership in _____ organizations is involuntary.
    a. normative
    b. anomic
    c. coercive
    d. utilitarian

12. According to Max Weber, rationality refers to
    a. the process by which bureaucracy is gradually replaced by alternative types of organization, such as quality circles.
    b. the process by which traditional methods of social organization are gradually replaced by bureaucracy.
    c. the level of sanity (or insanity) of people in an organization.
    d. the logic used by organizational leaders in decision making.

108

13. All of the following are ideal type characteristics of bureaucratic organizations, as specified by Max Weber, **except**:
    a. coercive leadership
    b. impersonality
    c. hierarchy of authority
    d. division of labor

14. In higher education, student handbooks, catalogs, and course syllabi are examples of:
    a. informal structures in bureaucracy
    b. impersonality
    c. hierarchy of authority
    d. rules and regulations

15. _____ theories posit that individuals weigh the costs and benefits when choosing a course of action, taking into account the resources available to them.
    a. Rational choice
    b. Social exchange
    c. Social resource
    d. Bureaucratic

16. The _____ perspective argues that groups involve power relationships that may not serve member's needs equally.
    a. functionalist
    b. conflict
    c. symbolic interactionist
    d. postmodern

17. All of the following are shortcomings of bureaucracy, **except**:
    a. division of labor
    b. inefficiency and rigidity
    c. resistance to change
    d. perpetuation of race, class, and gender inequalities

18. George Ritzer's "McDonaldization" includes all of the following **except**:
    a. efficiency
    b. predictability
    c. emphasis on quality over quantity
    d. control through nonhuman technologies

19. People join _____ organizations when they want to pursue a common interest.
    a. normative
    b. global
    c. coercive
    d. utilitarian

20. The _____ approach views informal networks as adaptive behaviors that should be nurtured.
    a. human relations
    b. oligarchy
    c. traditional management
    d. bureaucratic

21. Sociologist Charles H. Cooley used the term _____ to describe a small, less specialized group in which members engage in face-to-face, emotion-based interactions over an extended period of time.
    a. secondary group
    b. significant group
    c. tertiary group
    d. primary group

22. We have primary relationships with other individuals in our primary groups—that is, with our _____, who frequently serve as role models.
    a. generalized others
    b. significant others
    c. familial others
    d. fictive kin

23. People who are the same age, race, and gender, and who share the same educational level constitute a(n)
    a. category
    b. aggregate
    c. formal organization
    d. social group

24. _____ conducted famous studies on obedience.
    a. Solomon Asch
    b. Charles H. Cooley
    c. Georg Simmel
    d. Stanley Milgram

25. Bureaucracies require that all individuals be treated the same. This statement reflects which component of bureaucracies?
    a. impersonality
    b. division of labor
    c. qualifications
    d. hierarchy of authority

26. A _____ is an organizational model characterized by a hierarchy of authority, a clear division of labor, explicit rules and procedures, and impersonality in personnel matters.
    a. democracy
    b. bureaucracy
    c. monarchy
    d. formal organization

27. _____ distinguished between primary and secondary groups.
    a. Irving Janis
    b. Charles Cooley
    c. Karl Marx
    d. Max Weber

28. For a person who strongly believes in the value of human rights and equal opportunity, which of the following would **not** constitute a reference group?
    a. the Ku Klux Klan
    b. the American Civil Liberties Union
    c. the National Organization for Women
    d. the Council on Racial Equality

29. A(n) _____ is a web of social relationships that links one person with other people and, through them, with other people they know.
    a. informal organization
    b. network
    c. primary group
    d. reference group

30. According to functionalists, _____ needs are particularly important for self-expression and support from family, friends, and peers.
    a. expressive
    b. instrumental
    c. group
    d. individual

31. The text points out that ingroup and outgroup distinctions
    a. may encourage social cohesion among group members.
    b. serve to prevent racism, sexism, and ageism.
    c. discourage feelings of group superiority.
    d. are less likely to exist in urban societies than in rural societies.

32. _____ size refers to the number of potential members a group has.
    a. Relative
    b. Power
    c. Absolute
    d. Potential

33. A(n) _____ is a group to which a person does not belong and toward which the person may feel a sense of competitiveness or hostility.
    a. ingroup
    b. outgroup
    c. reference group
    d. secondary group

111

34. _____ leadership is goal- or task-oriented.
    a. Instrumental
    b. Expressive
    c. Primary
    d. Secondary

35. _____ leaders encourage group discussion and decision making through consensus building.
    a. Democratic
    b. Authoritarian
    c. Laissez-faire
    d. Socialistic

36. _____ is a form of compliance in which people follow direct orders from someone in a position of authority.
    a. Cohesion
    b. Obedience
    c. Learning
    d. Socialization

37. Thorstein Veblen used the term _____ to characterize situations in which workers have become so highly specialized, or have been given such fragmented jobs to do, that they are unable to come up with creative solutions to problems.
    a. trained incapacity
    b. bureaucratic personality
    c. bureaucratic formation
    d. organization men

38. The tendency of a bureaucracy to be ruled by a very limited number of people is referred to as the:
    a. iron cage
    b. Peter Principle
    c. iron law of oligarchy
    d. absolute regime

39. Eric works for the All-Star Baseball Card Company. Biweekly, Eric and about twelve of his co-workers meet with the shift supervisor to discuss ways to improve the work setting and the company's product. This group is an example of a(n):
    a. quality circle
    b. primary group
    c. bureaucratic subdivision
    d. assessment committee

40. John Pryor's research on sexual harassment showed that
    a. individuals today are less likely to engage in group conformity than in the past.
    b. individuals are more likely to engage in group conformity in primary groups than in secondary groups.
    c. individuals today still show high levels of group conformity.
    d. sexual harassment no longer occurs in the workplace.

TRUE/FALSE QUESTIONS

1. People in aggregates share a common purpose.
   T   F

2. Although formal organizations are secondary groups, they also contain many primary groups within them.
   T   F

3. According to functionalists, people form groups to meet instrumental and expressive needs.
   T   F

4. If one member withdraws from a dyad, the group ceases to exist.
   T   F

5. Laissez-faire leaders encourage group discussion and decision making through consensus building.
   T   F

6. Stanley Milgram's research focused on people's obedience to authority.
   T   F

7. The research of Solomon Asch demonstrated that group conformity is more likely in dyads than in groups of four or more.
   T   F

8. Recent research on sexual harassment by psychologist John Pryor suggests that there is a relationship between group conformity and harassment.
   T   F

9. The explosion of the space shuttle *Columbia* has been cited as an example of groupthink.
   T   F

10. A group that strongly influences a person's behaviors and attitudes regardless of whether the person is actually a member is referred to as an ingroup.
    T   F

11. Over the past century, the number of formal organizations increased dramatically in the United States.
    T   F

12. An ideal type is an abstract model that describes the recurring characteristics of some phenomenon.
T   F

13. George Ritzer's theory of "McDonaldization" is a contemporary application of Max Weber's theory of rationality and bureaucracy.

T   F

14. The "grapevine" informal structure of a bureaucracy spreads information much slower than the "official", formal channels of communication.
T   F

15. Women and people of color who are employed in positions traditionally held by white men often experience categorical exclusion from the informal structures.
T   F

16. A category is a number of people who may never have met one another, but share a similar characteristic.
T   F

17. According to Box 6.1, a fast-food restaurant can require all employees under the age of twenty-one to submit to periodic, unannounced drug testing.
T   F

18. According to Box 6.2, "Framing Community in the Media: Virtual Communities on the Internet," some social scientists believe virtual communities established in cyberspace constitute true communities, whereas others argue such communities are not true communities because they lack geographic and social boundaries and do not promote close interpersonal ties.
T   F

19. According to Box 6.3, "Computer Privacy in the Workplace," employers have the right to monitor everything that their employees do on company-owned computers.
T   F

20. Box 6.4, "Developing Invisible (But Meaningful) Networks on the Web," suggests that establishing social spaces almost always leads to creating greater social and individual problems.
T   F

FILL-IN-THE-BLANK QUESTIONS

1. According to _____, small groups have distinctive interaction patterns that do not exist in larger groups.

2. Sociologist _____ coined the terms "ingroup" and "outgroup."

3. The term "primary group" was coined by _____.

4. A series of controversial experiments dealing with people's obedience to authority were conducted by _____.

5. The term "groupthink" was coined by _____.

6. The theory of a _____ has been developed to explain how social class distinctions are perpetuated through different types of employment.

7. The tendency for bureaucracies to become ruled only by a few people is referred to as _____.

8. Many U.S. corporations use _____, which are small groups of workers who meet with managers regularly to discuss working conditions and performance.

9. Officials in a bureaucracy exhibit _____ when they emphasize rules and procedures more than the achievement of goals.

10. The term _____ describes workers more concerned with following correct procedures than with getting the job done correctly.

11. In the early 1980s in the United States, there was a movement to _____ bureaucracy with the major goal being to establish an organizational environment that develops rather than impedes human resources.

12. For several decades, the _____ model of organization has been widely praised for its innovative structure.

13. _____ leaders encourage group discussion and consensus.

14. _____ theories are based upon the assumption that self-interest is the basic motivating factor in people's interactions.

15. _____ theorists argue that groups and organizations in contemporary society are characterized by superficiality and depthlessness in social relationships.

SHORT ANSWER/ESSAY QUESTIONS

1. According to the social exchange/rational choice theories, what might some men gain from sexually harassing workers? What are some possible losses or penalties? What do women gain or lose by reporting incidents of sexual harassment?
2. Explain the three leadership styles discussed in the chapter. What are some strengths and weaknesses of each? Describe a scenario in which each would work best.
3. What is bureaucracy? What are some of the strengths and weaknesses of bureaucracies?
4. What is groupthink? How might one affect or be affected by groupthink?
5. What were the insights gained from Stanley Milgram's research? Do these insights outweigh the personal costs to the subjects of the research?

STUDENT CLASS PROJECTS AND ACTIVITIES

1. Prior to taking office, the U.S. President-elect announces to the public his/her choices for all the cabinet and chief advisory appointments. These appointments

profoundly shape many of the policies and directions of the U.S. Federal Government, which is one of the largest bureaucracies in the country. Select ten of these positions. Describe the position and its purpose; the requirements, if any, for the position; and the length of the term of the position. Name and briefly describe the credentials of the person who occupies that office. Then analyze each position, and the system as a whole, in terms of Weber's conceptualization of bureaucracy. [Some examples of cabinet posts are: U.S. Attorney General, Agriculture Secretary, Secretary of the Interior, Transportation Secretary, Secretary of State, Secretary of Defense, Education Secretary, Energy Secretary, Secretary of Veterans Affairs, Commerce Secretary, Labor Secretary, Secretary of the Treasury, etc. Some examples of non-cabinet appointments are: Office of Management and Budget, White House Chief of Staff, White House Council of Economic Advisers, and Environmental Protection Agency Administrator.]

2. One example of a formal organization is your college or university. Gather information about the characteristics of the organization. Is it constructed according to the specific characteristics of a bureaucracy as described by Max Weber? Explain your response. You may gather information from the college catalog, available organizational charts, and/or interviews with personnel who work in the organization. Draw an organizational chart to demonstrate the characteristics.

3. When Hurricane Katrina (August 29, 2005) hit the Gulf Coast, the failure of local, state, and federal government agencies to respond immediately, and the subsequent dysfunctions that followed, created national and international debates and resultant programs of assistance. Research this disaster in various publications from that time. Provide a description of what happened, what areas and cities of the United States were affected, and the resulting effects of this devastating hurricane. Research the *present* situation of the people, the cities, and the areas affected. Provide statistical as well as sociocultural descriptions of the hurricane and its impact on the major social institutions: family, educational, religious, economic, and political institutions. Provide sketches of the people affected and their personal stories. Provide pictures, statistical tables, and any other information pertinent to this topic.

INTERNET ACTIVITIES

1. Go to **Verstehen**, **http://www.faculty.rsu.edu/~felwell/Theorists/Weber/Whome.htm**, the Max Weber web site, and read about bureaucracy, rationality, oligarchy and ideal types.

2. Here are two sites on **McDonaldization**. These and other web sites can be used to acquire more information about this theory. Hold a discussion group with other class members and relate ideas from these sites to what you have learned about groups, organizations, and Weber's theories:
   http://www.geocities.com/mcdonaldization/
   http://www.umsl.edu/~rkeel/010/mcdonsoc.html

3. The sites below provide information on the devastating impact of Hurricane Katrina. From the information at the sites, describe how the victims of the hurricane are coping. What has been the federal government's response to the crisis? Locate

additional current sites to gather the most recent information, and present your written report in class: **http://www.msnbc.msn.com/id/9107338/**; **http://www.hurricaneadvisories.com/**

INFOTRAC COLLEGE EDITION EXERCISES

Visit the **InfoTrac College Edition** web site at: **http://www.wadsworthmedia.com/webtutor/infotrac.htm**. You will arrive at a screen that enables you to search topics.

1. Look for articles on **social isolation** that compare the consequences of web interaction to face-to-face interaction in a group. Note that much of the research to date finds negative consequences (loneliness and other psychological effects). After reviewing this research, explain the importance of face-to-face interactions to human development.
2. Look for research on **conformity**. What types of situations seem to increase or decrease conformity? What types of group membership are more likely to increase conformity (e.g., gangs, cults, etc.)?
3. Search for the keyword **leadership style** in the title, abstract, or citation of articles. Record information on effective leadership styles and bring that information to a meeting with some of your classmates. Work in a group to create a forum on leadership. Use information from the articles you found and from this chapter to construct a brochure on "How to Be a Successful Leader."
4. There are 107 periodical references listed under the category of **bureaucracy**. Try to discover the meaning of this common word. Bring to class examples of these articles. From these articles, construct a definition of the concept, along with its essential dimensions.

SOLUTIONS

MULTIPLE CHOICE QUESTIONS

1. D, p. 172-173
2. A, p. 174
3. D, p. 175
4. B, p. 179
5. D, p. 181-182
6. C, p. 182
7. B, p. 183
8. C, p. 184
9. A, p. 186
10. B, p. 188-189
11. C, p. 188-189
12. B, p. 186
13. A, p. 189
14. D, p. 189

15. A, p. 186
16. B, p. 177
17. A, p. 192-194
18. C, p. 191
19. A, p. 188-189
20. A, p. 192
21. A, p. 174
22. B, p. 174
23. A, p. 174
24. D, p. 184
25. A, p. 189
26. B, p. 189
27. B, p. 174
28. A, p. 176

29. B, p. 177
30. A, p. 177
31. A, p. 175
32. A, p. 179
33. B, p. 175
34. A, p. 181
35. A, p. 182
36. B, p. 184
37. A, p. 194
38. C, p. 195
39. A, p. 196
40. C, p. 185

TRUE/FALSE QUESTIONS

1. T, p. 174
2. T, p. 174-175
3. T. p. 177
4. T. p. 179
5. F, p. 182
6. T, p. 184
7. F. p. 183

8. T, p. 185
9. T, p. 186
10. F, p. 175
11. T, p. 188
12. T, p. 190
13. T, p. 191
14. F, p. 191-192

15. T, p. 192
16. T, p. 174
17. T, p. 176
18. T, p. 178
19. T, p. 193
20. T, p. 197

FILL-IN-THE-BLANK QUESTIONS

1. Georg Simmel, p. 179
2. William Graham Sumner, p. 175
3. Charles H. Cooley, p. 174
4. Stanley Milgram, p. 184
5. Irving Janis, p. 186
6. dual labor market, p. 194
7. the iron law of oligarchy, p. 195
8. quality circles, p. 196
9. goal displacement, p. 194
10. bureaucratic personality, p. 194
11. humanize, p. 195
12. Japanese, p. 196
13. Democratic, p. 182
14. Social exchange, p. 186
15. Postmodern, p. 177

7

DEVIANCE AND CRIME

BRIEF CHAPTER OUTLINE

CHAPTER SUMMARY

Deviance is any behavior, belief, or condition that violates significant social norms in the society or group in which it occurs. **Crime** is a form of deviant behavior that violates criminal law and is punishable by fines, jail terms, and other sanctions. A subcategory, **juvenile delinquency**, refers to a violation of a law or the commission of a status offense by a young person. All societies create norms in order to define, reinforce, and teach acceptable behavior. They also have various mechanisms of **social control**, systematic practices developed by social groups to encourage conformity and discourage deviance. However, deviance is relative, so what is defined as deviant varies across societies and changes over time. Each theoretical perspective views deviance differently. Functionalists suggest that deviance is inevitable in all societies and serves several functions: it clarifies rules, unites groups, and promotes social change. Functionalists use **strain theory** and opportunity theory to analyze access to **illegitimate opportunity structures**; they also use these theories to argue that socialization into the core value of material success without the corresponding legitimate means to achieve that goal accounts for much of the crime committed by people from lower income backgrounds, especially when a person's ties to society are weakened or broken. Conflict theorists suggest that people with economic and political power define as criminal any behavior that threatens their own interests and are able to use the law to protect their own interests. Symbolic Interactionists use **differential association theory**, differential reinforcement theory, **rational choice theory of deviance**, **social bond theory**, and **labeling theory** to explain how a person's behavior is influenced and reinforced by others. Postmodernists examine the intertwining nature of knowledge, power, and technology on social control and discipline. While the law classifies crime into felonies and misdemeanors based on the seriousness of the crime, sociologists categorize crimes according to how they are committed and how society views them. Some general categories of crime include: **violent**, **property**, public order (also referred to as **victimless**), **occupational** and **corporate**, **organized**, **and political crime.** Studies show that many more crimes are committed than are reported in official crime statistics. In explaining terrorism and crime, the rational choice approach suggests that terrorists are rational actors who constantly calculate the gains and losses of participation in violent—and often suicidal—acts against others. Gender, age, class, and race are key factors in official crime statistics. The criminal justice system includes the police, the courts, **punishment**, and corrections. These agencies often have considerable discretion in dealing with deviance. The death penalty, or capital punishment, has been used in the United States as an appropriate and justifiable response to very serious crimes and continues to be controversial. As we move into the future, we need new approaches for dealing with crime and delinquency. Equal justice under the law needs to be guaranteed, regardless of race, class, gender, or age. Global crime—the networking of powerful criminal organizations and their associates in shared activities around the world—has expanded rapidly in the era of global communications and rapid transportation networks. Reducing global crime will require a global response, including the cooperation of law enforcement agencies around the world.

LEARNING OBJECTIVES

After reading Chapter 7, you should be able to:

1. Define deviance, explain what it means to say that deviance is relative, and describe its most common forms in the U.S.

2. Discuss the functions of deviance from a functionalist perspective and outline the principal features of strain, opportunity, and control theories.

3. Identify and distinguish the three varieties of feminist approaches to deviance and crime.

4. State the four functions of punishment and explain how disparate treatment of the poor, all people of color, and white women is evident in the U.S. prison system.

5. Describe the key components of differential association theory, differential reinforcement theory, social control theory, rational choice theory, and labeling theory.

6. Compare and contrast functionalist, conflict, symbolic interactionist, and postmodern views of deviance.

7. Distinguish between legal and scholarly classifications of crime.

8. List the different categories of crime and explain and provide an example of each.

9. Describe the underground economy and the ways it enables criminal networks.

10. Differentiate between occupational and corporate crime and explain "criminals."

11. Describe organized crime and political crime and explain how each may weaken social control in a society.

12. Compare the different ways of measuring crime and describe the strengths and weaknesses of each.

13. Explain why official crime statistics may not be an accurate reflection of actual crime.

14. Discuss the relationship between gender, age, social class, race, and crime.

15. Define the criminal justice system.

16. Explain how police, courts, and prisons practice considerable discretion in dealing with offenders.

17. Discuss controversial issues surrounding the death penalty and explain recent court cases.

18. Discuss the concept of global crime, including the major types of global crime.

KEY TERMS

(defined at page number shown and in glossary)

corporate crime, p. 220
crime, p. 203
criminology, p. 205
deviance, p. 202
differential association theory, p. 214
illegitimate opportunity structures, p. 208
juvenile delinquency, p. 203
labeling theory, p. 215
occupational (white-collar) crime, p. 220
organized crime, p. 221
political crime, p. 222
primary deviance, p. 216

property crime, p. 220
punishment, p. 231
rational choice theory of deviance, p. 214
secondary deviance, p. 216
social bond theory, p. 215
social control, p. 204
strain theory, p. 206
tertiary deviance, p. 216
victimless crime, p. 220
violent crime, p. 219

KEY PEOPLE

(identified at page number shown)

Howard Becker, p. 216
Richard Cloward and Lloyd Ohlin, p. 208-209
Kathleen Daly and Meda Chesney-Lind, p. 212
Emile Durkheim, p. 206
Michel Foucault, p. 217
Travis Hirschi, p. 215

Edwin Lemert, p. 216
Karl Marx, p. 211
Robert Merton, p. 207-208
Felix Padilla, p. 208
Richard Quinney, p. 211-212
Walter Reckless, p. 215
Edwin Sutherland, p. 213-214

CHAPTER OUTLINE

I. WHAT IS DEVIANCE?
    A. **Deviance** is any behavior, belief, or condition that violates significant social norms in the society or group in which it occurs.
        1. Deviance is relative and it varies in its degree of seriousness: some forms of deviant behavior are officially defined as a crime.
        2. **Crime** is a form of deviant behavior that violates criminal law and is punishable by fines, jail terms, and other negative sanctions.
        3. **Juvenile delinquency** refers to a violation of law or the commission of a status offense by young people.

B. What is social control?
   1. All societies have norms that govern acceptable behavior and mechanisms of **social control**, systematic practices developed by social groups to encourage conformity and discourage deviance.
   2. **Criminology** is the systematic study of crime and the criminal justice system, including police, courts, and prisons.

II. FUNCTIONALIST PERSPECTIVES ON DEVIANCE
  A. What causes deviance, and why is it functional for society?
    1. Emile Durkheim regarded deviance as a natural and inevitable part of all societies.
    2. Deviance is universal because it serves three important functions:
      a. Deviance clarifies rules.
      b. Deviance unites a group.
      c. Deviance promotes social change.
    3. Functionalists acknowledge that deviance also may be dysfunctional for society; if too many people violate the norms, everyday existence may become unpredictable, chaotic, and even violent.
  B. Strain Theory: Goals and Means to Achieve Them
    1. **Robert Merton** modified Durkheim's concept of *anomie* to develop strain theory.
    2. According to **strain theory**, people feel strain when they are exposed to cultural goals that they are unable to obtain because they do not have access to culturally-approved means of achieving those goals. Robert Merton identified five ways in which people adapt to cultural goals and approved ways of achieving them:
      a. Conformity
      b. Innovation
      c. Ritualism
      d. Retreatism
      e. Rebellion
  C. Opportunity Theory: Access to Illegitimate Opportunities
    1. **Richard Cloward and Lloyd Ohlin** expanded Merton's strain theory to develop opportunity theory.
    2. Opportunity theory states that for deviance to occur, people must have access to **illegitimate opportunity structures**—circumstances that provide an opportunity for people to acquire through illegitimate activities what they cannot achieve through legitimate channels.

III. CONFLICT PERSPECTIVES ON DEVIANCE
  A. Deviance and Power Relations
    1. According to conflict theorists, people in positions of power maintain their advantage by using the law to protect their own interests.
  B. Deviance and Capitalism
    1. According to the Marxist/critical approach, the way laws are made and enforced benefits the capitalist class by ensuring that individuals at the bottom of the social class structure do not infringe on the property or threaten the safety of those at the top.

C. Feminist Approaches
   1. While there is no single feminist perspective on deviance and crime, three schools of thought have emerged:
      a. The *liberal feminist approach* is based on the assumption that women's deviance and crime is a rational response to gender discrimination experienced in work, marriage, and interpersonal relationships.
      b. The *radical feminist approach* is based on the assumption that women's deviance and crime is related to patriarchy (male domination over females). This approach focuses on forces that may cause women to commit crimes and biases that lead to female arrests (e.g., higher arrest rates for female prostitutes than male customers).
      c. The *Marxist (socialist) feminist approach* is based on the assumption that women's deviance and crime is the result of women's exploitation by capitalism and patriarchy (e.g., their overrepresentation in relatively low-wage jobs and their lack of economic resources).
   2. Feminist scholars of color have pointed out that these schools of feminist thought do not include race, ethnicity, or other significant social statuses in their analyses.
D. Approaches Focusing on Race, Class, and Gender
   1. According Arnold's research, criminal behavior of women in her study was linked to class, gender, and racial oppression, which the women experienced daily in their families, and at work and school.
   2. According to feminist sociologists and criminologists, research on women as both victims and perpetrators is long overdue.
IV. SYMBOLIC INTERACTIONIST PERSPECTIVES ON DEVIANCE
   A. Differential Association Theory and Differential Reinforcement Theory
      1. **Differential association theory**, created by **Edwin Sutherland**, states that individuals have a greater tendency to deviate from societal norms when they frequently associate with persons who are more favorable toward deviance than conformity.
      2. *Differential reinforcement theory* suggests that both deviant behavior and conventional behavior are learned through the same social processes.
   B. Rational Choice Theory
      1. The **rational choice theory of deviance** states that deviant behavior occurs when a person weighs the costs and benefits of nonconventional or criminal behavior and determines that the benefits will outweigh the risks involved in such actions.
   C. Control Theory: Social Bonding
      1. **Social bond theory**, created by **Travis Hirshi**, holds that the probability of deviant behavior increases when a person's ties to society are weakened or broken.
   D. **Labeling Theory**
      1. **Labeling theory** states that deviants are those people who have been successfully labeled as such by others.
      2. **Primary deviance** is the initial act of rule breaking.

3. **Secondary deviance** occurs when a person who has been labeled a deviant accepts that new identity and continues the deviant behavior.
4. **Tertiary deviance** occurs when a person labeled a deviant seeks to normalize the behavior by relabeling it as nondeviant.

V. POSTMODERNIST PERSPECTIVES ON DEVIANCE
  A. The postmodern perspective on deviance argues that the powerful exert control over the powerless by taking away their free will to think and act as they might choose.
    1. One example is **Michel Foucault**'s work, which examines the intertwining nature of knowledge, power, and technology on social control and discipline.
  B. Technologies make widespread surveillance and disciplinary power possible in many settings.
    1. These settings include state-police networks, factories, schools, and hospitals.
    2. Discipline would not sweep uniformly through society due to opposing forces that would use their power to oppose such surveillance.

VI. CRIME CLASSIFICATIONS AND STATISTICS
  A. How the Law Classifies Crime
    1. Crimes are divided into *felonies* and *misdemeanors* based on the seriousness of the crime.
  B. Other Crime Categories
    1. The *Uniform Crime Reports* are the major source of information on crimes and is complied by the Federal Bureau of Investigation.
      a. **Violent crime** consists of actions—murder, forcible rape, robbery, and aggravated assault—involving force or the threat of force against others.
      b. **Property crime** includes burglary, motor vehicle theft, larceny theft, and arson.
      c. *Public order crime* involves an illegal action voluntarily engaged in by participants, such as prostitution or illegal gambling. Public order crimes are also referred to as **victimless crimes** because they involve a willing exchange of illegal goods or services among adults.
      d. **Occupational (white collar) crime** includes illegal activities committed by people during employment. **Corporate crime** occurs when illegal acts are committed by corporate employees on behalf of the corporation and with its support.
      e. **Organized crime** is a business operation that supplies illegal goods and services for profit.
      f. **Political crime** refers to illegal or unethical acts involving the usurpation of power by government officials, or illegal/unethical acts perpetrated against the government by outsiders seeking to make a political statement, undermine the government, or overthrow it.
  C. Crime Statistics
    1. Official crime statistics, such as those found in the *Uniform Crime Reports*, provide important information on crime; however, the data reflect only those crimes that have been reported to the police.

2. The *National Crime Victimization Survey* and anonymous self-reports of criminal behavior have made researchers aware that the incidence of some crimes, such as theft, is substantially higher than reported in the UCR.
3. Data also show class and race biases in criminal justice enforcement.

D. Terrorism and Crime
1. In the twenty-first century, the United States and other nations are confronted with world terrorism and crime.
2. The nebulous nature of the "enemy" and other problems have resulted in a "war on terror" launched by the government.
3. *Terrorism* is the calculated, unlawful use of physical force or threats of violence against persons or property in order to intimate or coerce a government, organization or individual for the purpose of gaining some political, religious, economic, or social objective.

E. Street Crimes and Criminals
1. Gender and Crime
   a. The majority of people arrested are male.
   b. The three most common arrest categories for both men and women are driving under the influence of alcohol or drugs (DUI), larceny, and minor or criminal mischief types of offenses.
   c. There is a proportionately greater involvement of men in major property crimes and violent crime.
2. Age and Crime
   a. Arrest rates for index crimes are highest for people between the ages of 13 and 25, with the peak being between ages 16 and 17.
   b. Rates of arrest remain higher for males than females at every age and for nearly all offenses.
3. Social Class and Crime
   a. Individuals from all social classes commit crimes; they simply commit different kinds of crime.
   b. Persons from lower socioeconomic backgrounds are more likely to be arrested for violent and property crimes; only a very small proportion of individuals who commit white-collar or elite crimes will ever be arrested or convicted.
4. Race and Crime
   a. In 2007, whites (including Latinos/as) accounted for almost 70 percent of all arrests; African Americans are disproportionately arrested, comprising 12 percent of the population but 28 percent of all arrests. Latinos make up 13 percent of arrests, Native Americans 1.3 percent, and Asian Americans/Pacific Islanders 0.8 percent.
   b. Arrest records tend to produce overgeneralizations about who commits crime because arrest statistics are not an accurate reflection of the crimes actually committed in our society; white-collar and elite criminals are not always arrested; police may use racial bias when choosing who to question and arrest; law enforcement tend to focus more on certain neighborhoods and certain types of crime.

126

      c. Arrests should not be equated with guilt; being arrested does not mean that a person is guilty of the crime.

    5. Crime Victims

      a. Men are more likely to be victimized by crime although women tend to be more fearful of crime, particularly those directed toward them, such as forcible rape.

      b. The elderly also tend to be more fearful of crime, but are the least likely to be victimized.

      c. Young men of color between the ages of 12 and 24 have the highest criminal victimization rates.

      d. The burden of robbery victimization falls more heavily on males than females, African Americans more than whites, and young people more than middle aged and older persons.

VII. THE CRIMINAL JUSTICE SYSTEM

  A. *The criminal justice system* includes the police, courts, and correctional facilities. This system is a collection of bureaucracies that has considerable discretion regarding whether and how to take action on a given situation.

  B. The police are responsible for crime control and maintenance of order.

    1. *Racial profiling*, the use of ethnic/racial background to identify criminal suspects, remains a highly charged issue.

    2. The growing emphasis on *community-oriented policing*, a law enforcement approach where officers maintain a presence in the community, may enhance the image of police departments in the future.

  C. The courts determine the guilt or innocence of those accused of committing a crime.

    1. About 90 percent of criminal cases are resolved by plea bargaining.

    2. More serious crimes are more likely to proceed to trail.

    3. Sentencing involves various sanctions, such as fines, probation, incarceration, house arrest, electronic monitoring, and capital punishment.

  D. **Punishment** is any action designed to deprive a person of things of value (including liberty) because of something the person is thought to have done.

    1. Historically, punishment has had four major goals: *retribution*, *general deterrence*, *incapacitation*, and *rehabilitation*.

    2. Newer approaches, such as *restoration*, promote peacemaking, rather than punishment for offenders.

    3. The term *corrections*—which includes various programs, services, facilities, and organizations responsible for managing accused or convicted criminals—is often used instead of punishment.

    4. Corrections is a major activity today; more than one out of twenty males, one out of one hundred females are under some form of correctional control.

    5. Some argue that a *determinate sentence* with *mandatory sentencing guidelines* would help reduce disparities in corrections based on race, ethnicity, and class.

  E. The Death Penalty

    1. For many years, *capital punishment*, or the death penalty, has been used in the United States as a justifiable response to serious crimes.

2. Scholars have documented race and class biases in the imposition of the death penalty in this country.
3. Questions that remain today regarding capital punishment include the execution of individuals who are innocent, insane, have not received effective legal counsel, or are victims of racial discrimination.
4. Two landmark Supreme Court cases ruled on issues of capital punishment:
   a. The mentally retarded cannot be executed (*Atkins v. Virginia*).
   b. Juries, not judges, must decide if a convicted murderer should receive the death sentence (*Ring v. Arizona*).

VIII. DEVIANCE AND CRIME IN THE UNITED STATES IN THE FUTURE
   A. Although many people in the United States agree that crime is one of the most important problems facing this country, they are divided over what to do about it.
   B. The best approach for reducing delinquency and crime is prevention: to work with young people before they become juvenile offenders to help them establish family relationships, build self-esteem, choose a career, and get an education, which will help them pursue that career, as well as work to eliminate social injustices such as racism, sexism, classism, and ageism.

IX. THE GLOBAL CRIMINAL ECONOMY
   A. Global Crime—the networking of powerful criminal organizations and their associates in shared activities around the world—is a relatively new phenomenon that has expanded rapidly in the era of global communications and rapid transportation networks.
   B. Global networking and strategic alliances allow participants to escape police control and live beyond the laws of any one nation.
   C. Reducing global crime will require a global response, including the cooperation of law enforcement agencies around the world.

CRITICAL THINKING QUESTIONS

1. Explain the concepts of deviance and crime. How do certain actions come to be defined as deviant? How do these definitions change over time?
2. Which theoretical perspective do you think is most useful for understanding deviance and crime? Why? What are the strengths and weaknesses of each perspective?
3. In what ways has crime changed over time? How have broader social changes such as advances in technology and globalization contributed to changes in crime?
4. Summarize the arguments for and against the death penalty. Evaluate the arguments on both sides. After considering both sides, do you think the death penalty should be used in the U.S.? Why or why not?
5. In what ways do you think deviance, crime, and the criminal justice system will change in the future?

PRACTICE TESTS

MULTIPLE CHOICE QUESTIONS

Select the response that best answers the question or completes the statement.

1. Deviance is defined as any
   a. behavior that is inherently wrong.
   b. behavior, belief, or condition that violates social norms.
   c. serious violation of consistent moral codes.
   d. incorrect or inaccurate belief.

2. _____ refer(s) to systematic practices developed by social groups to encourage conformity and discourage deviance.
   a. Law
   b. Folkways
   c. Mores
   d. Social control

3. According to functionalists such as Emile Durkheim, deviance serves all of the following functions, **except**:
   a. deviance reduces crime
   b. deviance clarifies rules
   c. deviance promotes social change
   d. deviance unites a group

4. According to _____ theory, people are sometimes exposed to cultural goals that they are unable to obtain because they do not have access to culturally approved means of achieving those goals.
   a. containment
   b. status inaccessibility
   c. strain
   d. conflict

5. All of the following are included in Robert Merton's modes of adaptation to cultural goals and approved ways of achieving them, **except**:
   a. retribution
   b. ritualism
   c. retreatism
   d. rebellion

6. According to _____ theory, a teenager living in a poverty-ridden area of a central city is unlikely to become wealthy through a Harvard education, but some of his desires may be met through behaviors such as theft, drug dealing, and robbery.
   a. deviance management
   b. control
   c. illegitimate opportunity structures
   d. critical

129

7. Which theory would suggest that people must be taught that the risks of engaging in criminal behavior far outweigh any benefits gained from their actions?
   a. postmodern
   b. critical
   c. social bonding
   d. rational choice

8. _____ theory suggests that the probability of delinquency increases when a person's social bonds are weak and when peers promote antisocial values and violent behavior.
   a. Deviance management
   b. Social bond
   c. Illegitimate opportunity
   d. Critical

9. According to Edwin Lemert's typology, _____ deviance is exemplified by a person under the legal drinking age who orders an alcoholic beverage at a local bar but is not "caught" and labeled a deviant.
   a. primary
   b. secondary
   c. tertiary
   d. adolescent

10. The _____ approach argues that criminal law protects the interests of the affluent and powerful.
    a. functionalist
    b. liberal feminist
    c. interactionist
    d. conflict

11. _____ feminism explains women's deviance and crime as a rational response to gender discrimination experienced in work, marriage, and interpersonal relationships.
    a. Radical
    b. Communist
    c. Liberal
    d. Marxist

12. Marxist/socialist feminists argue that women's deviance and crime occurs because
    a. women are exploited by other women.
    b. women are exploited by capitalism and patriarchy.
    c. women are more likely to provide care for children.
    d. women are consumers and tend to purchase more than they can afford.

13. All of the following theorists are correctly matched with their theory, **except**:
    a. Howard Becker—labeling theory
    b. Robert Merton—strain theory
    c. Travis Hirschi—differential association
    d. Meda Chesney Lind—feminist approach

14. The _____ are the major source of information on crimes reported in the United States.
    a. Presidential Crime Reports
    b. Crime Statistical Lab Reports
    c. Civil Code Reports
    d. Uniform Crime Reports

15. Punishment for a _____ typically ranges from more than a year's imprisonment to death.
    a. felony
    b. misdemeanor
    c. parole violation
    d. moral crime

16. Much _____ crime is a violation of positions of trust at the expense of the general public.
    a. occupational or corporate
    b. street
    c. organized
    d. conventional

17. Drug trafficking, prostitution, loan sharking, and money laundering are examples of _____ crime.
    a. occupational/white-collar
    b. street
    c. organized
    d. conventional

18. All of the following are examples of political crime, **except**:
    a. unethical or illegal use of government authority for the purpose of material gain
    b. money laundering
    c. engaging in graft through bribery, kickbacks, or "insider" deals
    d. dubious use of public funds and public property

19. According to the text, rates of arrest
    a. are about the same for males and females at every age group and for most offenses.
    b. are slightly higher for females than males in the younger age levels and for violent crimes.
    c. are about the same for males and females for prostitution due to more stringent enforcement of criminal laws pertaining to male customers.
    d. remain higher for males than females at every age and for nearly all offenses.

131

20. All of the following are functions of punishment, **except**:
    a. deterrence
    b. retribution
    c. rehabilitation
    d. execution

21. Topically, sociologists and criminologists define a _____ as a group of people, usually young, who band together for purposes generally considered to be deviant or criminal by the larger society.
    a. gang
    b. peer formation
    c. criminal conspiracy
    d. juvenile coalition

22. For many years, capital punishment (the death penalty) has been used in the United States as an appropriate and justifiable response to very serious crime. In 2008, _____ inmates were executed and more than 3,300 people awaited executions, having received the death penalty under federal law or the law of one of the 38 states that have the death penalty.
    a. 12
    b. 37
    c. 76
    d. 197

23. _____ imposes a penalty on the offender and is based on the premise that the punishment should fit the crime. The greater the degree of social harm, the more the offender should be punished.
    a. Punishment
    b. Rehabilitation
    c. Restoration
    d. Retribution

24. African Americans are overrepresented in arrest data. In 2007, African Americans made up about 12 percent of the U.S. population but accounted for _____ percent of all arrests.
    a. 28
    b. 15
    c. 50
    d. 60

25. All of the following crimes are considered to be felonies, **except**:
    a. homicide
    b. aggravated assault
    c. larceny
    d. rape

26. The racial/ethnic group that accounts for the majority of all arrests is:
    a. whites
    b. African Americans
    c. Native Americans
    d. Asian Americans

27. According to research conducted by sociologists Richard Cloward and Llyod Ohlin, _____ gangs emerge in communities that do not provide either legitimate or illegitimate opportunities.
    a. retreatist
    b. criminal
    c. conflict
    d. street

28. A minister who is opposed to war conducts a nonviolent protest at a local military installation, thus committing a trespassing violation. This person has engaged in:
    a. conformity
    b. innovation
    c. rebellion
    d. retribution

29. Sociologist _____ believed that deviance is rooted in societal factors such as rapid social change and lack of social integration among people. As social integration (bonding and community involvement) decreases, deviance and crime increases.
    a. Emile Durkheim
    b. Karl Marx
    c. Edwin Sutherland
    d. Robert Merton

30. _____ is the systematic study of crime and the criminal justice system, including the police, courts, and prison.
    a. Criminology
    b. Survey
    c. Sociology
    d. Ethnomethodology

31. _____ social control takes place through the socialization process. Individuals internalize societal norms and values that prescribe how people should behave and then follow those norms and values in their everyday lives.
    a. Internal
    b. Interior
    c. External
    d. Exterior

32. We are most familiar with _____ deviance, based on a person's intentional or inadvertent actions. For example, a person may engage in intentional deviance by drinking too much or robbing a bank; or in inadvertent deviance by losing money in a Las Vegas casino or laughing at a funeral.
    a. intentional
    b. behavioral
    c. cultural
    d. social

33. Members of _____ seek to acquire a "rep" (reputation) by fighting over "turf" (territory) and adopting a value system of toughness, courage, and similar qualities.
    a. retreatist gangs
    b. criminal gangs
    c. street gangs
    d. conflict gangs

34. Sociologist Travis Hirschi's _____ theory holds that the probability of deviant behavior increases when a person's ties to society are weakened or broken.
    a. illegitimate opportunity
    b. social control
    c. labeling
    d. social bond

35. The theoretical perspective that examines the relationship between knowledge, power, and social control is:
    a. functionalism
    b. conflict theory
    c. symbolic interactionism
    d. postmodernism

36. John is a drug addict who believes that using marijuana or other illegal drugs is no more deviant that drinking alcoholic beverages and therefore should not be stigmatized. John would be in the _____ of labeling.
    a. career stage
    b. primary stage
    c. delinquent stage
    d. tertiary stage

37. Prostitution might be explained as a reflection of society's double standard, whereby it is acceptable for a man to pay for sex but unacceptable for a woman to accept money for such services. This explanation reflects a _____ feminist approach.
    a. Marxist
    b. liberal
    c. conventional
    d. radical

134

38. According to the text, the best way to solve the "crime problem" in the U.S. is to
   a. build more prisons.
   b. execute more criminals.
   c. create better education, job, and housing opportunities.
   d. increase racial profiling.

39. Of all factors associated with crime, the age of the offender is one of the most significant. Arrest rates for index crimes are highest for people between the ages of:
   a. 10 and 18
   b. 13 and 25
   c. 36 and 45
   d. 26 and 35

40. The National Crime Victimization Survey (NCVS) was developed by the Bureau of Justice Statistics as an alternate means of collecting crime statistics. The most recent NCVS indicates that _____ percent of violent crimes are not reported to the police.
   a. 30
   b. 40
   c. 50
   d. 80

TRUE/FALSE QUESTIONS

1. People may be regarded as deviant if they possess a certain condition, such as obesity or AIDS.
   T   F

2. Actions defined as deviant are relative, but crimes are inherently deviant.
   T   F

3. Deviance is found in all societies.
   T   F

4. A person who accepts culturally approved goals, but rejects the approved means of achieving them is an example of Merton's ritualism.
   T   F

5. According to differential association theory, people learn the necessary techniques and motivation for deviant behavior from people with whom they associate.
   T   F

6. Circumstances that provide individuals with an opportunity to acquire goods and services through illegal activities are referred to as illegitimate opportunity structures.
   T   F

7. Opportunity theory and strain theory are most useful for enhancing our understanding of the relationship between certain forms of deviance and gender.
   T   F

8. Michel Foucault argued that power and social control are exerted over individuals through the knowledge that exists in a society.
   T   F

9. Howard Becker's labeling theory posits that acts become deviant when they are defined that way.
   T   F

10. The radical feminist approach suggests that women's deviance and crime are a rational response to gender discrimination that women experience in families and the workplace.
    T   F

11. Political crime is a business operation that supplies illegal goods and services for a profit.
    T   F

12. The death penalty is used in all states.
    T   F

13. Moral crimes are often referred to as victimless crimes.
    T   F

14. According to the rational choice approach, terrorists are rational actors who constantly calculate the gains and losses of participation in violent and sometimes suicidal acts against others.
    T   F

15. A recent study by the justice department found that Native Americans are more likely to be victims of violent crimes than are members of any other racial/ethic group.
    T   F

16. According to Box 7.1, "Sociology and Everyday Life," gangs are only a problem in urban areas.
    T   F

17. Various measures of crime statistics show that about 85 percent of crimes committed are reported to police.
    T   F

18. Although corporate crimes are often more costly to society in terms of money and lives lost than street crimes, they are typically punished less harshly.
    T   F

19. In Box 7.4, "The Global Reach of Russian Organized Crime," analysts have documented that more than thirty Russian organized crime groups currently operate in the U.S.
    T   F

20. African Americans are eight to ten times more likely to be sentenced to death for homicidal rape than non-Latino/a whites who have committed the same crime.
T F

FILL-IN-THE-BLANK QUESTIONS

1. _____ is the systematic study of crime and the criminal justice system.

2. According to _____, criminal behavior is learned within intimate personal groups.

3. _____ occurs when a person accepts the label of "deviant" and continues to engage in the behavior that produced the label.

4. The _____ views the cause of women's crime as originating in patriarchy.

5. _____ crime comprises illegal activities committed by people in the course of their employment or financial affairs.

6. _____ crimes are illegal acts committed by corporate employees on behalf of the corporation.

7. According to the research of _____ and _____, the three basic gang types: criminal, conflict, and retreatist, emerge on the basis of what types of structures of illegitimate opportunities are available.

8. According to sociologist _____, people with economic and political power define behaviors that threaten their own interest as criminal.

9. _____ theory holds that the probability of deviant behavior increases when a person's ties to society are weakened or broken.

10. _____ theory refers to the systematic practices that social groups develop in order to encourage conformity to norms, rules, and laws and to discourage deviance.

11. A process in which the prosecution negotiates a reduced sentence for the accused in exchange for a guilty plea is know as _____.

12. The _____ suggests that both deviant behavior and conventional behavior are learned through the same social processes.

13. The _____ is based on the assumption that women are exploited by both capitalism and patriarchy.

14. The Supreme Court decision *Ring v. Arizona* ruled that _____ must decide whether a convicted murderer should receive the death penalty.

15. The practice of _____ is the use of ethnic or racial background as a means of identifying criminal suspects and remains a highly charged issue in the criminal justice system.

SHORT ANSWER/ESSAY QUESTIONS

1. What is meant by deviant behavior? What are some factors that help determine whether an act will be viewed as deviant?
2. Compare and contrast the main points of the functionalist, conflict, and symbolic interactionist perspectives of deviance.
3. What are the primary assumptions of conflict and feminist perspectives on deviance? What are some of the gaps in research that these two perspectives have attempted to fill?
4. What is meant by the global criminal economy? What are some examples? How can global crime be reduced?
5. How are age, race, class, and gender related to crime statistics?

STUDENT CLASS PROJECTS AND ACTIVITIES

1. The organized crime syndicate the Mafia has a long and interesting history. Research the origin, purpose, and growth of the Mafia in Italy and Sicily, noting some similarities and differences of each. Then explain the growth of the Mafia in the United States. Describe some of the original neighborhood programs that the Mafia sponsored. Explain why you think it is difficult to eradicate this crime syndicate.
2. Research the topic of crime rates in a cross-cultural perspective, with specific emphasis on comparing the rates of the U.S. with those of other industrialized countries. Summarize your findings, provide some explanations for your findings, and included a bibliography of your references.
3. As indicated in your textbook, global crime—the networking of powerful criminal organizations and their associates in shared activities around the world—is a relatively new phenomenon, although these organizations have existed for many years in their country of origin. 1) Select five global criminal organizations; 2) explain the origin, purpose, and growth of the organization; 3) describe some of their illegal activities; 4) examine and explain some key factors in the success of the expansion of their criminal activity; 5) provide some possible solutions for reducing global crime; and 6) critique your solutions. What would it take for your solutions to work?
4. Research the identification and expansion of world terrorism. What nations of the world today support and/or sponsor world terrorism? Why? In your research, identify various diverse "cells" of terrorists. Who are their leaders? What theoretical approaches best explain the existence of terrorists? What are some possible solutions for eradicating world terrorism and crime? Provide bibliographic references.

INTERNET ACTIVITIES

1. **"Defining Deviancy Down,"** **http://www.albany.edu/scj/jcjpc/vol2is5/deviancy.html**, explains how Senator Moynihan's misleading statement about criminal justice is being incorporated into popular culture. Read how this occurs at the site above.

138

2. Videos available from **CNN**, **http://www.cnn.com/CRIME/**, on recent trials or specific topics in the criminal justice system typically are well researched, objectively presented, and of great interest to students. Study specific court cases and follow the progress of each case from this web site. You can view video clips on your computer.

3. The Internet has become a highly useful source for obtaining statistical data and current information about crime in the United States. Search these sites and bring back to class important crime facts.
   http://www.ojp.usdoj.gov/bjs/pubalp2.htm
   http://www.ojp.usdoj.gov/bjs/welcome.html
   http://www.ojp.usdoj.gov/bjs/cvict.htm
   http://www.ojp.usdoj.gov/bjs/drugs.htm
   http://www.ncjrs.gov

INFOTRAC COLLEGE EDITION EXERCISES

Visit the **InfoTrac College Edition** web site at:
http://www.wadsworthmedia.com/webtutor/infotrac.htm. You will arrive at a screen that enables you to search topics.

1. Examine the **Juvenile Crime Bill of 1999** (draft). What kind of picture does it tend to paint about juvenile justice in America?

2. Search through the articles on **youth gangs**. Create your own questionnaire in order to conduct a survey based on factual information you find during your search. Have students on your campus participate in your survey and write up the results.

3. Search for information on **peer groups**. Type the phrase "age groups" in the search box. Go to the periodical listings. Compose a list of research findings from these articles.

4. There are numerous subdivisions under **deviant behavior**. Work in small groups and select a category of interest; then present your group's findings to the class. To enhance your class presentation, create a poster with facts you find within these categories.

5. Using the periodical references on **deviant behavior**, bring to class those which focus on the theories of deviance from this chapter. (You are not allowed to use articles with the word "theory" in the title!)

6. Look up the word **terrorism** and compare the various pieces of legislation that have been enacted to address this expanding problem.

SOLUTIONS

MULTIPLE CHOICE QUESTIONS

1. B, p. 202	15. A, p. 219	29. A, p. 206
2. D, p. 204	16. A, p. 220	30. A, p. 205
3. A, p. 206	17. C, p. 221	31. A, p. 204
4. C, p. 206	18. B, p. 222	32. B, p. 202
5. A, p. 207-208	19. D, p. 224	33. D, p. 210
6. C, p. 208	20. D, p. 231	34. D, p. 215
7. D, p. 214	21. A, p. 201	35. D, p. 217
8. B, p. 215	22. B, p. 232	36. D, p. 216
9. A, p. 216	23. D, p. 231	37. D, p. 212
10. D, p. 212	24. A, p. 226	38. C, p. 234
11. C, p. 212	25. C, p. 219	39. B, p. 224
12. B, p. 212	26. A, p. 226	40. C, p. 226-227
13. C, p. 218	27. C, p. 208-209	
14. D, p. 219	28. C, p. 206	

TRUE/FALSE QUESTIONS

1. T, p. 202	8. T, p. 217	15. T, p. 227
2. F, p. 203	9. T, p. 216	16. T, p. 204
3. T, p. 206	10. F, p. 212	17. F, p. 223
4. F, p. 208	11. F, p. 222	18. T, p. 221
5. T, p, 214	12. F, p. 232	19. T, p. 235
6. T, p. 208	13. T, p. 220	20. T, p. 232
7. F, p. 208	14. T, p. 214	

FILL-IN-THE-BLANK QUESTIONS

1. Criminology, p. 205
2. differential association theory, p. 214
3. Secondary deviance, p. 216
4. radical feminist approach, p. 212
5. Occupational (white-collar), p. 220
6. Corporate, p. 220
7. Cloward, p. 208; Ohlin, p. 208
8. Richard Quinney, p. 211
9. Social bond, p. 215
10. Social control, p. 204
11. plea bargaining, p. 229-230
12. differential reinforcement theory, p. 214
13. Marxist (socialist), p. 212 feminist approach, p. 212
14. juries, p. 233
15. racial profiling, p. 228

8

CLASS AND STRATIFICATION IN THE UNITED STATES

BRIEF CHAPTER OUTLINE

WHAT IS SOCIAL STRATIFICATION?
SYSTEMS OF STRATIFICATION
   Slavery
   The Caste System
   The Class System
CLASSICAL PERSPECTIVES ON SOCIAL CLASS
   Karl Marx: Relationship to the Means of Production
   Max Weber: Wealth, Prestige, and Power
CONTEMPORARY SOCIOLOGICAL MODELS OF THE U.S. CLASS STRUCTURE
   The Weberian Model of the U.S. Class Structure
   The Marxian Model of the U.S. Class Structure
INEQUALITY IN THE UNITED STATES
   Distribution of Income and Wealth
   Consequences of Inequality
POVERTY IN THE UNITED STATES
   Who Are the Poor?
   Economic and Structural Sources of Poverty
   Solving the Poverty Problem
SOCIOLOGICAL EXPLANATIONS OF SOCIAL INEQUALITY IN THE UNITED STATES
   Functionalist Perspectives
   Conflict Perspectives
   Symbolic Interactionist Perspectives
U.S. STRATIFICATION IN THE FUTURE

CHAPTER SUMMARY

Social stratification is the hierarchical arrangement of large social groups based on their control over basic resources. A key characteristic of systems of stratification is the extent to which the structure is flexible. **Social mobility** is the movement of individuals or groups from one stratification level to another. **Slavery**, a form of stratification in which people are owned by others, is a closed system. In a **caste system**, people's status is determined at birth based on their parents' position in society. The **class system**, which exists in the United States, is a type of stratification based on ownership of resources and the type of work people do. **Karl Marx** and **Max Weber** viewed class as a key determinant of social inequality and social change. According to Marx, capitalistic societies are comprised of two classes—the capitalists, who own the means of production, and the workers, who sell their labor to the owners. By contrast, Weber developed a multidimensional concept that focuses on the interplay of **wealth**, **prestige**, and **power**. Sociologists have developed several models of the class structure: one is the broadly based Weberian approach; the second is based on a Marxian approach. Throughout human history, people have argued about the distribution of scarce resources in society. Money, in the form of both income and wealth, is unevenly distributed in the U.S. Among the most prosperous nations in today's world, the United States has the highest degree of inequality of income distribution. **Income**—the economic gain derived from wages, salaries, income transfers, or ownership of property—and **wealth**—the value of all of a person's or family's economic assets—is unevenly distributed in the United States. The stratification of society into different social groups results in wide discrepancies in income, wealth, and access to available goods and services (including physical health, mental health, nutrition, housing, education, and safety). Sociologists distinguish between **absolute poverty**, which exists when people do not have the means to secure the basic necessities of life, and **relative poverty**, which exists when people may be able to afford basic necessities but are still unable to maintain an average standard of living. Poverty is highly concentrated according to age, gender, and race/ethnicity. There are both economic and structural sources of poverty. Low wages are a key problem, as are unemployment and underemployment. The United States has attempted to solve the poverty problem, with the most enduring program referred to as *social welfare*. Functionalist perspectives on the U.S. class structure view classes as broad groupings of people who share similar levels of privilege based on their roles in the occupational structure. According to the Davis-Moore thesis, inequality is necessary because positions that are most important within a society, requiring the most talent and training, must be highly rewarded. Conflict perspectives are based on the assumption that social stratification is created and maintained by one group in order to enhance and protect its own economic interests. Symbolic interactionists focus on a microlevel analysis, such as the impact of social class on people's everyday activities and interactions. As the gap between rich and poor, employed and unemployed widens, social inequality will continue to increase in the future if society does not address this problem. Given that the well-being of all people is linked, some analysts are urging a joint effort to regain the American Dream by attacking poverty.

LEARNING OBJECTIVES

After reading Chapter 8, you should be able to:

1. Define social stratification and social mobility.

2. Summarize the different types of social mobility and explain the reality of social mobility in the U.S.

3. Describe the different systems of stratification that exist throughout the world and summarize the main characteristics of each.

4. Discuss slavery and its relationship to global poverty.

5. Describe Marx's perspective on class position and class relationships.

6. Outline Weber's multidimensional approach to social stratification and explain how people are ranked on all three dimensions.

7. Describe contemporary sociological models of the U.S. class structure that are based on the ideas of Weber and Marx.

8. Compare socioeconomic status and social class.

9. Define income and wealth and describe their relation to social class.

10. Describe the ways that income is distributed in the U.S.

11. Explain and provide examples of the relationship between SES and life chances.

12. Discuss the distribution of income and wealth in the U.S. and describe how this distribution affects life chances.

13. Summarize the most important consequences of inequality in the U.S.

14. Describe current statistics about the poor in the U.S.

15. Distinguish between absolute and relative poverty and describe the characteristics and lifestyle of those who live in poverty in the U.S.

16. Compare functionalist and conflict approaches to explaining social inequality in the U.S.

17. Describe the contributions of the symbolic interactionist perspective to understanding social inequality.

KEY TERMS

(defined at page number shown and in glossary)

absolute poverty, p. 268
alienation, p. 248
capitalist class (or bourgeoisie), p. 248
caste system, p. 245
class conflict, p. 249
class system, p. 247
feminization of poverty, p. 270
income, p. 260
intergenerational mobility, p. 242
intragenerational mobility, p. 242
job deskilling, p. 271
life chances, p. 243

meritocracy, p. 272
pink-collar occupations, p. 255
power, p. 252
prestige, p. 252
relative poverty, p. 268
slavery, p. 245
social mobility, p. 244
social stratification, p. 243
socioeconomic status (SES), p. 252
wealth, p. 251
working class (or proletariat), p. 248

KEY PEOPLE

(identified at page number shown)

Peter Blau and Otis Duncan, p. 252-253
Kingsley Davis and Wilbert Moore, p. 273
Dennis Gilbert and Joseph A. Kahl, p. 254

Karl Marx, p. 248
Max Weber, p. 250-251
Erik Olin Wright, p. 257

CHAPTER OUTLINE

I. WHAT IS SOCIAL STRATIFICATION?
   A. **Social stratification** is the hierarchical arrangement of large social groups based on their control over basic resources.
   B. **Max Weber**'s term **life chances** describes the extent to which individuals within a particular layer of stratification have access to important scarce resources.
II. SYSTEMS OF STRATIFICATION
   A. Systems of stratification may be open or closed based on the availability of **social mobility**—the movement of individuals or groups from one level in a stratification system to another.
      1. **Intergenerational mobility** is the social movement experienced by family members from one generation to the next.
      2. **Intragenerational mobility** is the social movement of individuals within their own lifetime.
   B. **Slavery**, a closed system, is an extreme form of stratification in which some people are owned by others.

1. Some analysts suggest that, throughout recorded history, only five societies have been slave societies.
2. Some believe that slavery will not be totally abolished until debt bondage, child labor, contract labor, and coerced work cease to exist throughout the world.

C. A **caste system** is a system of social inequality in which people's status is permanently determined at birth based on their parents' ascribed characteristics.
   1. In India, caste is based in part on occupation; in South Africa it was based on racial classification.
   2. Cultural beliefs sustain caste systems.
   3. Caste systems grow weaker as societies industrialize; people start to focus on the type of skills needed for industrialization.

D. The **class system** is a type of stratification based on the ownership and control of resources and on the type of work people do.
   1. A class system is more open than a caste; boundaries between classes are less distinct.
   2. Status comes in part through achievement, rather than entirely by ascription.

III. CLASSICAL PERSPECTIVES ON SOCIAL CLASS
   A. **Karl Marx**: Relationships to the Means of Production
      1. According to Marx, class position in capitalistic societies is determined by people's work situation, or relationship to the means of production.
         a. The **bourgeoisie** or **capitalist class** consists of those who privately own the means of production; the **proletariat**, or **working class**, must sell their labor power to the owners in order to earn enough money to survive.
         b. Class relationships involve inequality and exploitation; workers are exploited as capitalists expropriate a surplus value from their labor; continual exploitation results in workers' **alienation**, a feeling of powerlessness and estrangement from other people and from oneself.
      2. The **capitalist class** maintains its position by control of the society's *superstructure*—comprised of the government, schools, and other social institutions that produce and disseminate ideas perpetuating the existing system; this exploitation of workers ultimately results in **class conflict**—the struggle between the capitalist class and working class.
   B. **Max Weber**: **Wealth**, **Prestige**, and **Power**
      1. Weber's multidimensional approach to stratification examines the interplay among wealth, prestige, and power as being necessary in determining a person's class position.
         a. Weber placed people who have a similar level of **wealth**—the value of all of a person's or family's economic assets, including income, personal property, and income-producing property—and income in the same class.
         b. **Prestige** is the respect or regard with which a person or status position is regarded by others, and those who share similar levels of social prestige belong to the same status group regardless of their level of wealth.
         c. **Power**—the ability of people or groups to carry out their own goals despite opposition from others—gives some people the ability to shape society in accordance with their own interests and direct the actions of others.

145

2. Wealth, prestige, and power are separate continuums on which people can be ranked from high to low; individuals may be high on one dimension while low on another.

3. **Socioeconomic status (SES)**—a combined measure that attempts to classify individuals, families, or households in terms of indicators such as income, occupation, and education—is used to determine class location.

4. **Peter Blau and Otis Duncan** measured the three dimensions from Weber's theory and found that occupational position is not identical to either economic class or prestige, but is closely related to both.

IV. CONTEMPORARY SOCIOLOGICAL MODELS OF THE U.S. CLASS STRUCTURE

A. The Weberian Model of the Class Structure: **Dennis Gilbert and Joseph A. Kahl** developed a model of social classes based on education, occupation, and family income. Their classification scheme generated the following groups:

1. The *upper (or capitalist) class* is the wealthiest and most powerful class, comprised of people who own substantial income producing assets. About 1 percent of the population is included in this class.

2. The *upper-middle class* is based on a combination of three factors: university degrees, authority and independence on the job, and high income. Examples of occupations for this class are highly educated professionals such as physicians, stockbrokers, or corporate managers. About 14 percent of the U. S. population is in this category.

3. The *middle class* has been traditionally characterized by a minimum of a high school diploma. Today, an entry-level requirement for employment in many middle-class occupations is a two- or four-year college degree. About 30 percent of the U. S. population is in this class, even though most people in this country think of themselves as middle class.

4. The *working class* is comprised of semiskilled machine operatives, clerks, and salespeople in routine, mechanized jobs, and workers in **pink-collar occupations**—relatively low paying, nonmanual, semiskilled positions primarily held by women. An estimated 30 percent of the U. S. population is in this class.

5. The *working poor* account for about 20 percent of the U. S. population and live from just above to just below the poverty line; they hold unskilled jobs, seasonal migrant employment in agriculture, lower-paid factory jobs, and service jobs (e.g., such as counter help at restaurants).

6. The *underclass* includes people who are poor, seldom employed, and caught in long-term deprivation resulting from low levels of education and income and high rates of unemployment. About 3 to 5 percent of the U. S. population is in this category.

B. The Marxian Model of the U.S. Class Structure

1. Contemporary Marxian (or conflict) models view classes based on people's relationships to each other in the production process.

   a. **Erik Olin Wright** outlined four criteria for placement in the class structure: (a) ownership of the means of production; (b) purchase of the labor of others (employing others); (c) control of the labor of others (supervising others on the job); and (d) sale of one's own labor (being employed by someone else).

   b. Wright used these criteria to identify four classes: capitalist, managerial, business, and working.

2. The *capitalist class* is composed of those who have inherited fortunes, own major corporations, or are top corporate executives who own extensive amounts of stock or control company investments.

3. The *managerial class* includes upper-level managers—supervisors and professionals who typically do not participate in company-wide decisions—and lower-level managers who may be given some control over employment practices, such as the hiring and firing of workers.

4. The *small-business class* consists of small business owners, craftspeople, and some doctors and lawyers who may hire a small number of employees but largely do their own work.

5. The *working class* is made up of blue-collar workers, including skilled workers (e.g., electricians, plumbers, and carpenters), unskilled blue-collar workers (e.g., laundry and restaurant workers), and white-collar workers who do not own the means of production, do not control the work of others, and are relatively powerless in the workplace. The working class constitutes about half of all employees in the U.S.

V. INEQUALITY IN THE UNITED STATES
   A. Distribution of Income and Wealth
      1. Income and wealth are very unevenly distributed in the United States. Among prosperous nations, the United States is number one in inequality of income distribution.
      2. Income Inequality
         a. Income is the economic gain derived from wages, salaries, income transfers, and ownership of property.
         b. In the last two decades of the twentieth century, the gulf between the rich and the poor widened in the United States. Income distribution varies by race/ethnicity, class, and gender.
      3. Wealth Inequality
         a. Wealth includes property and other assets such as bank accounts, corporate stocks, bonds, and insurances policies.
         b. An analysis of the Forbes 400 list of the wealthiest U.S. citizens found that 42 percent of the people on the list had inherited enough wealth to be on the list. Disparities in wealth are more pronounced across racial/ethnic and gender lines.
   B. Consequences of Inequality
      1. Physical and Mental Health and Nutrition
         a. As people's economic status increases, so does their health status; the poor have shorter life expectancies and are at greater risk for chronic illnesses and infectious diseases. About 45.7 million people in the United States are without health insurance coverage.
         b. Good health and adequate nutrition are essential to good life chances; low-income families are unable to provide adequate food for their children.

2. Housing
    a. Homelessness is a major problem in the U.S.
    b. Lack of affordable housing and substandard housing are also central problems brought about by economic inequality.
3. Education
    a. Education and life chance are directly linked; while functionalists view education as an "elevator" for social mobility, conflict theorists stress that schools are agencies for reproducing the capitalist class system and perpetuating inequality in society.
    b. Today there are vast disparities in the distribution of educational resources.
    c. Poverty affects the ability of many young people to finish high school, much less enter college.
4. Crime and Lack of Safety
    a. Both crime and lack of safety on the streets and at home are consequences of inequality.
    b. Poverty and violence are linked; street violence is often not random at all—but a response to profound social inequalities in the inner city.

VI. POVERTY IN THE UNITED STATES
    A. Although some people living in poverty are unemployed, many hardworking people with full-time jobs also live in poverty.
    B. The *official poverty line* is based on what is considered to be the minimum amount of money required for living at a subsistence level.
    C. Sociologists distinguish between **absolute poverty**—when people do not have the means to secure the most basic necessities of life—and **relative poverty**—when people may be able to afford basic necessities but still are unable to maintain an average standard of living.
    D. Who Are the Poor?
        1. Age: Children are more likely to be poor than older persons; older women are twice as likely to be poor as older men; older African Americans and Latino/as are much more likely to live below the poverty line than are non-Latino/a whites.
        2. Gender: About two-thirds of all adults living in poverty are women; this problem is described as the **feminization of poverty**—the trend in which women are disproportionately represented among individuals living in poverty.
        3. Race and Ethnicity: White Americans (non-Latino/as) account for approximately two-thirds of those below the official poverty line; however, a disproportionate percentage of the poverty population is made up of African Americans, Latino/as, and Native Americans, with Native Americans among the most severely disadvantaged.
    E. Economic and Structural Sources of Poverty
        1. Economic sources of poverty include unemployment and low wages. Poor economic times result in increased unemployment; others are poor due to low wages; half of all families living in poverty are headed by someone who is employed, and one-third of those family heads work full time.

148

2. Poverty also is exacerbated by structural problems such as (a) *deindustrialization*—millions of U.S. workers have lost jobs as corporations have disinvested here and opened facilities in other countries where "cheap labor" exists—and (b) **job deskilling**—a reduction in the proficiency needed to perform a specific job, which leads to a corresponding reduction in the wages paid for that job.

F. Solving the Poverty Problem
   1. The United States has attempted to solve the poverty problem with social welfare programs; however, the primary beneficiaries are not poor.

VII. SOCIOLOGICAL EXPLANATIONS OF SOCIAL INEQUALITY IN THE UNITED STATES

A. Functionalist Perspectives
   1. According to the Davis-Moore thesis created by **Kingsley Davis and Wilbert Moore**:
      a. Inequality is necessary for the smooth functioning of society.
      b. The most important positions in a society must be filled by the most qualified people, and therefore must provide the greatest rewards in order to attract the most talented and qualified people.
   2. This thesis assumes that social stratification results in **meritocracy**—a hierarchy in which all positions are rewarded based on people's ability and credentials.

B. Conflict Perspectives
   1. From a conflict perspective, inequality does not serve as a source of motivation for people; powerful individuals and groups use ideology to maintain their favored positions at the expense of others.
   2. Core values, laws, and informal social norms support inequality in the United States (e.g., legalized segregation and discrimination produce higher levels of economic inequality).

C. Symbolic Interactionist Perspectives
   1. Symbolic interactionists focus on microlevel concerns such as the effects of wealth and poverty on people's lives.
   2. Some studies focus on the social and psychological factors that influence the rich to contribute to charitable and arts organizations.
   3. Other researchers have examined the social and psychological aspects of life in the middle class; a few have examined social interactions between people from vastly different social classes.

VIII. U.S. STRATIFICATION IN THE FUTURE

A. The U.S. is facing one of the greatest economic challenges since the Great Depression.
B. According to some social scientists, wealth will become more concentrated at the top of the U.S. class structure; as the rich have grown richer, more people have found themselves among the ranks of the poor.
C. The gap between the earnings of workers and the income of managers and top executives in the U.S. has widened.

149

D. Structural sources of upward mobility are shrinking, while the rate of downward mobility has increased; the persistence of economic inequality is related to profound global economic changes.
E. Some call for a united effort to regain the American Dream by attacking poverty.

CRITICAL THINKING QUESTIONS

1. Compare the different systems of stratification discussed in the chapter. What are their similarities and differences?
2. Compare and contrast Karl Marx's and Max Weber's perspectives on social class. Which perspective do you think is most useful today? Why?
3. What factor(s) do you think is/are most important in determining social class in the U.S.? Why? Which model of the U.S. class structure do you think is most accurate? Why?
4. Which theoretical perspective do you think is most useful for understanding social class inequality in the U.S.? What are the strengths and weaknesses of each perspective?
5. What kinds of things could be done to reduce poverty? What could be done to reduce social class inequality?
6. What is the appropriate role of the government in reducing or eliminating poverty? Why?

PRACTICE TESTS

MULTIPLE CHOICE QUESTIONS

Select the response that best answers the question or completes the statement.

1. The term _____ refers to the hierarchical arrangements of large social groups based on their control over basic resources.
    a. social stratification
    b. social layering
    c. social distinction
    d. social accumulation

2. The extent to which individuals have access to important societal resources is known as:
    a. relative poverty
    b. absolute poverty
    c. social mobility
    d. life chances

150

3. A young woman's father is a carpenter; she graduates from college with a degree in accounting, becomes a CPA, and has a starting salary that represents more money than her father ever made in one year. This illustrates _____ mobility.
    a. intragenerational
    b. intergenerational
    c. horizontal
    d. subjective

4. All of the following are true statements about slavery, **except**:
    a. Slavery is a closed system in which slaves are treated as property.
    b. Slaves were forcibly imported to the United States as a source of cheap labor.
    c. Slavery has ended throughout the world.
    d. Some people have been enslaved because of unpaid debts, criminal behavior, or war and conquest.

5. A _____ system is a system of social inequality in which people's status is permanently determined at birth based on their parents' ascribed characteristics.
    a. class
    b. slavery
    c. capitalist
    d. caste

6. A young woman who comes from an impoverished background works at two full-time jobs in order to save enough money to attend college. Ultimately, she earns a degree, attends law school, graduates with highest honors, and is hired by a firm at a starting salary of $75,000. This person has experienced _____ mobility.
    a. vertical, intragenerational
    b. vertical, intergenerational
    c. horizontal, class
    d. direct, vertical

7. The _____ system is a type of stratification based on the ownership and control of resources and the type of work people do.
    a. class
    b. caste
    c. bureaucratic
    d. administrative

8. According to _____'s theory of class relations _____.
    a. Weber; the bourgeoisie consists of those who own the means of production.
    b. Marx; the proletariat consists of those who own the means of production.
    c. Marx; class relationships involve inequality and exploitation.
    d. Durkheim; wealth, prestige, and power are all important in determining a person's class position.

9. According to Karl Marx, _____ is a feeling of powerlessness and estrangement from other people and from oneself.
   a. alienation
   b. meritocracy
   c. class conflict
   d. classism

10. When workers overthrow the capitalists, according to Marx, they will eventually create a(n) _____ society.
    a. class
    b. caste
    c. egalitarian
    d. stratified

11. According to the _____ perspective, the explanation of social inequality in the U.S. is _____.
    a. conflict; that the beliefs and actions of people reflect their class location in society.
    b. symbolic interactionist; that powerful groups use ideology to maintain their favored position at the expense of others.
    c. functionalist; that some degree of social inequality is necessary for society to flow smoothly.
    d. absolute poverty; perception of one's status in comparison with the social status of others.

12. All of the following statements regarding Marx's analyses of class are correct, **except**:
    a. Class relationships involve inequality and exploitation.
    b. The exploitation of workers by the capitalist class ultimately will lead to the destruction of capitalism.
    c. The capitalist class maintains its position by control of the society's superstructure.
    d. Wealth, prestige, and power are separate continuums on which people can be ranked from high to low.

13. According to Max Weber, _____ is the respect or regard with which a person or status position is regarded by others.
    a. admiration
    b. power
    c. prestige
    d. rank

14. Entrepreneurs, as identified by Weber, are those who
    a. do not have to work.
    b. work for wages.
    c. are a privileged commercial class.
    d. live off their investments.

15. In 2007, ___ people in the United States were without health insurance coverage.
    a. 15.0 million
    b. 20.1 million
    c. 26.3 million
    d. 45.7 million

16. According to the Weberian model of the U.S. class structure, members of the _____ class have earned most of their money in their own lifetime as entrepreneurs, presidents of corporations, top level professionals, and so forth.
    a. upper-upper
    b. lower-upper
    c. upper-middle
    d. middle

17. A combination of three factors qualifies people for the upper-middle class. Which of the following is not one of these factors?
    a. university degrees
    b. authority and independence on the job
    c. inherited wealth
    d. high income

18. Over the past fifty years, Asian Americans, Latino/as, and African Americans have placed great emphasis on _____ as a means of attaining the American Dream.
    a. education
    b. affirmative action
    c. unemployment compensation
    d. vocational training

19. Women employed in pink-collar occupations are mainly classified in the:
    a. working class
    b. working poor
    c. middle class
    d. upper-middle class

20. In the Weberian Model of the U.S. class structure, medical technicians, nurses, lower-level managers, and semi-professionals make up the _____ class.
    a. upper
    b. middle
    c. working
    d. lower

21. The trend in which women disproportionately are represented among individuals living in poverty is referred to as:
    a. absolute poverty
    b. relative poverty
    c. situational poverty
    d. the feminization of poverty

153

22. All of the following are components of wealth, **except**:
    a. income
    b. a position on the local school board
    c. bank accounts
    d. insurance policies

23. In examining the unequal distribution of income and wealth in the United States,
    a. the wealthiest 20 percent of households receive almost 50 percent of the total income "pie."
    b. the poorest 20 percent of households receive about 30 percent of the total income "pie."
    c. income inequalities between the rich and the poor narrowed greatly in 2007.
    d. Differences in median income of married couples and female-headed households narrowed by the year 2005.

24. In recent years, the gulf between the rich and the poor has
    a. greatly fluctuated.
    b. decreased dramatically.
    c. widened.
    d. stayed the same.

25. According to the functionalist explanation of social inequality,
    a. caste systems are better for societies than class systems.
    b. the most important positions must offer the greatest rewards to attract the most qualified people.
    c. class conflict will eventually lead to the overthrow of capitalism.
    d. interpersonal interactions are influenced by social class.

26. According to Erik Olin Wright's model of social classes, doctors, lawyers, and accountants are part of the _____ class.
    a. managerial
    b. capitalist
    c. small-business
    d. working

27. According to Karl Marx, the _____ (capitalist class) consists of those who own the means of production—the land and capital necessary for factories and mines.
    a. proletariat
    b. bourgeoisie
    c. entrepreneurs
    d. rentiers

28. Prestige rankings have become the foundation for _____ research, which uses sophisticated statistical measurements to assess the influence of family background and education on people's occupational mobility and success.
    a. status attainment
    b. social
    c. class structure
    d. occupation position

154

29. According to the social class model developed by sociologists Dennis Gilbert and Joseph Kahl, members of the _____ class have earned most of their money in their own lifetime as entrepreneurs, presidents of major corporations, sports or entertainment celebrities, or top-level professionals.
    a. upper-middle
    b. lower-upper
    c. upper-lower
    d. upper-upper

30. According to your text, all of the following are factors that have eroded the American Dream for the middle class, **except**:
    a. higher housing prices
    b. increasing unemployment and layoffs
    c. negative media exposure
    d. cost of living squeeze

31. _____ attempt to determine what degree of control workers have over the decision-making process and the extent to which they are able to plan and implement their own work.
    a. Conflict theorists
    b. Feminist theorists
    c. Functionalist theorists
    d. Postmodern theorists

32. According to Forbes magazine's 2004 list of the richest people in the world, Bill Gates (co-founder of Microsoft Corporation) was the wealthiest, with a fortune of nearly $46.6 billion. According to sociologist Erik Wright's Marxian model, Bill Gates would be in the _____ class.
    a. capitalist
    b. entrepreneur
    c. working
    d. small-business

33. According to sociologist Dennis Gilbert, the top 5 percent of households alone received more than 20 percent of all income—an amount greater than that received by the bottom _____ percent of all households.
    a. 20
    b. 60
    c. 40
    d. 80

34. _____ theorists stress that schools are agencies for reproducing the capitalist class system and perpetuating inequality in society. Parents with limited income are not able to provide the same educational opportunities for their children as families with greater financial resources.
    a. Conflict
    b. Symbolic interactionist
    c. Postmodern
    d. Functionalist

35. Kingsley Davis and Wilbert Moore (Davis-Moore thesis) contend that social inequality
    a. assures that all necessary work in a society gets done.
    b. is the result of some people's poor self-esteem and self-concept.
    c. results from one group exploiting another.
    d. impacts the ways individuals interact in everyday life.

36. About two-thirds of all adults living in poverty are women. In 2007, single-parent families headed by women had a _____ percent poverty rate, as compared with a 5 percent poverty rate for two-parent families.
    a. 48
    b. 38
    c. 28
    d. 18

37. All of the following are reasons why social inequality may increase in the United States in this century, **except**:
    a. The purchasing power of the dollar has stagnated or declined since the early 1970s.
    b. Federal tax laws in recent years have benefited corporations and wealthy families at the expense of middle- and lower-income families.
    c. People living in poverty are demanding a larger share of the U.S. budget.
    d. Structural sources of upward mobility are shrinking, while the rate of downward mobility has increased.

38. Sociologists view poverty as mostly a result of:
    a. economic and structural causes
    b. individual intellectual deficiencies
    c. lack of ambition
    d. occupational competition

39. _____ examine(s) the impact of social class on everyday interactions.
    a. Feminists
    b. Symbolic interactionists
    c. Conflict
    d. Functionalists

40. All of the following statements regarding poverty and children in the United States are true, **except**:
    a. Poor children are mainly from two-parent households.
    b. Children in single-parent households headed by a woman are more likely to live in poverty.
    c. Many governmental programs established to alleviate childhood poverty have been cut back or eliminated.
    d. Most (55 percent) of white children under 18 in female-headed households live below the poverty line.

TRUE/FALSE QUESTIONS

1. The system of stratification in the U.S. is meritocracy.
    T   F

2. People no longer believe in the "American Dream."
    T   F

3. Both Karl Marx and Max Weber viewed class as an important determinant of
    social inequality.
    T   F

4. SES reflects Weber's multidimensional approach to determine social class.
    T   F

5. Most people in the U.S. identify themselves as the middle class.
    T   F

6. Of all the class categories, according to the Weberian Model, the one most
    shaped by formal education is the upper class.
    T   F

7. Accounting to the Weberian model of social class, an estimated 30 percent of the
    U.S. population is the working class.
    T   F

8. The working poor account for about 10 percent of the U.S. population.
    T   F

9. About 3 to 5 percent of the U.S. population make up the underclass.
    T   F

10. Peter Blau and Otis Duncan found that occupational position can be used
    interchangeably with economic class or prestige.
    T   F

11. The Marxian model suggests that the working class consists of about 50 percent
    of all employees of the U.S.
    T   F

12. Relative poverty exists when people do not have the means to secure the most
    basic necessities of life.
    T   F

13. About two-thirds of all adults living in poverty are men.
    T   F

14. The age group that has the highest risk of living in poverty is the elderly.
    T   F

15. The Davis-Moore thesis assumes that social stratification results in meritocracy.
    T   F

16. According to Box 8.2, "Framing Class in the Media," the television program "Deal or No Deal" is "framed" to highlight the idea that getting rich in the United States is very difficult.
T   F

17. According to Box 8.3, "Should Our Laws Guarantee People a Living Wage?," the high compensation packages received by many companies' chief executive officers is one of the major causes of social inequality in the United States.
T   F

18. Most poor people and welfare recipients are racial minorities.
T   F

19. According to the photo essay, "What Keeps the American Dream Alive," millions of people around the world see the United States and believe in the American Dream.
T   F

20. In recent years, federal tax laws have benefited corporations and wealthy families at the expense of middle- and low-income families.
T   F

FILL-IN-THE-BLANK QUESTIONS

1. _____ is the value of all of a person's or family's economic assets.

2. _____ is the ability of people or groups to achieve their goals despite opposition from others.

3. _____ is a combined measure that attempts to classify individuals, families, or households in terms of indicators such as income, occupation, and education to determine class.

4. Karl Marx identified _____ as the struggle between the capitalist class and the working class.

5. A reduction in the proficiency needed to perform a specific job that leads to a corresponding reduction in the wages for that job is known as _____.

6. Occupations that are relatively low-paying, nonmanual, semiskilled positions primarily held by women are called _____ occupations.

7. _____ revised Marx's analysis of the class structure to fit the occupations found in advanced capitalist societies.

8. _____ poverty exists when people may be able to afford basic necessities but are still unable to maintain an average standard of living.

9. The trend in which women are disproportionately represented among individuals living in poverty is known as _____.

10. A hierarchy in which all positions are rewarded based upon people's ability and credentials is termed _____.

11. The economic gain derived from wages, salaries, income, transfers, and ownership of property is identified as _____.

12. The movement of individuals or groups from one level in a stratification system to another is known as _____.

13. According to Marx, the working class, also known as the _____, are those who must sell their labor to the owners in order to survive.

14. According to the _____ thesis, inequality is necessary for the smooth functioning of society.

15. The _____ perspective focuses on the relationship between social inequality and people's everyday interactions.

SHORT ANSWER/ESSAY QUESTIONS

1. What is the "American Dream?" What assumption is it based upon? What are some of the facts about the "American Dream?"
2. What are some of the characteristics of the "new class society"? What are the two largest classes in the new class, and why do they possess fundamentally different and opposed interests?
3. What are some of the consequences of inequality in the United States?
4. Who are the poor in the United States? Why does poverty persist in the United States? What are some economic and structural sources of poverty?
5. What do you think are your own social mobility chances? Will you be better or worse off than your parents? Your grandparents? Why?

STUDENT CLASS PROJECTS AND ACTIVITIES

1. Using theories, concepts, facts, and discussions found in this chapter, as well as from other sources, write a proposal for a long-term plan that would reduce income inequality in our nation. Write your proposal as if you were going to present it to Congress. Document your facts and figures, and present compelling evidence and information supporting your proposal. You must give sufficient goals and objectives in your proposal and provide proof that your plan is feasible.
2. Examine social stratification in either your own hometown or the town/city in which your now reside. Conduct a driving route that will take you into a variety of neighborhoods that represent the six social classes, as discussed on pages 280-283 of the text. You can probably secure census track data that will be particularly helpful for this project. Drive through the various neighborhoods recording the characteristics of the neighborhoods, the type of housing, probable lifestyles, the streets and the condition of the streets, the landscaping, any vehicles, and other signs that would indicate social class. Look for symbols of wealth or poverty, such as fences, recreational facilities, statues, flagpoles, water fountains, trees, and noise level, or the absence of such. Record anything unique or special about each neighborhood. Look for children in the neighborhoods and describe the children and their activities. Look for signs of conspicuous consumption and/or conspicuous

waste. Submit a five-page paper describing each neighborhood using the above information as specific guidelines for the descriptions. Conclude with a summary of what you learned and how you felt about conducting this project.

3. Research the historical creation of the "War on Poverty" programs. What book, written by sociologist Michael Harrington, served as a reader to President John F. Kennedy and documented the vast existence of U.S. poverty? What were some of that book's startling statistics? Which U.S. president declared war on poverty? When? Where? Why? How did the name of the program—the War on Poverty—originate? What was the country's first response to this call to fight poverty? What major programs were set up as a part of this "war"? Name and describe the programs. What eventually happened to the "War on Poverty"? What programs still exist? Write a three- to five- page paper summarizing your findings.

INTERNET ACTIVITIES

1. To read about welfare reform, go to the **Urban Institute**, **http://www.urban.org/**, a nonpartisan economic and social policy research organization. Once there, look for the Issues in Focus, under Resources, and click on Welfare Reform to find current research on this topic.

2. The Bureau of Labor Statistics provides a wide range of economic data. Examine the **Economy at a Glance** section, **http://www.bls.gov/eag/**, which provides a regional and state-by-state breakdown of economic data. Then read the section on Industry at a Glance, which provides the same kind of information on specific industries. Click on your state or the region in which you live to get current local information.

3. Research Harvard University's **Inequality and Social Policy** web site think tank links: **http://www.ksg.harvard.edu/inequality/Focus/links.htm**. Explore some of the ideas that people are talking and "thinking" about in regards to inequality and social policy.

4. Use these web sites to collect facts about poverty in the United States. Use critical thinking skills to construct reasoned judgments about the question, "who are the poor?"
   The U.S. Census Bureau Income Statistics:
   http://www.census.gov/ftp/pub/hhes/www/income.html
   The U.S. Census Bureau Poverty Statistics:
   http://www.census.gov/ftp/pub/hhes/www/poverty.html
   The 2001 Health Human Services 2001 Poverty Guidelines:
   http://aspe.hhs.gov/poverty/01poverty.htm

INFOTRAC COLLEGE EDITION EXERCISES

Visit the **InfoTrac College Edition** web site at:
http://www.wadsworthmedia.com/webtutor/infotrac.htm. You will arrive at a screen that enables you to search topics.

1. Look up **social class**. Search for articles that address the plight of poor children. Bring a summary of the articles to class. Collectively construct a picture of the problems facing children in poverty.
2. Do a keyword search for the **American Dream**. What are the different ways that this term is used? What articles use the term to address social class issues? How often is this term used in a positive way? How has it become a negative symbol in our culture? Why do you think we use the word *dream* when we talk about our own social condition?
3. Look up **Karl Marx**. There are a number of references to his work *The Communist Manifesto*. You may want to actually read portions of this historic document. Write a paper that summarizes Marx's vision of communism. How does this ideal compare to the kinds of communism that have existed in countries such as Cuba and the Soviet Union?
4. Research five articles using InfoTrac. These articles must include all of the following terms: **caste system**, **class system**, **slavery,** and **poverty**. State the journal, author, and title of each article. Once you complete this, write a brief synopsis on each of the 5 articles. Make certain that your paper meets your professor's requirements.
5. Examine the relationship between **social networks** and **occupational prestige**. An excellent source is the article, "Social Networks and Prestige Attainment: New Empirical Findings" (*The American Journal of Economics and Sociology*, Oct. 1999).

SOLUTIONS

MULTIPLE CHOICE QUESTIONS

1. A, p. 243
2. D, p. 243
3. B, p. 242
4. C, p. 245
5. D, p. 245
6. B, p. 242
7. A, p. 247
8. C, p. 248
9. A, p. 248
10. C, p. 248
11. C, p. 272
12. D, p. 252
13. C, p. 252
14. C, p. 252

15. D, p. 262
16. B, p. 254
17. C, p. 254-255
18. A, p. 255
19. A, p. 255
20. B, p. 255
21. D, p. 270
22. B, p. 251
23. A, p. 260
24. C, p. 260
25. B, p. 272
26. A, p. 257
27. B, p. 248
28. A, p. 253

29. B, p. 254
30. C, p. 254
31. A, p. 273-274
32. A, p. 258
33. C, p. 254
34. A, p. 274
35. A, p. 273
36. B, p. 269-270
37. C, p. 275
38. A, p. 271
39. B, p. 274
40. D, p. 268

TRUE/FALSE QUESTIONS

1. F, p. 272
2. F, p. 244
3. T, p. 248
4. T, p. 252
5. T, p. 254
6. F, p. 254
7. T, p. 255

8. F, p. 255
9. T, p. 256
10. F, p. 253
11. T, p. 256
12. F, p. 268
13. F, p. 270
14. F, p. 278

15. T, p. 273
16. F, p. 250
17. T, p. 269
18. F, p. 271
19. T, p. 264
20. T, p. 275

FILL-IN-THE-BLANK QUESTIONS

1. Wealth, p. 251
2. Power, p. 252
3. Socioeconomic status, p. 252
4. class conflict, p. 249
5. job deskilling, p. 271
6. pink-collar, p. 255
7. Erik Olin Wright, p. 257
8. Relative, p. 268
9. feminization of poverty, p. 270
10. meritocracy, p. 272
11. income, p. 260
12. social mobility, p. 244
13. proletariat, p. 248
14. David-Moore, p. 273
15. symbolic interactionist, p. 274

9

GLOBAL STRATIFICATION

BRIEF CHAPTER OUTLINE

WEALTH AND POVERTY IN GLOBAL PERSPECTIVE
PROBLEMS IN STUDYING GLOBAL INEQUALITY
    The "Three Worlds" Approach
    The Levels of Development Approach
CLASSIFICATION OF ECONOMIES BY INCOME
    Low-Income Economies
    Middle-Income Economies
    High-Income Economies
MEASURING GLOBAL WEALTH AND POVERTY
    Absolute, Relative, and Subjective Poverty
    The Gini Coefficient and Global Quality of Life Issues
GLOBAL POVERTY AND HUMAN DEVELOPMENT ISSUES
    Life expectancy
    Health
    Education and Literacy
    Persistent Gaps in Human Development
THEORIES OF GLOBAL INEQUALITY
    Development and Modernization Theory
    Dependency Theory
    World Systems Theory
    The New International Division of Labor Theory
GLOBAL INEQUALITY IN THE FUTURE

CHAPTER SUMMARY

Global stratification refers to the unequal distribution of wealth, power, and prestige on a global basis, resulting in people having vastly different lifestyles and life chances, both within and among the nations of the world. The social and economic gaps between *developed nations* and *developing nations* of the world are much more pronounced than gaps in the United States. Since World War II, a common approach to defining global stratification is through the "three worlds" approach. *First World* nations are the rich, industrialized countries having primarily capitalistic economies and democratic political systems. *Second World* nations are those countries having a moderate level of economic development and a moderate standard of living. *Third World* countries are the poorest countries, with little or no industrialization and the lowest standards of living, shortest life expectancy, and highest mortality. Recently, the term *Fourth World* was created to refer to people who are barred from access to positions that would enable an autonomous livelihood due to **social exclusion.** Closely linked to the "three worlds" concept is the levels of development approach, which uses the terms *developed nations, developing nations, less-developed nations*, and *underdeveloped nations*. A more recent approach used by the World Bank classifies nations into three economic categories: *low-income economies, middle-income economies*, and *high-income economies*. The *gross domestic product* is now a means of measuring wealth and power on a global basis. Global poverty is sometimes defined in terms of *absolute, relative*, and *subjective poverty*. The World Bank uses the Gini coefficient as its measure of income inequality. Using the Human Development Index, the United Nations Development Program has established criteria for measuring the level of development in a country: life expectancy, education, and living standards. Overall, the gap between the poorest nations and the middle-income nations has continued to widen. Social scientists use four primary theoretical perspectives to explain the persistence of global inequality: (1) development and **modernization theory**, (2) **dependency theory**, (3) world systems theory, and (4) the new international division of labor theory. The future prospects of global inequality range from more to less optimistic predictions. Most analysts agree a nation of people can enjoy global prosperity only by ensuring that other people around the world have the opportunity to survive and thrive in their own surroundings.

LEARNING OBJECTIVES

After reading Chapter 9, you should be able to:
1. Define and describe global stratification.
2. Discuss problems in studying global inequality.
3. Define and describe the "three worlds" approach used to classify nations of the world.
4. Explain the levels of development approach used for describing global stratification.
5. Classify and describe nations of the world by the three economic categories.
6. List and explain different measures of global wealth and poverty.
7. Distinguish among absolute, relative, and subjective poverty.
8. Identify and explain the use of the Gini coefficient.
9. Discuss global poverty and its effects upon human development.
10. Describe some of the important ways that international aid has helped to fight global poverty and disease.
11. Compare and contrast the four major theories of global inequality, including development and modernization theory, dependency theory, world systems theory, and international division of labor theory.
12. Describe the contributions of the World Health Organization in addressing problems associated with global stratification.
13. Describe the future prospects of global inequality.

KEY TERMS

(defined at page number shown and in glossary)

core nations, p. 299
dependency theory, p. 297
modernization theory, p. 296
peripheral nations, p. 300
semiperipheral nations, p. 299
social exclusion, p. 285

KEY PEOPLE

(identified at page number shown)

Manuel Castells, p. 285
Walt Rostow, p. 296
Harry S. Truman, p. 286
Immanuel Wallerstein, p. 299

CHAPTER OUTLINE

I. WEALTH AND POVERTY IN GLOBAL PERSPECTIVE
    A. *Global stratification* refers to the unequal distribution of wealth, power, and prestige on a global basis.
    B. Global stratification results in people having vastly different lifestyles and life chances both within and among the nations of the world.
    C. The world is divided into unequal segments characterized by *high-income countries*, *middle-income countries*, and *low-income countries*. However, economic inequality is not the only dimension of global stratification; social inequality may result from factors such as discrimination based on race, ethnicity, gender, or religion.
II. PROBLEMS IN STUDYING GLOBAL INEQUALITY
    A. One of the problems is determining what terminology should be used to refer to the distribution of resources in various nations. The "three worlds" and levels of development approaches were used shortly after WWII, but are no longer used today.
    B. After World War II, the "three worlds" approach was utilized to distinguish among nations based on their economic development and their standard of living.
        1. *First World* nations consist of the rich, industrialized nations that primarily have capitalist economic systems and democratic political systems.
        2. *Second World* nations consist of the countries with at least a moderate level of economic development and a moderate standard of living.
        3. *Third World* nations consist of the poorest countries, with little or no industrialization, the lowest standards of living, the shortest life expectancies, and higher mortality rates.

4. Recently, Manuel Castells suggested the term *Fourth World* to describe the "multiple black holes of **social exclusion**" of people in wide-ranging areas of the world, including those from sub-Saharan Africa to U.S. inner-city ghettos.
5. **Social exclusion** occurs when certain individuals and groups are systematically denied access to positions that would enable them to maintain the standards and values of their society.

C. The levels of development approach includes concepts such as developed nations, developing nations, less-developed nations, and underdeveloped nations.
1. Leaders of the developed nations argued that economic development and growth was the primary way to solve poverty problems of the underdeveloped nations; if nations could increase their *gross national income* (GNI), then social and economic inequality among their citizens could be reduced.
2. This viewpoint requires that people in the less-developed nations accept the beliefs and values of people in the developed nations.
3. However, improving a country's GNI did not tend to reduce the poverty of the poorest people in that country, and inequality increased even with greater economic development.

III. CLASSIFICATION OF ECONOMIES BY INCOME
A. The World Bank classifies nations into three economic categories.
B. *Low-income economies* are nations that had a GNI per capita of $935 or less in 2007. About half of the world's population lives in the forty-nine low-income economies, with more women around the world tending to be more impoverished than men—a situation known as *the global feminization of poverty*.
C. Middle-income economies are nations that had a GNI per capita of between $936 and $11,455 in 2007. About one-third of the world's population resides in the ninety-five nations with middle-income economies. These nations typically have a higher standard of living and export diverse goods and services, ranging from manufactured goods to raw materials and fuels.
D. High-income economies are found in sixty-five nations that had a GNI per capita of more than $11,456 in 2007 and continue to dominate the world economy, despite *capital flight* and *deindustrialization*.

IV. MEASURING GLOBAL WEALTH AND POVERTY
A. In recent years, the United Nations and World Bank have begun to use the *gross domestic product* measurement—all the goods and services produced within a country's economy during a given year.
B. Global poverty is sometimes defined in terms of *absolute*, *relative*, and *subjective* poverty.
C. The Gini Coefficient and Global Quality of Life Issues
1. The World Bank uses as its measure of income inequality the *Gini coefficient*, which ranges from 0 (everyone has the same income) to 100 (one person receives all the income).
2. Using this measure, the World Bank concluded that inequality has increased in some nations, such as Bulgaria, the Baltic countries, and the Slavic countries of the former Soviet Union, to levels similar to those found in less-equal economies like the U.S.

166

V. GLOBAL POVERTY AND HUMAN DEVELOPMENT ISSUES
  A. Since the 1970s, the United Nations has focused on *human development*—expanding choices people have to lead a fulfilling and dignified life—as a crucial factor fighting poverty. The Human Development Index measures life expectancy, education, and living standards.
  B. Average life expectancy has increased by about one-third in the past three decades and is now more than 70 years in 95 countries. Life expectancy in low-income nations is often thirty years less than in high-income countries. Low life expectancy results from high rates of infant mortality, illness, disease, malnutrition, and poor health care.
  C. *Health* is defined by the World Health Organization as a stage of complete physical, mental, and social well-being and not merely the absence of disease or infirmity. Many people in low-income nations suffer from infectious and other diseases. New diseases have recently emerged in countries all over the world, while some middle-income countries are experiencing rapid growth in degenerative diseases such as cancer and coronary heart diseases.
  D. Education is fundamental to increasing literacy.
    1. The adult literacy rate in the low-income countries is about half of that of high-income countries, and is even lower for women.
  E. Persistent gaps in human development paint an overall dismal picture for the world's poorest people.
    1. The gap between the poorest nations and the middle-income nations has continued to widen, and women are particularly vulnerable to poverty due to increases in single-person and single-parent households and low-wage work.
VI. THEORIES OF GLOBAL INEQUALITY
  A. The most widely known development theory is **modernization theory**.
    1. Modernization theory links global inequality to different levels of economic development and suggests that low-income economies can move to middle- and high-income economies by achieving self-sustained economic growth.
    2. The theory states that changes in people's beliefs, values, and attitudes toward work are necessary for economic development.
    3. Walt Rostow suggested that all countries go through four stages of economic development: the *traditional stage*, *take-off stage*, *technological maturity*, and *high mass consumption*. In this fourth and final state the country reaches a high standard of living.
  B. **Dependency theory** states that global poverty partially results from exploitation of low-income countries by high-income countries.
    1. Dependency theory disputes modernization theory, arguing that poorer nations are trapped in a cycle of structural dependency on richer nations.
    2. Dependency theory has been more often applied to the newly industrializing countries (NICs) of Latin America, and may be less useful for understanding economic growth and development in NICs of East Asia.
  C. *World systems theory*, created by Immanuel Wallerstein, suggests that under capitalism, a global system is held together by economic ties.
    1. The capitalist world-economy is a global system divided into a hierarchy of three major types of nations: core, semiperipheral, and peripheral.

167

2. Core nations are dominant capitalist centers characterized by high levels of industrialization and urbanization, such as the United States, Germany, and Japan.
3. **Semiperipheral nations** are more developed than peripheral nations, but less developed than core nations, such as Taiwan, South Korea, Mexico, Brazil, India, Nigeria, and South Africa.
4. **Peripheral nations** are dependent on core nations for capital, have little or no industrialization (other than brought in by core nations), and have uneven patterns of urbanization. Most low-income countries in Africa, South America, and the Caribbean are peripheral nations.

D. According to the *new international division of labor theory*, commodity production is being split into fragments that can be assigned to whichever part of the world can provide the most profitable combination of capital and labor.
1. The global nature of these activities is referred to as *global commodity chains*, a complex pattern of international labor and production.
2. *Producer-driven commodity chains* describe industries in which transnational corporations play a central part in controlling the production process.
3. *Buyer-driven commodity chains* refers to industries wherein large retailers, brand-named merchandisers, and trading companies set up decentralized production networks in various middle- and low-income counties.

E. Social scientists describe an optimistic or a pessimistic scenario for the future depending upon the theoretical framework they apply in studying global inequality.
1. Some analysts highlight the human rights issues embedded in global inequality; others focus primarily on an economic framework.
2. In the future, continued population growth, urbanization, and environmental degradation threaten the living conditions of those residing in low-, middle-, and high-income countries; the quality of life diminishes as natural resources are depleted.
3. A more optimistic scenario suggests that, with modern technology and worldwide economic growth, it might be possible to reduce absolute poverty and increase people's opportunities, ensuring that people all over the world have the opportunity to survive and thrive in their own surroundings.

CRITICAL THINKING QUESTIONS

1. Explain the main differences between low-, middle-, and high-income countries. In what ways do you think your life would be different if you lived your whole life in a middle- or low-income country?
2. Explain the relationship between high-, middle-, and low-income countries. In what ways does your lifestyle depend on individuals in middle- and low-income countries?
3. What could be done to reduce global inequality? Do you think steps should be taken to reduce global inequality? Why or why not?
4. Which theory of global inequality do you think is most useful? Why? What are the strengths and weaknesses of each?
5. Compare and contrast pessimistic and optimistic views of global inequality in the future. Which view do you think is more accurate? Why?

PRACTICE TESTS

MULTIPLE CHOICE QUESTIONS

Select the response that best answers the question or completes the statement.

1. The unequal distribution of wealth, power, and prestige on a global basis is referred to as:
   a. global stratification
   b. global layering
   c. global distinction
   d. global accumulation

2. The income gap between the richest and the poorest 20 percent of the world population
   a. continues to widen.
   b. is greater in urban than in rural areas.
   c. has significantly decreased in the last decade.
   d. is slowly beginning to decline.

3. Between the 1950s and the 1970s, organizations such as the United Nations and the World Bank worked to increase the gross national income of developing countries. These efforts showed that increasing a country's GNI
   a. corresponded with increasing global poverty and inequality.
   b. resulted in decreasing global poverty, but not inequality.
   c. resulted in decreasing global poverty and inequality.
   d. was not possible to achieve.

4. The _____ approach was introduced by social analysts to distinguish among nations on the basis of their levels of economic development and the standard of living of their citizens.
   a. levels of development
   b. three worlds
   c. classification of economies
   d. global distinction

5. China and Cuba are classified as _____ nations.
   a. Developed World
   b. First World
   c. Second World
   d. Third World

6. First World nations are best represented by the countries of:
   a. Japan, the United States, China
   b. New Zealand, the United States, Great Britain
   c. Canada, China, Korea
   d. Russia, Australia, Japan

7. _____ occurs when individuals and/or groups are systematically barred from access to positions that would enable them to have an autonomous livelihood in keeping with the social standards and values of a given social context.
   a. Social isolation
   b. Social exclusion
   c. Social decay
   d. Social stigma

8. The ____ Plan, named after a U. S. Secretary of State, provided massive sums of money in direct aid and loans to rebuild countries destroyed during World War II.
   a. Albright
   b. Powell
   c. Marshall
   d. Rice

9. The term _____ refers to all the goods and services produced in a country in a given year.
   a. World Health Organization
   b. Gross Domestic Product
   c. Net Domestic Income
   d. Gross National Income

10. Low-income economies are primarily found in _____ nations, where half of the world's population resides.
    a. Asian and South American
    b. Asian and African
    c. Eastern European and African
    d. South American and Asian

11. The situation wherein women around the world tend to be more impoverished than men is referred to as:
    a. global feminization stratification
    b. global feminization
    c. global feminization of poverty
    d. global feminization of labor

12. Lower-middle-income economies include the nations:
    a. Bolivia, Colombia, El Salvador
    b. Nigeria, Mali, Mexico
    c. Ireland, Italy, Portugal
    d. Australia, Canada, Germany

13. High-income economies are found in _____ nations.
    a. 65
    b. 15
    c. 38
    d. 25

170

14. The movement of jobs and economic resources from one nation to another is defined as:
    a. capital flight
    b. deindustrialization
    c. transnational flight
    d. economic mobility

15. The closing of plants and factories because of their obsolescence or employment of cheaper workers in other nations is known as:
    a. capital flight
    b. capital destabilization
    c. deindustrialization
    d. reindustrialization

16. The World Bank uses the term _____ to indicate all of the goods and services produced within a country's economy during a given year:
    a. Gross Domestic Product
    b. Good and Service Production
    c. Economic Production
    d. Gross Product Output

17. _____ is a condition in which people do not have the means to secure the most basic necessities of life.
    a. Subjective poverty
    b. Relative poverty
    c. Absolute poverty
    d. Standard poverty

18. The World Bank uses the term _____ as its measure of income inequality.
    a. Gross National Product
    b. Gini coefficient
    c. relative poverty
    d. Mrydal coefficient

19. According to the _____ theory, the capitalist world-economy is a global system divided into a hierarchy of three major types of nations.
    a. dependency
    b. development and modernization
    c. new international division of labor
    d. world systems

20. According to the _____ theory, commodity production is split into fragments that can be assigned to the part of the world that can provide the most profitable combination of capital and labor.
    a. world systems
    b. new international division of labor
    c. dependency
    d. development and modernization

171

21. The idea of _____ has become the primary means used in attempts to reduce social and economic inequalities and alleviate the worst effects of poverty in the less industrialized nations of the world.
    a. social equity
    b. development
    c. evolution
    d. degeneration

22. One of the primary problems encountered by social scientists studying global stratification is
    a. what terminology should be used to refer to the distribution of resources in various nations.
    b. how to measure global inequality.
    c. how to interact with women from different societies.
    d. how to determine standards of health in different nations.

23. After _____, the terms "First World," "Second World," and "Third World" were introduced by social scientists.
    a. World War I
    b. World War II
    c. the Korean War
    d. the Vietnam War

24. The _____ was introduced by social analysts to distinguish among nations on the basis of their levels of economic development and the standard of living of their citizens.
    a. levels of development approach
    b. three worlds approach
    c. world systems approach
    d. global distinction approach

25. People living in _____ countries have little industrialization, high levels of poverty, and the shortest life expectancies.
    a. low-income
    b. lower-middle-income
    c. upper-middle-income
    d. high-income

26. According to sociologist Manuel Castells, the _____ has disintegrated with the advent of the Information Age, a period in which industrial capitalism is being superseded by global informational capitalism.
    a. First World
    b. Second World
    c. Third World
    d. Fourth World

27. Today the preferred terminology for referring to the distribution of resources across nations is reflected in the _____ approach.
    a. three worlds
    b. levels of development
    c. order of civilization
    d. classification of economies by income

28. Sociologist Manuel Castells stated that as informational capitalism grows around the world, the lack of regular work as a source of income is increasingly a key mechanism in:
    a. social isolation
    b. social inclusion
    c. social exclusion
    d. social exporting

29. According to Figure 9.1, the income gap between the richest and poorest people in the world _____ between 1960 and 2000.
    a. was less pronounced
    b. was most pronounced in the United States
    c. continued to decline
    d. continued to grow

30. The term _____ refers to material well-being that can be measured by the quality of goods and services that may be purchased by the per capita national income.
    a. standard of living
    b. standard of development
    c. standard of the economy
    d. standard of building

31. _____ are most affected by poverty in low-income economies.
    a. Women and children
    b. Adult men and women
    c. Aged men and women
    d. Aged women

32. By the year 2000, the wealthiest 20 percent of the world population had almost _____ times the income of the poorest 20 percent.
    a. thirty
    b. fifty
    c. eighty
    d. ninety

33. The recent financial crisis in the U.S. is expected to have the biggest _____ impact on _____.
    a. negative; the U.S.
    b. positive; middle-income nations
    c. positive; low-income nations
    d. negative; low-income nations

34. According to the _____ theory, the traditional caste system becomes obsolete as industrialization progresses.
    a. world systems
    b. new international division of labor
    c. dependency
    d. development and modernization

35. Complex patterns of international labor and production processes that result in a finished commodity ready for sale in the marketplace are referred to as:
    a. global buyer-driven commodities
    b. global labor-intensive chains
    c. global commodity chains
    d. global production chains

36. Sociologist Immanuel Wallerstein stated that most low-income countries in Africa and South America are _____ nations that are dependent on other nations for capital, have little or no industrialization, and have uneven patterns of urbanization.
    a. peripheral
    b. core
    c. semiperipheral
    d. tertiary

37. The _____ theory argues that commodity production is being divided into fragments assigned to whatever part of the world can provide the most profitable capital and labor.
    a. international division of labor
    b. world systems
    c. three worlds
    d. dependency

38. Health is defined in the Constitution of the World Health Organization as:
    a. a state of complete physical, mental, and social well-being
    b. absence of disease or infirmity
    c. a state of optimum health
    d. a state of mental happiness

39. Social scientists who hold a pessimistic view of global inequality are primarily concerned about
    a. education and literacy.
    b. population growth, urbanization and environmental degradation.
    c. religion and family values.
    d. the continued production of consumer goods.

40. Athletic footwear companies such as Nike and Reebok and clothing companies like Gap and Liz Claiborne are examples of _____ chains.
    a. individually-driven commodity
    b. producer-driven commodity
    c. international-driven commodity
    d. buyer-driven commodity

174

TRUE/FALSE QUESTIONS

1. Income disparities between rich and poor countries are greater than income disparities *within* countries.
   T   F

2. The classification of economies by income has been replaced in recent years by the three worlds approach.
   T   F

3. The Fourth World is a term *originally* designated for indigenous people who are descended from a country's aboriginal population.
   T   F

4. Castells uses the term *Fourth World* to refer to the new developing countries in the continent of Africa.
   T   F

5. Dependency theory argues that global poverty results from the exploitation of low-income countries by high-income countries, creating a cycle of structural dependency.
   T   F

6. Ideas regarding underdevelopment were popularized by U.S. President Harry S. Truman.
   T   F

7. The international division of labor theory suggests that global inequality involves a complex world system in which high-income nations benefit from other nations and exploit their citizens.
   T   F

8. High rates of population growth taking place in the underdeveloped nations have led to an increase in economic development of those countries.
   T   F

9. About half of the world's population lives in low-income economies, according to the World Bank classification.
   T   F

10. According to the classification of economies by income, many countries referred to as "middle income" have very few of their population living in poverty.
    T   F

11. Subjective poverty is defined as a condition in which people do not have the means to secure the most basic necessities.
    T   F

12. Since the 1970s, the United Nations has more actively focused on human development, as measured by the HDI, as a crucial factor for fighting poverty.
    T   F

13. More than 1 billion people worldwide live below the international poverty line, earning less than $1.25 per day.
    T   F

14. The assets of the wealthiest 200 people equal more than the income of over 40 percent of the world's population.
    T   F

15. According to UNESCO, women in low-income countries comprise about two-thirds of those who are illiterate.
    T   F

16. According to Box 9.1, the world's ten richest people are U.S. citizens.
    T   F

17. The life expectancy of people in low-income nations is often as much as 30 years less than that of people in high-income countries.
    T   F

18. According to "Marginal Migration" (Box 9.2), migration of individuals in poor nations to slightly less poor nations for work is referred to as "south-to-south" migration.
    T   F

19. According to Box 9.3, the boycotts and public pressure to reduce or eliminate exploitative and dangerous child labor always produce good results.
    T   F

20. According to Box 9.4, CARE assists the world's poor in their efforts to achieve social and economic well-being.
    T   F

FILL-IN-THE-BLANK QUESTIONS

1. The unequal distribution of wealth, power, and prestige on a global basis is known as _____.

2. According to Manuel Castells, _____ is the process by which certain individuals and groups are systematically barred from access to positions that would enable them to have an autonomous livelihood.

3. The United Nations defines _____ as the process of expanding people's choices to "lead a life to its full potential and in dignity, through expanding capabilities and people taking action themselves to improve their lives."

4. The concepts of underdevelopment and underdeveloped nations emerged out of the _____ following World War II.

5. The _____ is defined as all the goods and services produced within a country's economy during a given year.

176

6. A state in which people may be able to afford basic necessities, but are still unable to maintain an average standard of living, measured by comparing one's person's income with the incomes of others, is known as _____.

7. According to _____ theory, global poverty is partially caused by the exploitation of low-income countries by high-income countries.

8. Sociologist _____ is closely associated with world systems theory.

9. According to world systems theory, _____ nations are dominant capitalist centers characterized by high levels of industrialization and urbanization.

10. Countries such as the U.S. and Japan used to be referred to as _____ but now are referred to as _____.

11. The levels of development approach classifies countries based on their _____.

12. Manuel Castells uses the term _____ to describe the "multiple black holes of social exclusion" existing throughout the planet.

13. Ideas regarding underdevelopment were popularized by President _____ in his inaugural address.

14. The World Bank uses the _____ to measure income inequality.

15. According to the modernization theory of Walt W. Rostow, societies in the _____ stage are slow to change because the people hold a fatalistic value system, do not subscribe to the work ethic, and save very little money.

SHORT ANSWER/ESSAY QUESTIONS

1. What do you think are the future prospects for greater equality across and within nations? Cite at least one theoretical framework from this chapter that provides an optimistic scenario and one that provides a pessimistic scenario for the future. Which do you agree with more? Why?
2. What is global stratification, and what effects does it have upon economic inequality?
3. How would the new international division of labor theory explain global stratification? What is meant by global commodity chains, producer-driven commodity chains and buyer-driven commodity chains?
4. What is literacy, and why is it important for human development? Why is it especially important for women?
5. Why is the majority of the world's population growing richer while the poorest one-fifth remains poor? How are global poverty and human development related? What steps could be taken to reduce global poverty?

STUDENT CLASS PROJECTS AND ACTIVITIES

1. Drawing upon world systems theory, most closely associated with the work of Immanuel Wallerstein, select three countries that are members of ASEAN (Association of South East Asian Nations) who best represent the three major types in the hierarchical position of nations in the global economy. Prepare a research report on the economic conditions of each of the three you selected, including information such as (1) the status of the major infrastructures of each country, such as their transportation systems, power plants, and telecommunications systems; (2) the status of their financial capital; and (3) the status of their human capital. Provide substantial explanations of why and how you selected the countries representing each specific category. Include an analysis of the resources and obstacles characterizing each specific country. Are there any global cities (or command posts) in the countries you selected? Provide any other information that supports your selections.

2. Research the lives of the world's ten richest people, according to the latest Forbes magazine poll. (1) Provide a minimum one-paragraph biographical sketch on each of the "top ten." (2) Explain how each of them obtained their wealth. (3) What countries do they represent? (4) Who did they replace from the previous year's poll? (5) What are some similarities and differences among the top ten?

3. Select ten countries throughout the world that represent different income categories. Look up information on each of the countries, including their (1) gross national income, (2) average life expectancy, (3) infant mortality rates, and (4) literacy rates. Then investigate the major sources of income for each of the countries, and provide suggestions on how to increase the incomes of the low-income countries you selected.

4. Many corporations in high-income countries move factories to low-income countries to reduce costs through low wages and relaxed labor and environmental regulations. Select and investigate one U.S. corporation that engages in this practice, which is referred to as outsourcing. Find and summarize whatever information you can regarding (1) when the corporation first developed factories overseas; (2) which countries the corporation developed factories in; (3) the advantages to the company for outsourcing; (4) the impact of this corporation's outsourcing on U.S. workers and cities; and (5) the impact of outsourcing on the host countries. After you have completed and summarized your investigation, explain the relationship between outsourcing and global inequality, describing whether you believe outsourcing increases or decreases global inequality and why.

INTERNET ACTIVITIES

1. Visit the United Nations **Development** web site: **http://www.un.org/esa/**. This site has links to information about crime, trade, drugs, population, and other topics from a global perspective.
2. Another organization to visit is the **World Bank Group**: **http://www.worldbank.org/**. To read about proposals to reduce or end poverty, search "Poverty Reduction Strategy Papers."
3. **UNESCO** stands for the United Nations Educational, Scientific, and Cultural Organization: **http://www.unesco.org/**. What is the mission of UNESCO? How will reducing global illiteracy help the United States and other countries?
4. A site devoted to international issues of women's health is the **International Women's Health Coalition**: http://www.iwhc.org/. Read about their population policies and efforts to increase AIDS awareness for women around the world.
5. From the **Global Issues** site, **http://www.globalissues.org/TradeRelated/Poverty.asp**, you can gain information about stratification, poverty, and conflict.

INFOTRAC COLLEGE EDITION EXERCISES

Visit the **InfoTrac College Edition** web site at: **http://www.wadsworthmedia.com/webtutor/infotrac.htm**. You will arrive at a screen that enables you to search topics.

1. Examine **human rights violations** by searching for articles that address the human rights violations in Serbia, Russia, China, Saudi Arabia, and Sierra Leone, to name a few.
2. Search the subject **Gini coefficient** and see how this measure of income inequality is used in actual research articles.
3. There are a number of articles under the subject **women and poverty**. Find articles related to global poverty. What cultural values are related in some of the issues women face worldwide?
4. There are a number of subdivisions under the subject search for **developing countries**. The subdivision **environmental aspects** contains information regarding the interrelationship between the social structure and the physical environment.

SOLUTIONS

MULTIPLE CHOICE QUESTIONS

1. A, p. 282
2. A, p. 282
3. A, p. 286
4. B, p. 285
5. C, p. 285
6. B, p. 285
7. B, p. 285
8. C, p. 286
9. D, p. 286
10. B, p. 287
11. C, p. 288
12. A, p. 288
13. A, p. 290
14. A, p. 290

15. C, p. 290
16. A, p. 290
17. C, p. 291
18. B, p. 291
19. D, p. 299
20. B, p. 300
21. B, p. 283
22. A, p. 284
23. B, p. 285
24. B, p. 285
25. A, p. 285
26. B, p. 285
27. D, p. 287
28. C, p. 285

29. D, p. 282
30. A, p. 286
31. A, p. 289
32. C, p. 283
33. D, p. 283
34. D, p. 296
35. C, p. 301
36. A, p. 299
37. A, p. 300
38. A, p. 292
39. B, p. 302
40. D, p. 301

TRUE/FALSE QUESITONS

1. F, p. 282
2. F, p. 287
3. T, p. 285
4. F, p. 285
5. T, p. 297
6. T, p. 296
7. F, p. 300
8. F, p. 287
9. T, p. 287
10. F, p. 288

11. F, p. 291
12. T, p. 292
13. T, p. 284
14. T, p. 284
15. T, p. 294
16. F, p. 284
17. T, p. 292
18. T, p. 289
19. F, p. 294
20. T, p. 304

FILL-IN-THE-BLANK QUESTIONS

1. global stratification, p. 282
2. social exclusion, p. 285
3. human development, p. 292
4. Marshall Plan, p. 286
5. gross domestic product, p. 290
6. relative poverty, p. 291
7. dependency, p. 297
8. Immanuel Wallerstein, p. 299
9. core, p. 299
10. first-world countries; high-income economies, p. 290
11. gross national income, p. 286

12. Fourth World, p. 285
13. Harry Truman, p. 286
14. Gini coefficient, p. 291
15. traditional, p. 296

10
RACE AND ETHNICITY

BRIEF CHAPTER OUTLINE

RACE AND ETHNICITY
    Social Significance of Race and Ethnicity
    Racial Classification and the Meaning of Race
    Dominant and Subordinant Groups
PREJUDICE
    Stereotypes
    Racism
    Theories of Prejudice
    Measuring Prejudice
DISCRIMINATION
SOCIOLOGICAL PERSPECTIVES ON RACE AND ETHNIC RELATIONS
    Symbolic Interactionist Perspectives
    Functionalist Perspectives
    Conflict Perspectives
    An Alternative Perspective: Critical Race Theory
RACIAL AND ETHNIC GROUPS IN THE UNITED STATES
    Native Americans
    White Anglo-Saxon Protestants (British Americans)
    African Americans
    White Ethnic Americans
    Asian Americans
    Latino/as (Hispanic Americans)
    Middle Eastern Americans
GLOBAL RACIAL AND ETHNIC INEQUALITY IN THE FUTURE
    Worldwide Racial and Ethnic Struggles
    Growing Racial and Ethnic Diversity in the United States

CHAPTER SUMMARY

Issues of race and ethnicity permeate all levels of interaction in the United States. A **race** is a category of people who have been singled out as inferior or superior, often on the basis of physical characteristics such as skin color, hair texture, and eye shape. An **ethnic group** is a collection of people distinguished, by others or themselves, primarily on the basis of cultural or nationality characteristics. Race and ethnicity are ingrained in our consciousness, and often form the basis of hierarchical ranking and determine who gets what resources. A **dominant group** is one that is advantaged and has superior resources and rights in a society; a **subordinate group** is one whose members are disadvantaged and subjected to unequal treatment by the dominant group, and who regard themselves as objects of collective discrimination. **Prejudice** is a negative attitude based on faulty generalizations about the members of selected racial and ethnic groups and is rooted in **stereotypes** and *ethnocentrism*. **Racism** is a set of attitudes, beliefs, and practices that is used to justify the superior treatment of one racial or ethnic group and the inferior treatment of another racial or ethnic group. **Discrimination**—actions or practices of dominant group members that have a harmful impact on members of a subordinate group—may be either **individual** or **institutional**. **Institutional discrimination** involves day-to-day practices of organizations and institutions that have a harmful impact on members of subordinate groups. According to the contact hypothesis, increased contact between people from divergent groups should lead to favorable attitudes and behavior when a specific set of criteria are met. Two functionalist perspectives—**assimilation** and **ethnic pluralism**—focus on how members of subordinate groups become a part of the mainstream. Conflict theories analyze economic stratification and access to power in race and ethnic relations: caste and class perspectives, **internal colonialism**, **split labor market** theory, gendered racism, and racial formation theory. Critical race theory derives its foundation from the U.S. civil rights tradition and the writings of several civil rights leaders. The unique experiences of Native Americans, White Anglo-Saxon Protestants/British Americans, African Americans, White Ethnics, Asian Americans, Latino/as (Hispanic Americans), and Middle Easterners are discussed, and the increasing racial and ethnic diversity of the United States is examined. Globally, many racial and ethnic groups seek self-determination, resulting in ethnic wars in some areas. In the United States, racial and ethnic diversity is increasing; several possibilities of this increase remain. It is predicted that by 2056, the roots of the average U.S. resident will not be white Europe. Some analysts argue that to eliminate racism, schools and work places must equalize opportunities for all people.

LEARNING OBJECTIVES

After reading Chapter 10, you should be able to:

1. Define race and ethnic group and explain their social significance.

2. Discuss the social significance of race and ethnicity and the changes in racial classifications in the U.S.

3. Define and provide examples of the terms dominant group and subordinate group.

4. Define prejudice, stereotypes, and racism and outline the major theories of prejudice.

5. Describe Robert Merton's typology of the relationship between prejudice and discrimination and be able to give examples of each.

6. Discuss discrimination and distinguish between individual and institutional discrimination.

7. Distinguish between assimilation and ethnic pluralism.

8. Describe symbolic interactionist perspectives on racial and ethnic relations.

9. Describe the ways that language can be used to perpetuate racial and ethnic stereotypes.

10. Explain functionalist views of race and ethnicity, including the concepts of assimilation, ethnic pluralism, and segregation.

11. Explain the main ideas of conflict perspectives on racial and ethnic inequality, including the caste perspective, class perspectives, internal colonialism, and split labor market.

12. Compare and contrast the experiences of racial and ethnic subordinate groups in the United States.

13. Explain how the experiences of Native Americans have been different from those of other racial and ethnic groups in the United States.

14. Describe how the African American experience in the United States has been unique when compared with other groups.

15. Discuss racial and ethnic struggles from a global perspective.

KEY TERMS

(defined at page number shown and in glossary)

assimilation, p. 320
authoritarian personality, p. 317
discrimination, p. 318
dominant group, p. 314
ethnic group, p. 311
ethnic pluralism, p. 322
genocide, p. 319
individual discrimination, p. 319
institutional discrimination, p. 319
internal colonialism, p. 325

prejudice, p. 314
race, p. 311
racism, p. 315
scapegoat, p. 316
segregation, p. 322
social distance, p. 318
split labor market, p. 326
stereotypes, p. 314
subordinate group, p. 314

KEY PEOPLE

(identified at page number shown)

Theodor W. Adorno, p. 317
Joe R. Feagin, p. 319
Robert Merton, p. 318

Michael Omi and Howard Winant, p. 326
William Julius Wilson, p. 323

CHAPTER OUTLINE

I. RACE AND ETHNICITY
    A. A **race** is a category of people who have been singled out as inferior or superior, often on the basis of physical characteristics such as skin color, hair texture, and eye shape.
    B. An **ethnic group** is a collection of people distinguished, by others or by themselves, primarily on the basis of cultural or nationality characteristics.
    C. Social significance of race and ethnicity: Race and ethnicity impact the opportunities people have, how they are treated, and how long they live; race and ethnic stratification pervade all aspects of political, economic, and social life.
    D. Racial classifications in the U.S. census reflect how the meaning of race has continued to change over the past century in the U.S.
    E. A **dominant group** is one that is advantaged and has superior resources and rights in a society; a **subordinate group** is one whose members, because of physical or cultural characteristics, are disadvantaged and subjected to unequal treatment by the dominant group, and who regard themselves as objects of collective discrimination.

II. PREJUDICE
    A. **Prejudice** is a negative attitude based on faulty generalizations about members of selected racial and ethnic groups. Prejudice is often based on stereotypes.
    B. **Stereotypes** are overgeneralizations about the appearance, behavior, or other characteristics of all members of a category. *Ethnocentrism* is maintained and perpetuated by stereotypes.
    C. **Racism** is a set of attitudes, beliefs, and practices that is used to justify the superior treatment of one racial or ethnic group and the inferior treatment of another.
    D. Theories of Prejudice
        1. The frustration-aggression hypothesis states that people who are frustrated in their efforts to achieve a highly desired goal will respond with a pattern of aggression toward a **scapegoat**—a person or group that is incapable of offering resistance to the hostility or aggression of others.
        2. Social learning theory (a symbolic interactionist approach) states that prejudice is learned from observing and imitating significant others.
        3. Theodor W. Adorno argued prejudice resulted from an **authoritarian personality**, which is characterized by excessive conformity, submissiveness to authority, intolerance, insecurity, a high level of superstition, and rigid, stereotypic thinking.
    E. To measure prejudice, some sociologists use **social distance**, which is the extent to which people are willing to interact and establish relationships with members of racial and ethnic groups other than their own.

III. DISCRIMINATION
    A. **Discrimination** is defined as actions or practices of dominant group members that have a harmful impact on members of a subordinate group.
    B. Robert Merton identified four combinations of attitudes and responses:

184

1. *Unprejudiced nondiscriminators*—persons who are not personally prejudiced and do not discriminate against others.
2. *Unprejudiced discriminators*—persons who may have no personal prejudice but still engage in discriminatory behavior because of peer group prejudice or economic, political, or social interests.
3. *Prejudiced nondiscriminators*—persons who hold personal prejudices but do not discriminate due to peer pressure, legal demands, or a desire for profits.
4. *Prejudiced discriminators*—persons who hold personal prejudices and actively discriminate against others.

C. Discriminatory actions vary in severity from the use of derogatory labels to violence against individuals and groups.
1. **Genocide** is the deliberate, systematic killing of an entire people or nation.
2. More recently, the term *ethnic cleansing* has been used to define a policy of "cleansing" geographic areas (such as in Yugoslavia) by forcing persons of other races or religions to flee—or die.

D. Discrimination also varies in how it is carried out.
1. **Individual discrimination** consists of one-on-one acts by members of the dominant group that harm members of the subordinate group or their property.
2. **Institutional discrimination** is the day-to-day practices of organizations and institutions that have a harmful impact on members of subordinate groups.
3. Joe R. Feagin has identified four types of discrimination:
   a. *Isolate discrimination*—harmful action taken by one dominant group member against a subordinate group member; discrimination occurs without support of other dominant group members.
   b. *Small group discrimination*—harmful action taken by a small number of dominant group members against subordinate group members; discrimination occurs without the support of other group members.
   c. *Direct institutionalized discrimination*—action prescribed by an organization or community that intentionally has a differential and negative impact on members of subordinate groups.
   d. *Indirect institutionalized discrimination*—practices that have a harmful impact on subordinate group members even though organizationally or community-prescribed regulations were created with no intent to harm.

IV. SOCIOLOGICAL PERSPECTIVES ON RACE AND ETHNIC RELATIONS
A. Symbolic Interactionist Perspectives
1. The *contact hypothesis* suggests that contact between people from divergent groups should lead to favorable attitudes and behavior when members of each group have equal status, pursue the same goals, cooperate, and receive positive feedback when interacting in nondiscriminatory ways.
2. However, scholars have found that increasing contact may have little or no effect on existing prejudices.
B. Functionalist Perspectives
1. **Assimilation** is a process by which members of subordinate racial and ethnic groups become absorbed into the dominant culture.

2. **Ethnic pluralism** is the coexistence of a variety of distinct racial and ethnic groups within one society.
3. **Segregation** is the spatial and social separation of categories of people by races, ethnicity, class, gender, and/or religion.

C. Conflict Perspectives
1. The caste perspective views racial and ethnic inequality as a permanent feature of U.S. society.
2. Class perspectives emphasize the role of the capitalist class in racial exploitation.
   a. William Julius Wilson has suggested that race may be less significant than class in determining life chances of inner-city residents.
3. **Internal colonialism** occurs when members of a racial or ethnic group are conquered or colonized and forcibly placed under the economic and political control of the dominant group.
4. **Split labor market** refers to the division of the economy into two areas of employment: a primary sector composed of higher paid (usually dominant group) workers in more secure jobs, and a secondary sector comprised of lower paid (often subordinate group) workers in jobs with little security and frequently hazardous working conditions.
5. *Gendered racism* refers to the interactive effect of racism and sexism in the exploitation of women of color.
6. The *theory of racial formation*, influenced by Michael Omi and Howard Winant, states that actions of the government substantially define racial and ethnic relations in the United States.

D. An Alternative Perspective: Critical Race Theory
1. One premise is that racism is so ingrained in U.S. society that it appears to be ordinary and natural to many people.
2. *Interest convergence* is a crucial factor in bringing about social change.
3. Formal equality under the law does not necessarily equate to actual equality in society.
4. Ironies and contradictions exist in civil rights laws, which actually serve the dominant group.

V. RACIAL AND ETHNIC GROUPS IN THE UNITED STATES
A. Native Americans
1. Historically, Native Americans experienced the following kinds of treatment in the United States:
   a. Genocide
   b. Forced migration
   c. Forced assimilation
2. Today, about 2.5 million Native Americans live in the United States (primarily in the Southwest), and about one-third live on reservations.
3. Native Americans are the most disadvantaged racial or ethnic group in the United States in terms of income, employment, housing, nutrition, and health (especially among individuals living on reservations).

B. White Anglo-Saxon Protestants (British Americans)
 1. White Anglo-Saxon Protestants (WASPs) have been the most privileged group in the U.S.
 2. Although many English settlers initially were indentured servants or sent here as prisoners, they quickly emerged as the dominant group, creating a core culture to which all other groups were expected to adapt.
 3. Like other racial and ethnic groups, British Americans are not all alike; social class and gender affect their life chances and opportunities. Many do not think of themselves as having a race or ethnicity.
C. African Americans
 1. About 500,000 Africans were brought by force to the U.S. under the system of slavery, primarily to work on southern plantations.
 2. Informal practices in the north and *Jim Crow laws* in the south segregated African Americans in housing, employment, education, and all public accommodations; some slaves and whites engaged in active resistance that eventually led to the abolition of slavery.
 3. *Lynching*—a killing carried out by a group of vigilantes seeking revenge for an actual or imagined crime by the victim—was used by whites to intimidate African Americans into staying "in their place."
 4. During World Wars I and II, African Americans were a vital source of labor in war production industries; however, racial discrimination continued both on and off the job.
    a. After African Americans began to demand sweeping societal changes in the 1950s, racial segregation slowly was outlawed by the courts and the federal government.
    b. Civil rights legislation attempted to do away with discrimination in education, housing, employment, and health care; affirmative action policies attempt to assure equal education and employment opportunities.
 5. Today, African Americans make up about 13.4 percent of the U.S. population; many have made significant gains in education, employment, and income since the 1960s. However, other African Americans have not fared so well; for example, the African American unemployment rate remains twice as high as that of whites.
D. The term *white ethnic Americans* is used to identify immigrants who came from European countries other than England. Ireland, Poland, Italy, Greece, Germany, Yugoslavia, Russia, and other former Soviet republics are examples.
 1. White ethnic groups migrated to the U.S. in the late nineteenth and early twentieth centuries and experienced high levels of prejudice and discrimination.
E. The term *Asian Americans* refers to many diverse groups with roots in Asia.
 1. Chinese Americans
    a. The initial wave of Chinese immigration occurred between 1850 and 1880 when Chinese men came to the United States seeking gold in California and jobs constructing the transcontinental railroads.

b. Chinese Americans were subjected to extreme prejudice and stereotyping; the Chinese Exclusion Act of 1882 was passed because white workers feared for their jobs.

c. Today, one third of all Chinese Americans were born in the United States; as a group, they have enjoyed considerable upward mobility, but many Chinese Americans provide low-wage labor and live in poverty.

2. Japanese Americans

a. The earliest Japanese immigrants primarily were men who worked on sugar plantations in the Hawaiian Islands in the 1860s. The immigration of Japanese men was curbed in 1908; however, Japanese women were permitted to enter the U.S. for several more years because of the shortage of women.

b. During World War II, when the United States was at war with Japan, nearly 120,000 Japanese Americans were placed in internment camps because they were seen as a security threat; many Japanese Americans lost all that they owned during the internment. Four decades later, the U.S. government apologized and paid $20,000 to some who had been placed in internment camps.

3. Korean Americans

a. The first wave of Korean immigrants were male workers who arrived in Hawaii between 1903 and 1910; the second wave came to the mainland following the Korean War in 1954 (mostly the wives of servicemen and Korean children who had lost their parents in the war); the third wave arrived after the Immigration Act of 1965 permitted well-educated professionals to migrate to the U.S.

b. Korean Americans have helped each other open small businesses by pooling money through the *kye*—an association that grants members money on a rotating basis to gain access to more capital.

4. Filipino Americans

a. Most of the first Filipino immigrants were men who were employed in agriculture; following the Immigration Act of 1965, Filipino physicians, nurses, technical workers, and other professionals moved in large numbers to the U.S. mainland.

b. Unlike other Asian Americans, most Filipinos have not had the start-up capital necessary to open their own businesses, and workers generally have been employed in the low-wage sector of the dual labor market; however, average household income is relatively high due to high levels of employment and college degrees among women.

5. Indochinese Americans

a. Most Indochinese Americans (including people from Vietnam, Cambodia, Thailand, and Laos) have come to the U.S. in the past three decades.

b. Vietnamese refugees who had the resources to flee at the beginning of the Vietnam War were the first to arrive. Next came Cambodians and lowland Laotians, referred to as "boat people" by the media.

c. Today, most Indochinese Americans are foreign born; about half live in western states, especially California.

F. Latino/as (Hispanic Americans)
1. Mexican Americans or Chicano/as, who make up the largest segment (approximately two-thirds) of Latinos/as in the U.S., have experienced disproportionate poverty as a result of internal colonialism.
   a. More recently, Mexican Americans have been seen as cheap labor at the same time that they have been stereotyped as lazy.
   b. When anti-immigration sentiments are running high, Mexican Americans often are the objects of discrimination.
   c. Today, the families of many Mexican Americans have lived in the United States for four or five generations and have made significant contributions in many areas.
2. When Puerto Rico became a territory of the United States in 1917, Puerto Ricans acquired U.S. citizenship and the right to move freely to and from the mainland. While living conditions have improved substantially for some, others continue to live in poverty.
3. Cuban Americans have fared somewhat better than other Latinos/as. Early waves of Cuban immigrants were affluent business and professional people; the second wave of Cuban Americans in the 1970s fared worse—many had been released from prisons and mental hospitals in Cuba. More recent arrivals have developed their own ethnic and economic enclaves in Miami, and have become mainstream professionals and entrepreneurs.
G. Middle Eastern Americans
1. Since 1970, many immigrants have arrived in the United States from Middle Eastern countries such as Egypt, Syria, Lebanon, Saudi Arabia, Kuwait, Yemen, Iran, and Iraq.
2. Middle Eastern Americans speak a variety of languages and have diverse backgrounds.
3. Hate crimes and other discriminatory acts against Middle Eastern Americans escalated after the terrorist attacks on September 11, 2001.
VI. GLOBAL RACIAL AND ETHNIC INEQUALITY IN THE FUTURE
A. Worldwide Racial and Ethnic Struggles
1. The cost of self-determination—the right to choose one's own way of life—often results in the loss of life and property in ethnic warfare, such as in Yugoslavia, Spain, Britain, Romania, Russia, Moldova, Georgia, the Middle East, Africa, Asia, and Latin America.
2. Today, the struggle between the Israeli government and Palestinian factions has heightened tensions around the world.
B. Growing Racial and Ethnic Diversity in the United States
1. Racial and ethnic diversity is increasing in the United States; by 2056, the roots of the average U.S. resident will be Africa, Asia, Hispanic countries, the Pacific Islands, and the Middle East—not white Europe.
2. Conflict may become more overt if people continue to use *sincere fictions*—personal beliefs that are a reflection of larger societal mythologies, such as "I am not a racist"—even when these are inaccurate perceptions.
3. Some analysts believe that there is reason for cautious optimism—throughout U.S. history, subordinate racial and ethnic groups have struggled to gain the

189

freedom and rights which were previously withheld from them, and movements comprised of both whites and people of color will continue to oppose racism in everyday life, to aim at healing divisions among racial groups, and to teach children about racial tolerance.

CRITICAL THINKING QUESTIONS

1. What impact do you think your own race or ethnicity has had on your life? How would your life be different if you were a different race or ethnicity?
2. Compare and contrast the frustration-aggression hypothesis, authoritarian personality theory, and social learning theory. Which theory do you think is most useful for understanding prejudice?
3. Which theoretical perspective do you think is most useful for understanding race and ethnic relations? Why? What are the strengths and weaknesses of each theory?
4. Compare and contrast the experiences of each racial/ethnic group in the U.S. Compare the impact of sports on each group. What other institutions do you think have a significant impact on each group? Why?
5. For each racial/ethnic group, examine the relationship between their immigration experience and their current status in U.S. society.
6. In what ways do you think race relations will change in the future?

PRACTICE TESTS

MULTIPLE CHOICE QUESTIONS

Select the response that best answers the question or completes the statement.

1. _____ is a category of people who have been singled out as inferior or superior, often on the basis of real or alleged physical characteristics such as skin color, hair texture, eye shape, or other subjectively selected attributes.
   a. Ethnic group
   b. Race
   c. National group
   d. Subcultural group

2. All of the following are characteristics of ethnic groups, **except**:
   a. unique cultural traits
   b. a feeling of ethnocentrism
   c. territoriality
   d. the same skin color

3. According to the text, the terms "majority group" and "minority group" are
   a. accurate because "majority groups" always are larger in number than "minority groups."
   b. misleading because people who share ascribed racial or ethnic characteristics automatically constitute a group.
   c. less accurate terms than "dominant group" and "subordinate group," which more accurately reflect the importance of power in the relationships.
   d. no longer used because they are "politically incorrect."

4. _____ is a negative attitude based on faulty generalizations about members of selected racial and ethnic groups.
   a. Prejudice
   b. Discrimination
   c. Stereotyping
   d. Genocide

5. _____ is an overgeneralization about the characteristics of members of certain categories.
   a. Prejudice
   b. Stereotype
   c. Discrimination
   d. Amalgamation

6. According to the text, the use of Native American names, images, and mascots by sports teams is an example of:
   a. prejudice
   b. discrimination
   c. stereotyping
   d. genocide

7. _____ is often hidden from sight and more difficult to prove.
   a. Subtle racism
   b. Open racism
   c. Closed racism
   d. Obvious racism

8. Racism tends to intensify in times of:
   a. increased assimilation
   b. economic uncertainty
   c. low rates of immigration
   d. changing gender roles

9. All of the following are examples of overt racism, **except**:
   a. a police detective who repeatedly refers to African Americans in derogatory terms while discussing police department procedures with the author of a screenplay
   b. a statement that African Americans are naturally more athletic than other groups
   c. segregation of students by race/ethnicity in public schools
   d. hate crimes perpetrated against victims because of their race/ethnicity, sexual orientation, or religion

10. _____ states that people who are disappointed in their efforts to achieve a highly desired goal will respond with a pattern of assertiveness toward others.
    a. Social learning theory
    b. Authoritarian personality theory
    c. Frustration-aggression hypothesis
    d. Social convergence hypothesis

11. When members of subordinate racial and ethnic groups become substitutes for the actual source of frustration within a community, they become:
    a. authoritarian personalities
    b. ethnocentric
    c. racists
    d. scapegoats

12. According to symbolic interactionists, prejudice results from:
    a. social learning
    b. anger and frustration
    c. the authoritarian personality
    d. discrimination

13. Psychologist Theodore W. Adorno and his colleagues concluded that highly prejudiced individuals tend to have a(n) _____ personality.
    a. frustrated-aggressive
    b. authoritarian
    c. racist
    d. discriminatory

14. Michael Omi and Howard Winant created the theory of
    a. racial formation.
    b. authoritarian personality.
    c. social distance.
    d. social learning.

15. According to sociologist Robert Merton's typology, an umpire who is prejudiced against African Americans and deliberately makes official calls against them when they are at bat is an example of a(n):
    a. unprejudiced nondiscriminator
    b. unprejudiced discriminator
    c. prejudiced nondiscriminator
    d. prejudiced discriminator

192

16. Institutional discrimination consists of
    a. one-on-one acts by members of the dominant group that harm members of the subordinate group or their property.
    b. day-to-day practices of organizations and institutions that have a harmful impact on members of subordinate groups.
    c. the division of the economy into two areas of employment, a primary sector or upper tier, and a secondary sector or lower tier.
    d. the deliberate, systematic killing of an entire people or nation.

17. A prejudiced judge giving harsher sentences to all African American defendants without the support of the judicial system in that action, is an example of _____ discrimination.
    a. indirect institutionalized
    b. direct institutionalized
    c. small group
    d. isolate

18. _____ is the deliberate, systematic killing of an entire people or nation.
    a. Institutional discrimination
    b. Isolate discrimination
    c. Genocide
    d. Global prejudice

19. According to the contact hypothesis, contact between people from divergent groups should lead to favorable attitudes and behaviors when certain factors are present. Members of each group must have all of the following, **except**:
    a. equal status
    b. pursuit of the same goals
    c. being competitive with one another to achieve their goals
    d. positive feedback from interactions with each other

20. _____ is a process by which members of subordinate racial and ethnic groups become absorbed into the dominant culture.
    a. Assimilation
    b. Ethnic pluralism
    c. Accommodation
    d. Internal colonialism

21. _____ occurs when members of an ethnic group adopt dominant group traits, such as language, dress, values, religion, and food preferences.
    a. Acculturation
    b. Integration
    c. Amalgamation
    d. Miscegenation

22. _____ occurs when members of subordinate racial or ethnic groups gain acceptance in everyday social interaction with members of the dominant group.
    a. Amalgamation
    b. Miscegenation
    c. Integration
    d. Acculturation

23. _____ is the coexistence of a variety of distinct racial and ethnic groups within one society.
    a. Assimilation
    b. Ethnic pluralism
    c. Accommodation
    d. Internal colonialism

24. Five young white men target a young black man for physical violence because of his race. Based on Joe R. Feagin's identification of four types of discrimination, this scenario is an example of _____ discrimination.
    a. direct institutional
    b. isolate
    c. small group
    d. indirect institutional

25. Based on early theories of race relations by W.E.B. Du Bois, sociologist Oliver Cox suggested that African Americans were enslaved because
    a. of prejudice based on skin color.
    b. they were the cheapest and best workers that owners could find for heavy labor in the mines and on plantations.
    c. they allowed themselves to be dominated by wealthy people.
    d. they agreed to work for passage to the United States, believing that they would be released after a period of indentured servitude.

26. According to sociologist William Julius Wilson, _____ is the most important factor to understanding the state of African Americans in the United States.
    a. class
    b. race
    c. culture
    d. gender

27. According to _____, sports reflect the interests of the wealthy and powerful; athletes are exploited in order for coaches, managers, and owners to gain high levels of profit and prestige.
    a. interactionists
    b. sports theorists
    c. functionalists
    d. conflict theorists

28. The division of the economy into a primary sector, composed of higher-paid workers in more secure jobs, and a secondary sector, composed of lower-paid workers in jobs with little security and hazardous working conditions, is referred to as:
    a. the split labor market
    b. racial formation
    c. gendered racism
    d. stacking

29. The theory of _____ refers to the interactive effect of racism and sexism on the exploitation of women of color.
    a. gendered racism
    b. sexual racism
    c. colonial racism
    d. historical racial sexism

30. According to _____ theory, interest convergence is a critical factor in changing race and ethnic inequality.
    a. functionalist
    b. symbolic interactionist
    c. critical race
    d. ethnic study

31. The experiences of Native Americans in the United States have been characterized by:
    a. slavery and segregation
    b. restrictive immigration laws
    c. genocide, forced migration, and forced assimilation
    d. internment

32. The "Trail of Tears" was
    a. systematic discrimination against Latino/as.
    b. specific Jim Crow laws in the 1950s.
    c. segregationist policies against Asian Americans.
    d. the forced migration of Native Americans.

33. President _____ was the first to establish a committee with the goal of ending discrimination in employment in government and its contractors.
    a. Truman
    b. Kennedy
    c. Johnson
    d. Nixon

34. Nearly 120,000 _____ were placed in internment camps, experiencing one of the worst forms of discrimination sanctioned by U.S. laws.
    a. African Americans
    b. Native Americans
    c. Mexican Americans
    d. Japanese Americans

35. Today, _____ are the most disadvantaged racial or ethnic group in the U.S. in terms of income, employment, housing, nutrition, and health.
    a. African Americans
    b. Native Americans
    c. Mexican Americans
    d. Korean Americans

36. Many _____ have experienced disproportionate poverty as a result of internal colonialism.
    a. Mexican Americans
    b. African Americans
    c. Chinese Americans
    d. Italian Americans

37. _____ live primarily in the Southeast, especially Florida. Many were affluent professionals and businesspeople who fled their homeland after Fidel Castro's 1959 Marxist revolution.
    a. Puerto Rican Americans
    b. Mexican Americans
    c. Cuban Americans
    d. Central Americans

38. Middle Eastern women often experience discrimination for wearing a _____, or a "head-to-toe covering," showing only one's face.
    a. hijab
    b. burkah
    c. sari
    d. scarf

39. Following the September 11, 2001 terrorist attacks, the United States passed the _____—a law giving the federal government greater authority to engage in searches and surveillance with less judicial review.
    a. U.S. Homeland Security Act
    b. U.S. Patriot Act
    c. U.S. Immigration Act of 2001
    d. U.S. Protection Act

40. Throughout the world, many racial and ethnic groups seek _____, the right to choose their own way of life.
    a. self-determination
    b. freedom
    c. liberty
    d. civil disobedience

TRUE/FALSE QUESTIONS

1. Sociologists suggest that race is a socially constructed reality, not a biological one.
   T   F

2. Racial classifications in the U.S. census have remained unchanged for the past four decades.
   T   F

3. In order for an ethnic group to exist, people must identify themselves as a group or be identified as such by others.
   T   F

4. Analysts have found that whites that accept racial stereotypes have no greater desire for social distance from people of color than do whites that reject racial stereotypes.
   T   F

5. The extermination of 6 million European Jews by Nazi Germany is an example of genocide.
   T   F

6. Discrimination is an individual activity; it is never based on norms of an organization or community.
   T   F

7. Equalitarian pluralism is an ideal typology that has never actually been identified in any society.
   T   F

8. De facto segregation refers to racial separation and inequality enforced by custom.
   T   F

9. The internal colonialism model helps explain the continued exploitation of immigrant groups such as the Chinese, Filipinos, and Vietnamese.
   T   F

10. Racial classifications have remained stable throughout U.S. history.
    T   F

11. White Anglo-Saxon Protestants (WASPS) have been the most privileged group in the United States.
    T   F

12. The segregation of African Americans from whites occurred only in the southeastern United States.
    T   F

13. White ethnics have not experienced discrimination in the United States.
    T   F

197

14. All Middle Eastern Americans are of the Muslim faith.
    T   F

15. It is predicted that by 2056, the roots of the average U.S. resident will be in Africa, Asia, Hispanic countries, the Pacific Islands, and the Middle East.
    T   F

16. According to Box 10.1, racially linked genetic traits explain many of the differences among athletes.
    T   F

17. According to Box 10.2, "Sociology in Global Perspective," racism is an issue in European soccer just as it is an issue in U.S. sports.
    T   F

18. Amalgamation, or intermarriage among members of different racial/ethnic groups, is the most common form of assimilation in the U.S.
    T   F

19. According to Box 10.3, "Framing Race in the Media," multiracial scenes in television have no redeeming social values.
    T   F

20. According to Box 10.4, "You Can Make a Difference: Working for Racial Harmony," racial harmony objectives could be reached over time.
    T   F

FILL-IN-THE-BLANK QUESTIONS

1. A _____ group is one that is advantaged and has superior resources and rights in a society.

2. The term _____ refers to individuals who share some common cultural or national characteristics.

3. Sociologists emphasize that race is a _____, not a biological one.

4. _____ occurs when members of a racial or ethnic group are conquered or colonized and forcibly placed under the economic and political control of the dominant group.

5. _____ is an attitude that is used to justify the superior or inferior treatment of another racial or ethnic group.

6. Highly prejudiced individuals tend to have a _____, which is characterized by excessive conformity, intolerance, and rigid stereotypes.

7. _____ involves actions or practices of dominant group members that have a harmful impact on members of a subordinate group.

8. The day-to-day practices of organizations and institutions that have a harmful impact on members of subordinate groups are termed _____.

9. Segregation may be enforced by custom, _____ segregation, or by law, which is an example of _____ segregation.

10. The process by which members of subordinate racial and ethnic groups become absorbed into the dominant culture is known as_____.

11. Dr. Martin Luther King, Jr. and the civil rights movement used the term _____ to refer to nonviolent action seeking to change a policy or law by refusing to comply with it.

12. _____ programs were instituted in public and private sector organizations in order to bring about greater opportunities for previously excluded racial and ethnic groups.

13. The _____ Exclusion Act of 1882 brought this group's immigration into the United State to a halt.

14. _____ were made wards of the government and were subjected to forced assimilation.

15. The term _____ is preferred by people who have their origins to Haiti, Puerto Rico, or Jamaica.

SHORT ANSWER/ESSAY QUESTIONS

1. Select any two racial/ethnic groups and compare and contrast their historical and contemporary experiences in the U.S.
2. What are some social factors that have prevented African Americans from achieving full equality in the United States?
3. What is the fastest growing ethnic minority group in the United States today? Why?
4. Why is education a crucial issue for Latino/as? Which Latino/a group has fared the best? Why?
5. Why are race and ethnicity so socially significant? In your response, provide both historic and contemporary explanations.

STUDENT CLASS PROJECTS AND ACTIVITIES

1. Investigate pluralism on your campus by conducting a brief campus study. Record your information in a written paper. Ask the following questions and include any others you may want to ask: (1) What is the number and percentage of the different ethnic or racial groups enrolled on the campus? (2) What is the number and percentage of students from other countries? (3) What percentage of various ethnic groups holds important campus offices? Identify the office and the race or ethnicity of the office holder. (4) What percentage of racial or ethnic groups are in the campus sororities and fraternities and other organizations? This information may be available from the admissions or registrar's office, student activities, or athletic office. Write a paper that summarizes your findings and evaluate your campus on how pluralistic it appears from your findings.
2. Conduct an informal survey of 20 people who are ethnic or racial minorities in the U.S. Some general questions you may want to ask include: (1) The respondent's specific race or ethnicity. (2) Do you ever experience racism, bigotry, or discrimination in some form in our society? (3) If so, what kinds? (4) Can you remember how you learned that you were a member of a minority group? (5) How did it feel to know that you were considered a minority? (6) What do you think could be done to improve race/ethnic relations in the U.S? Provide any other questions

that you would like to ask and then provide a summary, conclusion, and evaluation of this project.

3. Examine your own ethnicity or racial heritage, if known. If your ethnicity or race is mixed, then select the one that you believe to be the most dominant heritage in your life. If your ethnicity or race is unclear, or not known, then examine your own socialization into an ethnic or racial group. Note that many people, because of adoption or other factors, may not know their original ethnic heritage, but may identify with a racial or ethnic heritage. Prepare a minimum five-page report on your experience as a member of your particular group. Gather as much background information as possible in exploring your own ethnic or racial heritage. If known, trace the ethnic history of your own family, such as the approximate date of your ancestors' migration to the United States, the obstacles they faced, their cultural practices and experiences from childhood, usual activities, and family experiences. Include activities with extended kin members, type of education received, if any, and church and neighborhood experiences. Record any other event or experience that influenced your identity and membership in the group.

INTERNET ACTIVITIES

1. Search the internet for general information about **race and ethnicity**. You can look through the sites below to collect relevant facts about racial and ethnic groups, and then conduct a search on your own.
   http://eserver.org/race, **http://www.trinity.edu/mkearl/race.html**
2. These are sites that are more specific to various racial and ethnic groups in the U.S. Students wishing to go more in-depth with a study of specific racial and ethnic groups should start here.
   African-American (http://www.aframnews.com/), **Latino (http://www.lif.org/)**, **Asian-American (http://goldsea.com/Air/Issues/issues.html)**, and, **Native American (http://www.hanksville.org/NAresources/)**.
3. Browse the **National Association for the Advancement of Colored People**'s web site to review current projects, events, and issues at: **http://www.naacp.org/**

INFOTRAC COLLEGE EDITION EXERCISES

Visit the **InfoTrac College Edition** web site at: **http://www.wadsworthmedia.com/webtutor/infotrac.htm**. You will arrive at a screen that enables you to search topics.

1. Scan the InfoTrac for articles dealing with **racism**. What are some common themes that emerge?
2. **Ethnic cleansing** remains a powerful tool for one group to eliminate another. There are two categories listed in InfoTrac: genocide and forced migration. Look for articles that describe where ethnic cleansing is taking place, and with what consequences.
3. Select one of the **ethnic minority groups** addressed in this chapter and research for information on that group, as found in the archives of InfoTrac. Answer the following questions: What are the key issues impacting this group? What kinds of suggestions do the articles offer to improve the status of this group?
4. Search for these theoretical concepts in articles contained in the archives of Infotrac: **scapegoat**, **contact hypothesis**, **assimilation**, **internal colonialism**, **split labor market**, and **critical race theory**.

SOLUTIONS

MULTIPLE CHOICE QUESTIONS

1. B, p. 311
2. D, p. 311
3. C, p. 314
4. A, p. 314
5. B, p. 314
6. C, p. 314
7. A, p. 315
8. B, p. 315
9. B, p. 315
10. C, p. 316
11. D, p. 316
12. A, p. 317
13. B, p. 317
14. A, p. 326

15. D, p. 318
16. B, p. 319
17. D, p. 319
18. C, p. 319
19. C, p. 320
20. A, p. 320
21. A, p. 320
22. C, p. 321
23. B, p. 322
24. B, p. 319
25. B, p. 323
26. B, p. 323
27. D, p. 323
28. A, p. 326

29. A, p. 326
30. C, p. 326
31. C, p. 328
32. D, p. 329
33. B, p. 332
34. D, p. 336
35. B, p. 329
36. A, p. 338
37. C, p. 338
38. A, p. 339
39. B, p. 339
40. A, p. 340

TRUE/FALSE QUESTIONS

1. T, p. 311
2. F, p. 313
3. T, p. 311
4. F, p. 318
5. T, p. 319
6. F, p. 318
7. F, p. 322

8. T, p. 322
9. F, p. 325
10. F, p. 313
11. T, p. 330
12. F, p. 331
13. F, p. 333
14. F, p. 339

15. T, p. 341
16. F, p. 312
17. T, p. 316
18. F, p. 322
19. F, p. 324
20. T, p. 341

FILL-IN-THE-BLANK QUESTIONS

1. dominant, p. 314
2. ethnicity, p. 311
3. socially constructed reality, p. 311
4. Internal colonialism, p. 325
5. Racism, p. 315
6. authoritarian personality, p. 317
7. Discrimination, p. 318
8. institutional discrimination, p. 319
9. de facto, p. 323; de jure, p. 322
10. assimilation, p. 320
11. civil disobedience, p. 331
12. Affirmative action, p. 332
13. Chinese, p. 336
14. Native Americans, p. 329
15. black, p. 331

11

SEX AND GENDER

BRIEF CHAPTER OUTLINE

CHAPTER SUMMARY

Sex refers to the biological and anatomical differences between females and males, whereas **gender** refers to the culturally and socially constructed differences between females and males found in the meanings, beliefs, and practices associated with "femininity" and "masculinity." Gender is socially significant because it leads to differential treatment of men and women; sexism (like racism) often is used to justify discriminatory treatment. Sex is not always clear-cut; a hormone imbalance before birth produces a **hermaphrodite**, while a **transsexual** identifies with the opposite gender.

The closest approximation of a third sex in Western societies is a **transvestite**, which is an individual who lives as the opposite sex but does not change their genitalia. **Sexual orientation** refers to an individual's preference for emotional-sexual relationships with members of the opposite sex, same sex, or both. Gender is a social construction with important consequences in everyday life. **Sexism** is linked to **patriarchy**, a hierarchical system in which cultural, political, and economic structures are male dominated. Gender relationships vary in different types of societies. In most hunting and gathering societies, fairly equitable relationships exist because neither sex has the ability to provide all of the food necessary for survival. In horticultural societies, hoe cultivation is compatible with child care, and a fair degree of gender equality exists because neither sex controls the food supply; in pastoral societies, women have relatively low status. In agrarian societies, male dominance is very apparent; tasks require more labor and physical strength, and women are seen as too weak or too tied to childrearing activities to perform these activities. In industrialized societies, a gap exists between unpaid work performed by women at home and paid work performed by men. In postindustrial societies, technology supports a service- and information-based economy; most adult women are in the labor force and have double responsibilities—those from the labor force and those from the family. The key agents of gender socialization are parents, peers, teachers and schools, sports, and the mass media, all of which tend to reinforce stereotypes of appropriate gender behavior. **Gender bias** and **sexual harassment** remain major problems in American society. Gender inequality results from the economic, political, and educational discrimination of women. In most workplaces, jobs are either gender segregated or the majority of employees are of the same gender. Gender-segregated occupations lead to a disparity, or **wage gap**, between women's and men's earnings. Even when women are employed in the same job as men, on average they do not receive the same pay. **Comparable worth** is the belief that wages should reflect the worth of a job, not the gender or race of the worker. Many women have a "second shift" because of their dual responsibilities for paid and unpaid work. According to functional analysts, husbands perform instrumental tasks of economic support and decision making, and wives assume expressive tasks of providing affection and emotional support for the family. Conflict analysts suggest that the gendered division of labor within families and the workplace results from male control and dominance over women and resources. **Feminism** is the belief that women and men are equal, and that they should be valued equally and have equal rights. Although feminist perspectives vary in their analyses of women's subordination, they all advocate social change to eradicate gender inequality.

LEARNING OBJECTIVES

After reading Chapter 11, you should be able to:

1. Distinguish between sex and gender and explain their sociological significance.

2. Distinguish between primary sex characteristics and secondary sex characteristics.

3. Differentiate between hermaphrodites, transsexuals, and transvestites.

4. Explain why definitions of sex are not always clear-cut.

5. Define sexual orientation and discuss issues and examples of homophobia.

6. Define gender roles, gender identity, and body consciousness and describe the relationships between the three concepts.

7. Explain some of the important ways that people "do" gender in their daily interactions.

8. Define sexism and explain how it is related to discrimination and patriarchy.

9. Describe gender stratification in hunting and gathering societies, horticultural societies, pastoral societies, agrarian societies, industrial societies, and postindustrial societies.

10. Describe the process of gender socialization and identify specific ways in which parents, peers, teachers, sports, and mass media contribute to the process.

11. Describe gender bias and explain how schools operate as a gendered institution.

12. Describe the most important ways that the mass media contributes to gender socialization.

13. Discuss the gendered division of paid work and explain its relationship to the issue of pay equity or comparable worth.

14. Trace changes in labor force participation by women and note how these changes have contributed to the "second shift."

15. Use examples to explain the feminist perspective on gender equality.

16. Describe functionalist and neoclassical economic perspectives on gender stratification and contrast them with conflict perspectives.

17. Outline the key assumptions of liberal, radical, socialist, and multicultural feminism.

KEY TERMS

(defined at page number shown and in glossary)

body consciousness, p. 350
comparable worth, p. 367
feminism, p. 373
gender, p. 350
gender bias, p. 359
gender identity, p. 350
gender role, p. 350
hermaphrodite, p. 348
homophobia, p. 349
matriarchy, p. 352

patriarchy, p. 352
primary sex characteristics, p. 347
secondary sex characteristics, p. 347
sex, p. 346
sexism, p. 352
sexual harassment, p. 360
sexual orientation, p. 349
transsexual, p. 348
transvestite p. 348
wage gap, p. 365

KEY PEOPLE

CHAPTER OUTLINE

I. SEX: THE BIOLOGICAL DIMENSION
    A. **Sex** refers to the biological and anatomical differences between females and males.
        1. **Primary sex characteristics** are the genitalia used in the reproductive process; **secondary sex characteristics** are the physical traits (other than reproductive organs) that identify an individual's sex.
    B. Hermaphrodites/transsexuals are examples of when sex is not always clear-cut.
        1. **Hermaphrodites** are born with ambiguous genitalia.
        2. **Transsexuals** feel they are the opposite sex from their sex organs.
    C. **Sexual orientation** is a preference for emotional sexual relationships with members of the opposite sex (heterosexuality), the same sex (homosexuality), or both (bisexuality).
        1. **Homophobia** refers to extreme prejudice directed at gays, lesbians, bisexuals, and others who are perceived as not being heterosexual.
II. GENDER: THE CULTURAL DIMENSION
    A. **Gender** refers to the culturally and socially constructed differences between females and males found in the meanings, beliefs, and practices associated with "femininity" and "masculinity."
        1. A microlevel analysis of gender focuses on how individuals learn **gender roles**—the attitudes, behavior, and activities that are socially defined as appropriate for each sex and are learned through the socialization process—and **gender identity**—a person's perception of the self as female or male. **Body consciousness**, how a person feels about his or her body, is a part of gender identity.
        2. A macrolevel analysis of gender examines structural features, external to the individual, which perpetuate gender inequality, including gendered institutions that are reinforced by a *gender belief system*, based on ideas regarding masculine and feminine attributes that are held to be valid in a society.
    B. The Social Significance of Gender
        1. Gender is a social construction with important consequences in everyday life.
        2. Gender stereotypes hold that men and women are inherently different in attitudes, behavior, and aspirations.

C. Sexism
   1. **Sexism**—the subordination of one sex, usually female, based on the assumed superiority of the other sex—is interwoven with **patriarchy**—a hierarchical system of social organization in which cultural, political, and economic structures are controlled by men.
   2. **Matriarchy** is a hierarchical system of social organization in which cultural, political, and economic structures are controlled by women; few (if any) societies have been organized in this manner.
III. GENDER STRATIFICATION IN HISTORICAL AND CONTEMPORARY PERSPECTIVE
   A. Preindustrial Societies
      1. The earliest known division of labor between women and men is in hunting and gathering societies; men hunt while women gather food. A relatively equitable relationship exists because both sexes are necessary for survival.
      2. In horticultural societies, women make an important contribution to food production because hoe cultivation is compatible with child care; a fairly high degree of gender equality exists because neither sex controls the food supply
      3. In pastoral societies, herding primarily is done by men; women contribute relatively little to subsistence production and thus have relatively low status.
      4. Gender inequality increases and becomes institutionalized in agrarian societies as men become more involved in food production. Though scholars cannot agree on why, some suggest private property is to blame.
   B. Industrial Societies
      1. In industrial societies—those in which factory or mechanized production has replaced agriculture as the major form of economic activity—the status of women tends to decline further.
      2. Gendered division of labor increases the economic and political subordination of women.
   C. Postindustrial Societies
      1. In postindustrial societies, technology supports a service- and information-based economy.
      2. Gender segregation in the workplace continues; today, more households are headed by women with no male present.
IV. GENDER AND SOCIALIZATION
   A. Parents and Gender Socialization
      1. From birth, parents act toward children on the basis of the child's sex; children's clothing and toys reflect their parents' gender expectations.
      2. Gender socialization may vary by race and ethnicity.
   B. Peers and Gender Socialization
      1. Peers help children learn prevailing gender role stereotypes, as well as gender appropriate and inappropriate behavior.
      2. During adolescence, peers often are stronger and more effective agents of gender socialization than are adults.
      3. Among college students, peers play an important role in career choices and the establishment of long-term, intimate relationships.
   C. Teachers, Schools and Gender Socialization

206

1. From kindergarten through college, schools operate as gendered institutions; teachers provide important messages about gender through both the formal content of classroom assignments and informal interactions with students.
2. Teachers may unintentionally demonstrate **gender bias**—the showing of favoritism toward one gender over the other—toward male students.
3. When girls complain of **sexual harassment**—unwanted sexual advances, requests for sexual favors, or other verbal or physical conduct of a sexual nature—their concerns are sometimes downplayed or overlooked by school personnel.

D. Sports and Gender Socialization
1. The type of game played differs with the child's sex; boys are socialized to play competitive, rule-oriented games with large numbers of participants, whereas girls are socialized to play noncompetitive games in groups of two or three.
2. For many males, sports participation and spectatorship is a training ground for masculinity; for females, sports still is tied to the male gender role, thus making it very difficult for girls and women to receive the full benefits of participating in such activities.

E. Mass Media and Gender Socialization
1. The media are powerful sources of gender stereotyping that have a unique ability to shape ideas.
2. On television, more male than female roles are shown, and male characters typically are more aggressive, constructive, and direct, while females are deferential toward others or use manipulation to get their way.
3. Advertising also plays an important role in gender socialization.

F. Adult Gender Socialization
1. Men and women are taught gender-appropriate conduct in education and the workplace.
2. As people enter "middle age," a double standard of aging views aging as positive for men and negative for women.

V. CONTEMPORARY GENDER INEQUALITY
A. Gendered Division of Paid Work
1. *Gender-segregated work* refers to the concentration of women and men in different occupations, jobs, and places of work.
2. *Labor market segmentation*—the division of jobs into categories with distinct working conditions—results in women having separate and unequal jobs that are lower paying, less prestigious, and have fewer opportunities for advancement.

B. Pay Equity (Comparable Worth)
1. Occupational segregation contributes to a **wage gap**—the disparity between women's and men's earnings.
2. **Comparable worth** is the belief that wages ought to reflect the worth of a job, not the gender or race of the worker.

207

C. Paid Work and Family Work
1. Although most married women now share the breadwinner role, family demands remain women's responsibility, even among women who hold full-time paid employment.
2. Many women have a "double day" or "second shift" because of their dual responsibilities for paid and unpaid work.

VI. PERSPECTIVES ON GENDER STRATIFICATION
A. Functionalists and neoclassical economic perspectives on the family view the division of family labor as ensuring that important societal tasks will be fulfilled.
1. According to functionalists, the gendered division of labor—where husbands fulfill *instrumental tasks* of providing money and decision making and wives fulfill *expressive tasks* of affection and emotional support—provides stability for families and ensures all societal tasks are completed.
2. According to the human capital model, individuals vary widely in the amount of education and job training they bring to the labor market; what individuals earn is the result of their own choices and labor market demand for certain kinds of workers at specific points in time.
3. Other neoclassical economic models attribute the wage gap to such factors as: (1) the different amounts of energy men and women expend on their work; (2) the occupational choices women make (choosing female-dominated occupations so that they can spend more time with their families); and (3) the crowding of too many women into some occupations (suppressing wages because the supply of workers exceeds demand).
B. According to conflict perspectives, the gendered division of labor within families and the workplace results from male control of and dominance over women and resources.
1. Although men's ability to use physical power to control women diminishes in industrial societies, they still remain the head of household, control the property, and hold more power through their predominance in the most highly paid and prestigious occupations and the highest elected offices.
2. Conflict theorists in the Marxist tradition assert that gender stratification results from private ownership of the means of production; some men not only gain control over property and the distribution of goods but also gain power over women.
C. Feminist Perspectives
1. **Feminism** refers to a belief that women and men are equal, and that they should be valued equally and have equal rights.
2. In liberal feminism, gender equality is equated with equality of opportunity, such as civil rights and education.
3. According to radical feminists, male domination causes all forms of human oppression, including racism and classism; childbearing and childrearing is viewed as the root of patriarchy.
4. Socialist feminists suggest that women's oppression results from their dual roles as paid and unpaid workers in a capitalist economy. In the workplace, women are exploited by capitalism; at home, they are exploited by patriarchy.

5. Multicultural feminism is based on the belief that women of color experience a different world than other people because of multilayered oppression based on race/ethnicity, gender, and class.

VII. GENDER ISSUES IN THE FUTURE
   A. During the twentieth century, women made significant progress in the labor force; laws were passed to prohibit sexual discrimination in the workplace and school; and affirmative action programs helped make women more visible in education, the government, and the professional world, and able to enter the political arena as candidates instead of volunteers.
   B. Many men have joined feminist movements not only to raise their consciousness about the need to eliminate sexism and gender bias, but also recognizing that what is harmful to women is harmful to men too.
   C. However, in the midst of these changes, many gender issues remain unsolved, such as (1) gender segregation and the wage gap in the labor force, (2) women's burden of the "second shift" at home, (3) eating disorders for both women and some men, (4) body image and obsession with physical fitness, and (5) women as possessions or sex objects.

CRITICAL THINKING QUESTIONS

1. In what ways do you think sex, gender, and sexuality impact each other?
2. Why do you think industrial and postindustrial societies have more gender inequality than other types of societies?
3. In what ways were you socialized into your gender? In what ways does your gender socialization continue today?
4. In what ways does gender inequality impact you? How might it impact you in the future?
5. Which theoretical perspective do you think is most useful for understanding gender stratification? What are the strengths and weaknesses of each?
6. In what ways do you think gender stratification will change or stay the same in the future?

PRACTICE TESTS

MULTIPLE CHOICE QUESTIONS

Select the response that best answers the question or completes the statement.

1. _____ is the process of treating people as if they were *things*, rather than human beings. This occurs when we judge people on the basis of their physical appearance rather than on the basis of their individual qualities or actions.
   a. Stereotyping
   b. Objectification
   c. Sexism
   d. Imaging

2. Social scientists classify individuals as gay or lesbian who
   a. self-identify as gay or lesbian.
   b. have engaged in a homosexual act.
   c. have a gay gene.
   d. are transgender.

3. Sociologists use the term _____ to refer to the biological attributes of men and women; _____ is used to refer to the distinctive qualities of men and women that are culturally created.
   a. gender; sex
   b. sex; gender
   c. primary sex characteristics; secondary sex characteristics
   d. secondary sex characteristics; primary sex characteristics

4. A(n) _____ is a person in whom sexual differentiation is ambiguous or incomplete.
   a. hermaphrodite
   b. transsexual
   c. transvestite
   d. homosexual

5. A person's perception of the self as female or male is referred to as one's:
   a. self-concept
   b. gender identity
   c. gender role
   d. gender belief system

6. According to historian Susan Bordo, the anorexic body and the muscled body
   a. illustrate that eating problems and bodybuilding are unrelated.
   b. are opposites.
   c. are both acquired through a rejection of fat.
   d. are not gendered experiences.

7. According to social psychologist Hilary Lips, an example of _____ directed against men is the mistaken idea that it is more harmful for female soldiers to be killed in combat than male soldiers.
   a. gender stereotypes
   b. matriarchy
   c. patriarchy
   d. sexism

8. All of the following statements are **correct** regarding sexism, **except**:
   a. sexism, like racism, is used to justify discriminatory treatment
   b. evidence of sexism is found in the underevaluation of women's work
   c. women may experience discrimination in leisure activities
   d. men cannot be the victims of sexist assumptions

9. Sociologist Virginia Cyrus suggests that under _____, men are seen as "natural" heads of the households, presidential candidates, corporate executives, and presidents, while women are seen as men's subordinates, playing supportive roles as housewives, mothers, nurses, and secretaries.
    a. patriarchy
    b. matriarchy
    c. polyarchy
    d. hierarchy

10. According to sociologists, _____ are the most important factors in defining what females and males are, what they should do, and what sorts of relations should exist between them.
    a. genetalia
    b. evolutionary processes
    c. social and cultural processes
    d. genes and chromosomes

11. In most hunting and gathering societies,
    a. a relatively equitable relationship exists because neither sex has the ability to provide all of the food necessary for survival.
    b. women are not full economic partners with men.
    c. relationships between women and men tend to be patriarchal in nature.
    d. menstrual taboos place women in subordinate positions by monthly segregation into menstrual huts.

12. Gender inequality and male dominance became institutionalized in _____ societies.
    a. hunting and gathering
    b. horticultural and pastoral
    c. agrarian
    d. industrial

13. All of the following are characteristics of agrarian societies that may contribute to gender inequality, **except**:
    a. private property
    b. male control over distribution of the surplus and the kinship system
    c. the cult of domesticity
    d. emphasis on legitimate heirs to inherit the surplus

14. In the industrial United States, the creation of the home as a private, personal sphere in which women were expected to create a haven for the family is referred to as:
    a. matriarchal theory
    b. cult of domesticity
    c. second shift
    d. theory of male dominance

15. The gendered division of labor in industrialized societies that assigned women domestic roles in the private sphere of the home and men breadwinner roles in the public sphere of paid labor contributed to
    a. the rise of multicultural feminism.
    b. increased economic and political subordination of women.
    c. higher rates of single motherhood.
    d. the association of slightly overweight bodies with social status.

16. All of the following statements regarding gender socialization are correct, **except**:
    a. virtually all gender roles have changed dramatically in recent years
    b. many parents prefer boys to girls because of stereotypical ideas about the relative importance of males and females to the future of the family and society
    c. parents tend to treat baby boys more roughly than baby girls
    d. parents tend to assign chores to children based on gender

17. In postindustrial societies
    a. technological developments have reduced boundaries in men's and women's work.
    b. the proportion of female-headed households with children decreases compared to industrial societies.
    c. gender stereotyping and segregation in the workforce remains remarkably stable.
    d. the extent to which women's social status depends on their husband is highest.

18. Teachers devoting more time, effort, and attention to boys than to girls in schools is an example of:
    a. gender socialization
    b. gender identity
    c. gender role differentiation
    d. gender bias

19. A comprehensive study of gender bias in schools suggested that girls' self-esteem is undermined in school through all of the following experiences, **except**:
    a. a relative lack of attention from teachers
    b. sexual harassment from male peers
    c. an assumption that girls are better in visual-spatial ability, as compared with verbal ability
    d. the stereotyping and invisibility of females in textbooks

20. The 1972 passage of Title IX requires equal
    a. pay for equal work.
    b. educational opportunities in science and math.
    c. opportunities in academic and athletic programs.
    d. treatment in the military.

21. Which of the following statements is **true** regarding men and women in higher education?
    a. Men are more likely than women to earn a college degree.
    b. Men tend to have higher grade point averages in college.
    c. Women typically have higher scores on standardized admissions tests like the Scholastic Assessment Test.
    d. Men constitute the majority of majors in architecture, engineering, computer technology, and physical science.

22. According to the text's discussion of gender and athletics,
    a. women who engage in activities that are assumed to be "masculine" (such as bodybuilding) may either ignore their critics or attempt to redefine the activity or its result as "feminine" or womanly.
    b. half of all high school and college athletes are female.
    c. girls are typically socialized from a young age to participate in team sports with many players.
    d. women bodybuilders are more likely to win competitions if they appear more "masculine."

23. Which of the following statements regarding gender issues and television are **true**?
    a. Television programs are sex-typed and white-male-oriented.
    b. In television today, an equal number of males and females are depicted in leading roles.
    c. Most of the characters in educational programs have female names and voices.
    d. Gender stereotyping no longer exists in the media.

24. Regardless of women's marital status, women have been viewed as _____ wage earners in a male-headed household.
    a. critical
    b. primary
    c. prestigious
    d. supplemental

25. African American professional women are most likely to be concentrated in
    a. public sector employment.
    b. managerial positions.
    c. sales.
    d. large corporations and private law firms.

26. _____ is the division of jobs into categories with distinct working conditions, which results in women having separate and unequal jobs.
    a. Gender bias
    b. Sexual division
    c. Labor market segmentation
    d. Gender disparity

213

27. Occupational segregation contributes to the _____—the disparity between women's and men's earnings.
    a. wage gap
    b. gendered employment
    c. comparable worth distinction
    d. gendered earnings

28. Overall, women make approximately _____ for every $1 earned by men.
    a. 80 cents
    b. 90 cents
    c. $1
    d. $1.20

29. The belief that wages ought to reflect the worth of a job, not the gender or race of the worker, is referred to as:
    a. the earnings ratio
    b. comparable worth
    c. equal pay for equal work
    d. the wage gap

30. According to the functionalist/human capital model,
    a. what women earn is the result of their own choices and the needs of the labor market.
    b. the gendered division of labor in the workplace results from male control of and dominance over women and resources.
    c. women's oppression results from their dual roles as paid and unpaid workers in a capitalist economy.
    d. society will function more smoothly when gender equality is achieved.

31. Critics of functionalist and neoclassical economic perspectives have pointed out that these perspectives
    a. exaggerate the problems inherent in traditional gender roles.
    b. fail to critically assess the structure of society that makes educational and occupational opportunities more available to some than others.
    c. overemphasize factors external to individuals that contribute to the oppression of white women and people of color.
    d. focus on differences between men and women without taking into account the commonalities they share.

32. According to the conflict perspective on gender stratification,
    a. male dominance over women accumulates steadily due to survival demands in hunting and gathering societies.
    b. the gendered division of labor within families and in the workplace results from male control of and dominance over women and resources.
    c. the gendered division of labor provides stability for the family and assures all of society's work will get done.
    d. gender stratification can be explained using the human capital model.

214

33. A women's group fights for better child care options, a woman's right to choose an abortion, and the elimination of sex discrimination in the workplace, arguing that the roots of women's oppression lie in women's lack of equal civil rights and educational opportunities. The platform of this group reflects:
    a. black (African American) feminism
    b. socialist feminism
    c. radical feminism
    d. liberal feminism

34. According to _____, male domination causes all forms of human oppression, including racism and classism.
    a. liberal feminism
    b. radical feminism
    c. socialist feminism
    d. multicultural feminism

35. _____ believe that the only way to achieve gender equality is to eliminate capitalism and develop an economy that would bring equal pay and equal rights to women.
    a. Multicultural feminists
    b. Socialist feminists
    c. Radical feminists
    d. Liberal feminists

36. According to _____, race, class, and gender all simultaneously oppress African American women.
    a. liberal feminism
    b. socialist feminism
    c. multicultural feminism
    d. radical feminism

37. _____ feminists often trace the roots of patriarchy to women's childbearing and childrearing responsibilities, which make women dependent on men.
    a. Liberal
    b. Multicultural
    c. Socialist
    d. Radical

38. According to multicultural feminism,
    a. women of different races and ethnicities are equal to each other, but oppressed by men.
    b. race, class, and gender combine to create a "double" or "triple" jeopardy for minority women.
    c. different characteristics may be more significant in some situations than others.
    d. all women benefit from "relational privilege."

215

39. Sociologist William Goode suggested that some men have felt "under attack" by women's demands for equality because
   a. they are being paid less than women in some specific jobs.
   b. they do not see themselves as responsible for patriarchy.
   c. they have tried to eliminate inequality.
   d. they view society now as female-dominated.

40. Which of the following is true regarding lower wages paid to women in the workforce?
   a. Women's lower wages in the labor force raise men's wages.
   b. The wage gap between men and women continues to widen.
   c. Women's lower wages in the labor force suppress men's wages as well.
   d. If women receive higher wages, they will begin to dominate world markets.

TRUE/FALSE QUESTIONS

1. Thinness has always been the "ideal" body for females in the U.S.
   T   F

2. Some societies recognized three sexes.
   T   F

3. Gendered belief systems generally do not change over time.
   T   F

4. Sexism is the subordination of one sex, usually female, based on the assumed superiority of the other sex.
   T   F

5. The earliest known division of labor between women and men is in agrarian societies.
   T   F

6. As societies industrialize, the status of women tends to decline further.
   T   F

7. The cult of true womanhood increased white women's dependence on men and became a source of discrimination against women of color.
   T   F

8. In postindustrial societies, gendered segregation in the workforce dramatically decreases.
   T   F

9. Unlike parents in many other countries, parents in the U.S. do not prefer boys to girls.
   T   F

10. Individuals who are transsexuals are always also homosexuals.
   T   F

11. In studies of parental gender socialization, gender-linked chore assignments occur less frequently in African American families, where sons and daughters are both socialized toward independence, employment, and child care.
    T   F

12. Female peer groups place more pressure on girls to do "feminine" things than male peer groups place on boys to do "masculine" things.
    T   F

13. By young adulthood, men and women no longer receive gender-related messages from peers.
    T   F

14. Research shows the females receive more praise from teachers for their contribution in the class than do males.
    T   F

15. The degree of gender segregation in the professional labor market has declined since the 1970s, but racial-ethnic segregation has remained deeply embedded in the social structures.
    T   F

16. In discussing body image and gender, recent studies show that up to 95 percent of men express dissatisfaction with some aspect of their bodies.
    T   F

17. Many societies today are considered matriarchies.
    T   F

18. In "Framing Gender in the Media," the news item states that Chinese teenagers are being given cosmetic surgery by their parents as a reward for their hard work in school.
    T   F

19. In Box 11.4, "You Can Make a Difference," many feminist organizations exist for women, but none for men.
    T   F

20. In the "Photo Essay," it is suggested that more men are assuming greater responsibility at home, doing household tasks and childrearing.
    T   F

FILL-IN-THE-BLANK QUESTIONS

1. The word _____ is used to refer to the distinctive qualities of men and women that are culturally created.

2. A person in whom the sex-related structures of the brain that define gender identity are opposite from the physical sex organs of the person's body is known as a(n) _____.

217

3. A(n) _____ is a male who lives as a woman or a female who lives as a man but does not alter the genitalia.

4. Extreme prejudice directed at gays, lesbians, bisexuals, and others who are perceived as not being heterosexual is termed _____.

5. A person's perceptions of self as female or male is one's _____.

6. _____ is how a person perceives and feels about his or her body.

7. Gender-appropriate behavior is learned through a process called _____.

8. _____ work refers to the concentration of women and men in different occupations.

9. Education, politics, and the workplace are examples of _____.

10. The disparity between women's and men's earnings is the _____.

11. The attitudes, behavior, and activities that are socially defined as appropriate for each sex and are learned through the socialization process are known as a(n) _____.

12. A hierarchial system of social organization in which cultural, political, and economic structures are controlled by women is defined as a(n) _____.

13. _____ is unwanted sexual advances, requests for sexual favors, or other verbal or physical conduct of a sexual nature.

14. _____ refers to an individual's preference for emotional-sexual relationships with members of the opposite sex, same sex, or both.

15. _____ is made up of negative attitudes, beliefs, and discrimination against women.

SHORT ANSWER/ESSAY QUESTIONS

1. Explain the difference between sex and gender and the significance of this difference.
2. Describe the process of gender socialization. Include in your answer a discussion of at least three agents of gender socialization.
3. What are some of the consequences of gender inequality in the United States?
4. How are gender relations in the labor force and in families different in postindustrial societies compared to other types of societies?
5. Explain the concept of feminism and the main ideas proposed by liberal, socialist, radical, and multicultural feminism.

STUDENT CLASS PROJECTS AND ACTIVITIES

1. Conduct an informal survey of twenty students on your campus about their opinions of the women's movement and feminism. Some suggested guideline questions to ask: (1) What is your age, gender, and college major? (2) Do you support women's rights? Why? Why not? (3) What is your opinion of the women's movement in our country? (4) Do you consider yourself a feminist? Why? Why not? (5) Has the women's movement improved your life in any way? If so, how? (6) Is there still a need for a strong women's movement? Why? Why not? (7) Does the present women's movement reflect the views of most women? Explain your answer. (8) Is a woman's place at home with her children, if it is economically feasible? (9) How do you think women's lives today are different from women's lives in the past? In writing up this project, provide the responses to each particular question and a summary of the responses. Conclude with a personal evaluation of your findings.

2. Research the history of sexual harassment in the United States. Look up and report information on: (1) how the term came to be defined; (2) how it became an issue in our society; (3) how it has affected women in the workplace in general; (4) some specific examples of how it has affected women in the workplace; (5) the prevalence of sexual harassment in the workplace; and (6) some court cases involving charges of sexual harassment and their outcomes. Include any other information relevant to this research project. Provide in your paper a bibliography and a summary of your findings.

3. Research the topic of oppression and resistance among women in Afghanistan. In the writing up of your paper, include the following information: (1) Describe the plight of women and girls before, during, and after the Taliban's takeover of the country, and after the forced exit of the Taliban. (2) Describe the situation today. Is the situation better or worse? Why or why not? (3) What problems remain for females in Afghanistan today? (4) Describe how the major social institutions, the family, education, political, economic, and religious institutions impact females today. (5) What specific rights do females have today and what specific rights are they denied? In the writing up of your paper, include any other information pertinent to this topic and provide a bibliography of your references.

4. Research the history of the feminist movement in the United States. Write a paper that summarizes each "wave" of feminism, and includes the following information about each wave: (1) Who were the prominent leaders of the movement? (2) What were the key issues or rights women were fighting for? (3) What techniques did women use to try to gain rights? (4) What gains did women make as a result of the movement? (5) What ended this "wave" of feminism? Include any other important information and provide a bibliography of your references.

INTERNET ACTIVITIES

1. Surf these web sites in order to compare similarities and differences in the ways that both sexes are defining the issues.
   a. **Men's Studies Web sites:**
      i. http://www.ncfm.org/
      ii. http://www.nomas.org/
      iii. http://www.xyonline.net/
   b. **Feminist Web sites:**
      i. http://vos.ucsb.edu/browse.asp?id=2711
      ii. http://bailiwick.lib.uiowa.edu/wstudies/
      iii. http://www.now.org/
2. Learn more about **feminist theory** at **http://www.cddc.vt.edu/feminism/**. This site provides a lot of information on different theorists, as well as references for different topics related to feminism, such as education, sports, etc.
3. Learn more about **gender issues** at this site, **http://www.trinity.edu/mkearl/gender.html**.

INFOTRAC COLLEGE EDITION EXERCISES

Visit the **InfoTrac College Edition** web site at: **http://www.wadsworthmedia.com/webtutor/infotrac.htm**. You will arrive at a screen that enables you to search topics.

1. Compare research on **transsexualism**, **hermaphroditism**, and **transvestitism**. Search for these terms in the media as well and see how each group is portrayed differently.
2. Search for articles on **body esteem**. Read at least three articles. Write a summary and discussion question for each article and bring them to class to discuss.
3. Go to InfoTrac and conduct a general search for the term **gender**. You will find a multitude of articles in numerous subfields. Come to class with an idea or issue that is new to you. This is a good way to start a class discussion or introduce the topic of gender.
4. Look up the keyword **fatherhood**. Make a list of current issues impacting fathers today.
5. There are over a hundred articles on **patriarchy** listed on InfoTrac. Write a brief essay on this subject and use at least three sources from this list. The essay should introduce the concept and then demonstrate an important dimension.

SOLUTIONS

MULTIPLE CHOICE QUESTIONS

1. B, p. 346	15. B, p. 356	29. D, p. 365
2. A, p. 349	16. A, p. 358	30. A, pp. 368-369
3. B, p. 346	17. C, p. 356	31. B, p. 369
4. A, p. 348	18. D, p. 359	32. B, p. 369
5. B, p. 350	19. C, p. 360	33. D, p. 373
6. C, p. 352	20. C, p. 361	34. B, p. 373
7. D, p. 352	21. D, p. 361	35. B, p. 373
8. D, p. 352	22. A, pp. 360-361	36. C, p. 373
9. A, p. 352	23. A, pp. 362-363	37. D, p. 373
10. C, p. 350	24. D, p. 365	38. C, p. 373
11. A, p. 354	25. A, p. 364	39. B, p. 375
12. C, p. 355	26. C, p. 364	40. C, p. 375
13. C, p. 355	27. A, p. 365	
14. B, p. 356	28. A, p. 365	

TRUE/FALSE QUESTIONS

1. F, p. 348	8. F, p. 356	15. T, p. 364
2. T, p. 348	9. F, p. 358	16. T, p. 348
3. F, p. 352	10. F, p. 349	17. F, p. 352
4. T, p. 352	11. T, p. 358	18. F, p. 362
5. F, p. 354	12. F, p. 359	19. F, p. 374
6. T, p. 355	13. F, p. 359	20. T, p. 370
7. T, p. 356	14. F, p. 360	

FILL-IN-THE-BLANK QUESTIONS

1. gender, p. 350
2. transsexual, p. 348
3. transvestite, p. 348
4. homophobia, p. 349
5. gender identity, p. 350
6. Body consciousness, p. 350
7. gender socialization, p. 357
8. Gender-segregated, p. 364
9. gendered institutions, p. 351
10. wage gap, p. 365
11. gender role, p. 350
12. matriarchy, p. 352
13. Sexual harassment, p. 360
14. Sexual orientation, p. 349
15. Sexism, p. 352

12

AGING AND INEQUALITY BASED ON AGE

BRIEF CHAPTER OUTLINE

CHAPTER SUMMARY

Aging is the physical, psychological, and social processes associated with growing older. Over the past two decades, the older U.S. population has been increasing, due to increased **life expectancy** and declining birth rates. As a result of these changing population trends, the discipline of **gerontology**—the study of aging and older people—has grown dramatically. People are assigned roles and positions based on the age structure and role structure in a particular society. In hunting and gathering societies, young people are valuable assets and older people may be perceived as being less productive. In horticultural, pastoral, and agrarian societies, people begin to live longer and may accumulate a surplus, which can provide prestige for older men. In industrial societies, living standards improve and advances in medicine contribute to greater longevity for more people. In postindustrial societies, a shift from a primarily young society to an older society brings about major changes in societal patterns and in the needs of the population. According to a case study, the aging of the Japanese population has taken only twenty-five years, whereas it took almost a century in North America and Europe. Age differentiation in the United States produces **age stratification**, the inequalities, differences, segregation, or conflict between age groups. In contemporary society, various strata of the life course present unique problems at each level. The strata range from infancy and childhood, adolescence, young adulthood, middle adulthood, and late adulthood. **Ageism**—prejudice and discrimination against people on the basis of age—is reinforced by stereotypes based on one-sided and exaggerated images of older people. Age, gender, and inequality are intertwined. Older women who retire from the workforce do not garner economic security for their retirement years at the same rate that many men do. Age, race/ethnicity, and economic inequality are also closely intertwined; the lower income status of older American ethnic minority group members can be traced to patterns of limited economic opportunities throughout their lives. Social Security, Medicare, Medicaid, and civil service pensions are the major **entitlements** for persons aged 65 and older. Older people in rural areas are more likely to have lower incomes and tend to be classified as poor. **Elder abuse** includes physical abuse, psychological abuse, financial exploitation, and medical abuse or neglect of people age 65 or older. In examining the living arrangements for older adults, relatives provide most of the care, although support services help older individuals who can afford services cope with their daily problems. Either by choice or necessity, some older adults move to smaller housing units or apartments. The most restrictive environment for older persons is the nursing home setting. Functionalist explanations of aging—such as **disengagement theory**—focus on how older persons adjust to their changing roles in society by detaching themselves from their social role and preparing for their eventual death. **Activity theory**— va symbolic interactionist perspective—states that people change in late middle age and find substitutes for previous statuses, roles, and activities. Conflict theorists link the loss of status and power experienced by many older persons to their lack of ability to produce and maintain wealth in a capitalist economy. The association of death with the aging process contributes to ageism in our society. Three well-known frameworks explain how people cope with the process of dying: the *stage-based approach*, the *dying trajectory*, and the *task-based approach*. How a person dies is shaped by many social and cultural factors. A **hospice** is an organization that provides a home-like facility or home-based care for people who are terminally ill. Over time, hospice care has moved toward hospital standards. In examining aging in the future, advances in medical technology may lead to a more positive outlook on aging; however with the increase of the aged population and the decrease of the birth rate, concerns over entitlements and other governmental assistance for the aged continue. Classism, racism, sexism, and ageism all restrict individual's access to valued goods and services in society. Older people continue to resist ageism through organizations such as the Grey Panthers, AARP, and the Older Women's League.

LEARNING OBJECTIVES

After reading Chapter 12, you should be able to:

1. Define aging and explain why research on aging has recently increased.

2. Distinguish between chronological age and functional age and note the social significance of each.

3. Explain the social significance of age.

4. Describe trends in aging and explain how life expectancy has changed in the United States during the twentieth century.

5. Compare the experiences and statuses of older people in preindustrial, industrial, and postindustrial societies.

6. Discuss the changes related to the increase in the older population in Japan.

7. Trace the process of aging through the life course, noting the social consequences of age at each stage.

8. Discuss ageism and describe the negative stereotypes associated with older persons.

9. Describe how age, gender, and poverty are intertwined.

10. Describe how age, race/ethnicity, and economic inequality are intertwined.

11. Describe the programs available to help older people with income and health care.

12. Explain the typical American living arrangements for older persons.

13. List and describe support services for older persons.

14. Describe and provide explanations for elder abuse.

15. Explain and apply functionalist, symbolic interactionist, and conflict perspectives on aging, including disengagement theory and activity theory.

16. Explain and apply the three major frameworks that explain how people cope with death and dying.

KEY TERMS

(defined at page number shown and in glossary)

activity theory, p. 401
age stratification, p. 386
ageism, p. 379
aging, p. 380
chronological age, p. 380
cohort, p. 381
disengagement theory, p. 401

elder abuse, p. 398
entitlements, p. 395
functional age, p. 380
gerontology, p. 382
hospice, p. 405
life expectancy, p. 381

KEY PEOPLE

(identified at page number shown)

Elaine C. Cumming and William E. Henry, p. 401
Elisabeth Kübler-Ross, p. 404

CHAPTER OUTLINE

I. THE SOCIAL SIGNIFICANCE OF AGE
    A. **Aging** is the physical, psychological, and social processes associated with growing older.
    B. **Chronological age** is a person's age based on date of birth; **functional age** is observable individual attributes (physical appearance, mobility, etc.) that are used to assign people to age categories.
    C. Trends in Aging
        1. The U.S. population has been aging over the past twenty-five years, partly due to Baby Boomers moving into middle age and partly due to longer life expectancy.
        2. **Life expectancy** is the average length of time a group of individuals of the same age will live.
        3. Life expectancy shows the average length of life of a **cohort**—a group of people born within a specific time period.
        4. **Gerontology** is the study of aging and older people; *Social gerontology* is the study of the social (nonphysical) aspects of aging.
        5. While young persons historically were considered "little adults" and were expected to do adult work, today the skills necessary for many roles are more complex, and the number of unskilled positions is more limited.
II. AGE IN GLOBAL PERSPECTIVE
    A. In hunting and gathering societies, young people are assets in the nomadic lifestyle involved in hunting and gathering food, whereas older people may be perceived as a liability.

B. In horticultural, pastoral, and agrarian societies, more people reach older ages, but life is still very hard.
  1. Accumulating a surplus becomes possible, and older individuals, especially men, are often the most privileged in society.
  2. In agrarian societies, food surplus, because of farming, makes it possible for more people to live to adulthood and old age; in these societies, most older people live with the other family members.
C. In industrial societies, living standards improve and advances in medicine contribute to greater longevity for more people, while in postindustrial societies, a shift from a primarily young society to an older society brings about major changes in societal patterns and in the needs of the population.
D. An increase in the older population in Japan has occurred over the past thirty years, whereas it took almost a century in North America and Europe.
  1. The sociocultural change and population shift may bring about gradual change in the social importance of the elderly in Japan.
  2. With over 60 percent of Japanese women in the labor force, greater pressure is being placed on Japanese policymakers to consider how the government can play a larger role in the care of the aging population, rather than having the responsibility placed completely on the family, especially women.

III. AGE AND THE LIFE COURSE IN CONTEMPORARY SOCIETY
A. Expectations for people's capabilities, responsibilities, and entitlement are largely based on age. These expectations are somewhat arbitrary and create **age stratification**—inequalities, differences, segregation, or conflict between age groups.
B. Infancy and childhood are typically thought of as carefree years; however, children are among the most powerless and vulnerable people in society.
  1. Early socialization plays a significant part in children's experiences and their quality of life.
  2. Two-thirds of all childhood deaths are caused by injury; far too many children lose their lives at an early age due to the abuse, neglect, or negligence of adults.
C. Adolescence roughly spans the teenage years; today it is a period in which young people are expected to continue their education and perhaps hold a part-time job.
  1. A variety of reports have labeled contemporary U.S. teenagers as a "generation at risk" because of many social problems, including crime and violence, teenage pregnancy, suicide, drug abuse, and peer pressure.
  2. One way to save the adolescent generation of today is to reduce poverty among them and their families and move away from age-based laws that limit adolescents' employment opportunities and freedom.
D. Young Adulthood follows adolescence and lasts to about age 39.
  1. During this time, people are expected to get married, have children, and get a job.
  2. People who are unable to earn income and pay into Social Security or other retirement plans in early adulthood become disadvantaged in later life.

E. Middle Adulthood—roughly the ages of 40 to 65—did not exist in the United States until fairly recently.
  1. The process of *senescence*, primary molecular and cellular changes in the body due to age, occurs at this time.
  2. Women undergo *menopause* (the cessation of the menstrual cycle); men undergo a *climacteric* in which the production of the "male" hormone decreases.
  3. For some people, middle adulthood represents the time period of having the highest levels of income and prestige.
F. Late adulthood is generally considered to begin at age 65, which formerly was known as the "normal" retirement age.
  1. *Retirement* is the institutionalized separation of an individual from an occupational position, with continuation of income through a retirement pension based on prior years of service.
  2. Some gerontologists subdivide late adulthood into three categories: (1) the "young-old" (ages 65-74); (2) the "old-old" (ages 75-85); and (3) the "oldest-old" (over age 85).
  3. The rate of biological and psychological changes in older persons may be as important as their chronological age in determining how they are perceived by themselves and others. A loss of height, impaired ability to function, heart attacks, strokes, and cancer are not uncommon at this stage of life; these changes can cause stress.
  4. Late adulthood may turn into a time of despair or one of integrity.
IV. INEQUALITIES RELATED TO AGING
  A. **Ageism**—prejudice and discrimination against people on the basis of age, particularly when they are older persons—is reinforced by stereotypes of older persons as cranky, sickly, and lacking in social value.
  B. If we compare wealth (all economic resources of value) with income (available money or its purchasing power), we find that older people tend to have more wealth but less income than younger people; among older people, there is a wider range of assets and income than in other age categories.
  C. Age, gender, and poverty are intertwined; although middle-aged and older women make up an increasing portion of the workforce, they are paid much less than men their age, receive raises at a slower rate, and still work largely in gender-segregated jobs.
    1. The percentage of people age 65 and older living below the poverty line decreased between 1980-2000 because of increasing benefit levels of entitlements.
      a. **Entitlements** are benefits paid by the government, including Social Security, Supplemental Security Income, Medicaid, Medicare, and civil service pensions.
      b. Ninety percent of all retired people in the U.S. draw Social Security benefits.
      c. Medicare is a nationwide health care program for persons age 65 and over who participate in Social Security or "buy into" the program.

D. Age, race/ethnicity, and economic inequality are closely intertwined.
   1. The lower income status of older American ethnic minority group members can be traced to patterns of limited economic opportunities throughout their lives.
E. Older people in rural areas are more likely to have lower incomes, be poor, and have less education.
   1. The rural elderly have lower lifetime earnings, limited savings, and fewer opportunities for part-time work.
   2. The homes of the rural elderly have lower value and are in greater need of repair.
F. **Elder abuse** refers to physical abuse, psychological abuse, financial exploitation, and medical abuse or neglect of people age 65 or over.

V. LIVING ARRANGEMENTS FOR OLDER ADULTS
A. Many older people live alone or in a family setting where care is provided informally by family or friends.
B. Support services, homemaker services, and day care centers help older people cope with the problems of daily care.
C. Alternative housing options for older people who can afford them include *retirement communities* and *assisted-living* facilities.
D. Nursing homes are the most restrictive environments for older persons.
   1. Financing of long-term care is a problem for most older persons.
   2. Medicare and most private insurance policies exclude long-term care.

VI. SOCIOLOGICAL PERSPECTIVES ON AGING
A. Functionalists examine how older persons adjust to their changing roles. According to **disengagement theory**, created by **Elaine C. Cumming** and **William E. Henry**, older persons make a normal and healthy adjustment to aging when they detach themselves from their social roles and prepare for their eventual death.
B. **Activity theory**—a symbolic interactionist perspective on aging—states that people tend to shift gears in later life and find substitutes for previous statuses, roles, and activities.
C. Conflict theorists note that, as people grow older, their power tends to diminish unless they are able to maintain their wealth. Consequently, those who have been disadvantaged in their younger years become even more so in late adulthood.

VII. DEATH AND DYING
A. In contemporary industrial societies, death is looked on as unnatural because it has been removed from everyday life: most deaths occur among older persons and in institutionalized settings.
B. Three well-known frameworks provide explanations of how people cope with the process of dying.
   1. The *stage-based approach* was popularized by Elisabeth Kübler-Ross, who proposed five stages in the dying process:
      a. Denial ("Not me!")
      b. Anger ("Why me?")
      c. Bargaining ("Yes me, but....")
      d. Depression and sense of loss
      e. Acceptance

2. The *dying trajectory* approach focuses on the perceived course of dying and the expected time of death, and involves three phases:
   a. Acute phase (maximum anxiety and fear)
   b. Chronic phase (decline in anxiety as the person faces reality)
   c. Terminal phase (withdrawal from others)
3. The *task-based approach* is based on the assumption that the dying person can and should go about daily activities and complete physical, social, or spiritual tasks.
C. A **hospice** is an organization that provides a homelike facility or home-based care for patients with terminal illnesses; over time, hospice care has moved toward hospital standards.
VIII. AGING IN THE FUTURE
   A. By the year 2050, there will be approximately 80 million persons age 65 and over, as compared with 35 million in 2000. More people will survive to age 85—or even 95 and over.
   B. A 1994 report warned that this increase (coupled with a decreasing birth rate) may result in entitlements consuming nearly all federal tax revenues by the year 2012, leaving the government with no money for anything else.
   C. New biomedical research discoveries may lead to more positive aging; however, many of the benefits and opportunities of living in a highly technological, affluent society are not available to all.

CRITICAL THINKING QUESTIONS

1. What are some ways that chronological age differs from functional age? What are some examples?
2. What challenges do you experience based on your current stage of the life course? In what ways would your experience of your current stage of the life course be different if you lived in a different type of society (e.g., hunting and gathering or pastoral)?
3. What sort of living arrangement would you prefer when you enter older age? How do you think your race, class, and gender will help or hinder your ability to have the living arrangement of your choice?
4. Which sociological perspective do you think is most useful for understanding aging in contemporary U.S. society? What are the strengths and weaknesses of each?
5. Which of the three frameworks—the stage-based approach, the dying trajectory, and the task-based approach—do you think is most useful for understanding death and dying?
6. What are some ways you think the U.S. can prepare for the expected increase in the older population in the future?

PRACTICE TESTS

MULTIPLE CHOICE QUESTIONS

Select the response that best answers the question or completes the statement.

1. _____ age refers to a person's age based on date of birth.
   a. Obvious
   b. Functional
   c. Subjective
   d. Chronological

2. _____ age refers to observable individual attributes such as physical appearance, mobility, strength, coordination, and mental capacity that are used to assign people to age categories.
   a. Subjective
   b. Obvious
   c. Functional
   d. Chronological

3. The median age of the U.S. population is
   a. increasing due to an increase in life expectancy combined with a decrease in birth rates.
   b. decreasing due to an increase in birth rates.
   c. roughly the same as it has been for most of the twentieth century due to stabilization of both birth and death rates.
   d. decreasing due to advances in medical technology.

4. Two hundred years ago, the age spectrum was divided into
   a. babyhood and adulthood.
   b. babyhood, a very short childhood, and adulthood.
   c. babyhood, childhood, adulthood, and old age.
   d. infancy, childhood, adolescence, adulthood, and old age.

5. The number of people of each age level in a society is referred to as the:
   a. age structure
   b. age pyramid
   c. role structure
   d. chronological age

6. In the U.S., most childhood deaths today are caused by:
   a. HIV/AIDS
   b. injuries
   c. communicable diseases
   d. birth defects

7. Middle adulthood represents the time when many people
   a. are likely to suffer from disabling illnesses.
   b. may have grandchildren who give them another tie to the future.
   c. have a decline in income as they approach their retirement years.
   d. get married, have children, and get a job.

8. Which of the following statements is **correct** regarding Alzheimer's disease?
   a. Most older people suffer from Alzheimer's disease.
   b. Most people with Alzheimer's disease have an extremely short life expectancy.
   c. About 55 percent of all organic mental disorders in the older population are caused by Alzheimer's disease.
   d. In 1995, researchers found a possible cure for Alzheimer's disease.

9. According to Erik Erikson, older people must resolve a tension of _____ versus _____ in their lives.
   a. integrity; despair
   b. activity; disengagement
   c. stability; change
   d. autonomy; entitlements

10. Ageism is
    a. prejudice and discrimination against people on the basis of age, particularly when they are younger persons.
    b. prejudice and discrimination against people on the basis of age, particularly when they are older persons.
    c. gradually being overcome by positive depictions of older persons in the media.
    d. experienced equally by both women and men.

11. In horticultural, pastoral, and agrarian societies,
    a. older people may be viewed as a liability because they are less able to travel and acquire food.
    b. it is possible to accumulate surplus, so older men are often more privileged due to wealth, power, and prestige.
    c. improved living standards and advances in medicine contribute to greater longevity.
    d. is characterized by a shift from a society that was primarily young to a society that is older.

12. Advertising typically portrays the elderly as
    a. far more attractive than many elderly people can actually achieve.
    b. disproportionately female.
    c. feeble-minded, wrinkled, and frail.
    d. important role models for young people.

13. All of the following statements are **correct** regarding elder abuse, **except**:
   a. Abuse and neglect of older persons has received increasing public attention in recent years.
   b. Financial exploitation is a form of elder abuse.
   c. More than 1.6 million older people in the United States are victims of elder abuse.
   d. Nursing home personnel are the most frequent abusers of older persons.

14. Studies have shown that _____ were the most frequent abusers of older persons.
   a. nursing home personnel
   b. younger siblings
   c. daughters
   d. sons

15. Functionalist perspectives of aging focus on
   a. how older persons adjust to their changing roles in society.
   b. the connection between personal satisfaction in a person's later years and a high level of activity.
   c. reasons why aging is especially problematic in contemporary capitalist societies.
   d. the ways aging differs based on gender, race, and class.

16. _____ theory states that people tend to shift gears in late middle age and find substitutes for previous statuses, roles, and activities.
   a. Continuity
   b. Disengagement
   c. Activity
   d. Engagement

17. _____ theorists analyze the lack of power individuals have as they grow older unless they are able to maintain wealth.
   a. Functionalist
   b. Activity
   c. Conflict
   d. Symbolic interactionist

18. The _____ framework focuses on the perceived course of dying and the expected time of death.
   a. dying trajectory
   b. stage-based approach
   c. task-based approach
   d. identity theory

19. All of the following about hospice care are true **except**:
   a. A hospice provides care for the terminally ill.
   b. Over time, hospice care has moved away from hospital standards.
   c. Hospice philosophy asserts that people should participate in their own care.
   d. The approach is family-based.

232

20. _____ theory posits that older persons make a normal and healthy adjustment to aging when they detach themselves from social roles and prepare for their eventual death.
    a. Continuity
    b. Disengagement
    c. Activity
    d. Engagement

21. The _____ approach to dying suggests that the dying person can and should go about daily activities and fulfill tasks that make the process of dying easier on family, friends, and the dying person.
    a. dying trajectory
    b. stage-based
    c. task-based
    d. identity theory

22. When people say, "Act your age," they are referring to _____ age—a person's age based on date of birth.
    a. functional
    b. normative
    c. chronological
    d. sequential

23. When feminist scholars state that functional age is subjective for men and women, they mean
    a. women and men age physically in a similar fashion.
    b. women and men are evaluated differently throughout the aging process.
    c. women age physically more slowly than men.
    d. men age physically more slowly than women.

24. As they age, men are believed to become more _____ and women are thought to become more _____.
    a. distinguished; grandmotherly
    b. helpless; powerful
    c. active; passive
    d. mentally depressed; physically active

25. The age at which half of the people in a population are younger and the other half are older is referred to as the _____ age of that population.
    a. mean
    b. modal
    c. median
    d. standard

26. Research indicates that many older adults buffer themselves against ageism by
    a. purposely conforming to stereotypes of older people.
    b. subjecting themselves to illness and early death.
    c. spending as much time as possible at home to avoid contact with the public.
    d. continuing to identify themselves as middle-aged.

233

27. The aging of the U. S. population resulted from
    a. increased life expectancy and increased birth rates.
    b. decreased life expectancy and decreased birth rates.
    c. decreased life expectancy and increased birth rates.
    d. increased life expectancy and decreased birth rates.

28. In 1900, about 4 percent of the U.S. population was over age 65. In 2005, approximately _____ percent of the population was 65 or over.
    a. 9
    b. 13
    c. 20
    d. 24

29. Older people tend to have _____ compared with younger people.
    a. less wealth and income
    b. more wealth and income
    c. more wealth and less income
    d. less wealth and more income

30. One of the fastest growing segments of the U.S. population is made up of _____. This cohort is expected to almost double in size between 2000 and 2025.
    a. teenagers
    b. those between ages 25-35
    c. those between ages 45-55
    d. those age 85 and over

31. The age pyramid that depicts the current distribution of the U.S. population can best be described as
    a. a perfect pyramid.
    b. a pyramid that is larger at the base because of an increase in birth rates.
    c. a pyramid that bulges in the middle because of the Baby Boomers and is smaller at the base because of declining birth rates among post-Baby Boomers.
    d. a pyramid that bulges at the very top because of the extraordinary growth of the old-old population.

32. _____ is the study of the nonphysical aspects of aging.
    a. Human gerontology
    b. Urban gerontology
    c. Gerontology
    d. Social gerontology

33. Research suggests that _____ is more directly related to poverty among older persons than other demographic characteristics.
    a. gender
    b. race/ethnicity
    c. education
    d. occupational status

234

34. _____ refers to physical, psychological, financial, and medical abuse or neglect of people age 65 or older.
    a. Age stratification
    b. Ageism
    c. Elder abuse
    d. Gerontological abuse

35. The group of elderly most likely to be physically abused is _____ aged 75-85 who are impaired.
    a. white middle-class men
    b. white middle-class women
    c. black working-class women
    d. black middle-class women

36. The text points out that adolescence is a time when individuals are
    a. still allowed to act "childish."
    b. expected to continue their education and perhaps hold a part-time job.
    c. expected to get married, have children, and get a job.
    d. accorded adult status.

37. Young adulthood follows adolescence and lasts to about age_____.
    a. 25
    b. 30
    c. 39
    d. 45

38. Middle adulthood refers to people between the ages of:
    a. 30 and 50
    b. 40 and 55
    c. 40 and 65
    d. 55 and 70

39. Elder abuse is most commonly perpetrated by
    a. strangers in public places.
    b. nursing home and hospital staff.
    c. family members on whom the elderly person is dependent.
    d. family members who are dependent on the elderly person.

40. _____ provide(s) in-home services, such as basic chores and meals, to older individuals who are not able to fully care for themselves.
    a. Homemaker services
    b. Day care
    c. Hospice
    d. Retirement communities

TRUE/FALSE QUESTIONS

1. A group of people born within a specific period of time is known as a cohort.
   T   F

2. As men age, they are believed to become less powerful when compared to women.
   T   F

3. Homemaker services, day care, retirement communities, and nursing home care is free for all elderly who need services.
   T   F

4. Gerontology is the study of the nonphysical aspects of aging.
   T   F

5. In hunting and gathering societies, the elderly have great prestige because they accumulate a surplus of goods.
   T   F

6. The graying of Japan has taken longer than in North America and Europe.
   T   F

7. Adolescence is a twentieth century concept.
   T   F

8. In the U.S., young adulthood lasts to about age 39.
   T   F

9. Scientific studies show that women age faster than men.
   T   F

10. Age, race/ethnicity, gender, and economic inequality are interrelated.
    T   F

11. Nearly half of all women over age 65 are widowed and living alone on fixed incomes.
    T   F

12. According to analysts, Social Security keeps more white men out of poverty than it does white women and people of color.
    T   F

13. Older Native Americans are among the most disadvantaged of all categories.
    T   F

14. Only about 5 percent of frail, older persons are currently in nursing homes.
    T   F

15. Over time, the hospice movement has moved away from hospital standards to more home-based care.
    T   F

16. According to Box 12.1, the rate of elder abuse in the United States has been greatly exaggerated by the media.
T   F

17. According to the article on driving while elderly, most states have policies regarding drivers over the age of 65.
T   F

18. In Box 12.3, the article suggests that many films today showing what it means to be older are depicting a positive image of aging and romance.
T   F

19. Most older people are economically secure today as a result of Social Security, Medicare, and retirement plans.
T   F

20. Most people lose their sex drive after age 45.
T   F

FILL-IN-THE-BLANK QUESTIONS

1. A person's age based on date of birth is one's _____ age, while observable individual attributes is one's _____ age.

2. The average number of years that a group of people born in the same year could be expected to live is known as _____.

3. The physical, psychological, and social processes associated with growing older are called _____.

4. The study of aging and older people is known as _____.

5. The inequalities, differences, segregation, or conflict between age groups is known as _____.

6. Prejudice and discrimination against people on the basis of age is termed _____.

7. _____ refers to physical abuse, psychological abuse, financial exploitation, medical abuse, and neglect of people age 65 and older.

8. A national health care program for persons aged 65 and older who are covered by Social Security or who buy into the program is known as _____.

9. The theory of dying that focuses on the perceived course of dying and expected time of death is known as the _____ approach.

10. A(n) _____ is an organization that provides home-based or a homelike facility care for terminally-ill people.

11. _____ states that people tend to shift gears in late middle age and find substitutes for previous statuses, roles, and activities.

12. The first stage in the dying process, according to the stage-based approach, is _____.

237

13. A group of people born in a specific time period is called a _____.

14. _____ states that older persons make a normal and healthy adjustment to aging when they detach themselves from their social roles and prepare for their eventual death.

15. The _____ approach is based upon the assumption that the dying person should go about daily activities.

SHORT ANSWER/ESSAY QUESTIONS

1. Is the United States aging? Support your answer, and describe any implications today and in the future.
2. How are race, class, gender, and aging related?
3. What living arrangements are common for older adults in the United States? What are the advantages and disadvantages of each?
4. What are some examples of ageism in the United States?
5. What are some examples of problems of the elderly, primarily problems of elderly women?

STUDENT CLASS PROJECTS AND ACTIVITIES

1. Conduct research on the history of the original Social Security legislation in the U.S. In your research, examine these issues and provide responses to the following questions: (1) What was the original purpose of the act, and what were the major provisions of the act? (2) Did most legislators support the act? Why or why not? (3) What groups were opponents of the act? What were their reasons for opposing it? (4) What was the political position of the President toward the act? (5) What year did the legislation become law? (6) Why? (7) Why was the United States one of the last industrialized nations in the world to provide its elderly citizens with old-age insurance? (8) What Western country was the first to enact old-age benefits? (9) List at least ten different countries that had some type of old-age insurance programs before the United States. (10) How was the program originally funded? (11) How did workers become eligible for full benefits? (12) Evaluate the effectiveness of this legislation. Did it meet its goals? Why? Why not? (13) Personally critique this act. Do you agree with the premises of Social Security? Why? Why not?
2. Conduct research on an old-age insurance program available in another country for the purpose of comparing the current state of its program with the current "crisis" of Social Security in the United States. Some possible choices include Germany, Denmark, Sweden, the United Kingdom, Austria, France, Ireland, Italy, Spain, Belgium, Greece, the Netherlands, New Zealand, Australia, or Japan. Examine the history of the legislation and answer the following questions: (1) What was the purpose of the act, and what were its major provisions? (2) What groups supported the act? Why? (3) What groups opposed the act? Why? (4) What was the political climate of the time; did the leader of the country support the act? Why? (5) When did the act become law? (6) How is the program financed? (7) How do people qualify for

the program? (8) What is its current status? Are there any indications of a financial crisis for the program? (9) Briefly compare this country's program with that of the Social Security program in the United States. Are there any similarities or differences? (10) Provide any other relevant information.

3. Choose at least two elderly people who are accessible to you (such as grandparents, neighbors, retired faculty, etc.) and interview them about their experiences in aging. If they do not object, you may use a tape recorder to record the interview. In the interview, ask them to describe some of the important events that occurred in their childhood, adolescence, adulthood, and, most recently, in the past several years. For consistency in your interviews and written papers, answer each of these questions: (1) How do they describe themselves? (2) What have been some of the most important events in their lives? Remember to cover childhood, adolescence, and adulthood. (3) How do members of their family, friends, and other people treat them? Is this different from previous times? How? (4) Do they think of themselves as old? Why? Why not? (5) What is one thing that they like about their lives today? (6) What was the most significant event in their life up to this point? (7) How do they think the majority of society views them today? How do they feel about this view? Summarize the major points of the interviews and present this information in a written paper.

4. The year 2005 was the midpoint of a decade that presented a unique picture of the world's population. Before 2000, "young people" always outnumbered "old people." From 2000 forward, "old people" began to and will continue to outnumber "young people." The graying of the world's population is not moving in unison around the globe. Search relevant resources to research and respond to these questions: (1) What are at least six major consequences of the "graying" the world?" (2) What countries represent this "graying" process? (3) Select two of these representative countries and compare and contrast their plans to address their ever-growing aging population concerns and needs, such as health care, illnesses and diseases associated with aging, the dependency ratio, and the sustainability of the elderly population. (4) Analyze the effectiveness of these plans and provide any alternatives you can think of. Summarize your responses in a paper.

INTERNET ACTIVITIES

1. To learn more about research being conducted on aging, go the site for the **National Institute on Aging**: http://www.nia.nih.gov/. Click on *Research* to read about current research on aging. Look for research on myths about aging and how drug costs affect the elderly.

2. A very powerful interest group is the **AARP** (The American Association of Retired Persons), **http://www.aarp.org/**. Click on AARP in your state and then select either a region or your state to read about the services that are provided by the AARP to members. Next, select some other states for purposes of comparison. For example, Florida and Arizona have large older populations. Does the AARP have a stronger presence in those states, and does it offer more services?

3. The **Administration on Aging**, **http://www.aoa.gov/**, provides extensive information on aging for professionals in the field of aging, service providers, and

policymakers, as well as older Americans themselves. Find out more about this agency at the web site.

4. You can calculate your own life expectancy from the **Living to 100 Life Expectancy Calculator** web site: **http://www.livingto100.com/**

5. To find health-related research, such as information on Alzheimer's, go to the **National Institutes of Health** web site, **http://www.nih.gov/**, and do a search for aging research.

INFOTRAC COLLEGE EDITION EXERCISES

Visit the **InfoTrac College Edition** web site at**: http://www.wadsworthmedia.com/webtutor/infotrac.htm**. You will arrive at a screen that enables you to search topics.

1. Look for newspaper articles on **age discrimination**. Compare these with research articles on the same topic. Does it appear that the general public really understands this issue?

2. Examine research on **elder abuse** in the home and in nursing homes. Take a look at the *California Elder Abuse and Dependent Adult Civil Protection Act*. What factors seem to contribute to elder abuse?

3. Write an essay on the **social aspects of aging**. There are a number of journal and news articles on this subfield.

4. Look for articles that explore **aging** in other countries, especially Asian countries.

SOLUTIONS

MULTIPLE CHOICE QUESTIONS

1. D, p. 379	15. A, p. 401	29. C, p. 394
2. C, p. 380	16. C, p. 401	30. D, p. 381
3. A, p. 380	17. C, p. 402	31. C, pp. 381-382
4. B, p. 383	18. A, p. 404	32. D, p. 382
5. A, p. 383	19. B, p. 405	33. A, p. 395
6. B, p. 386	20. B, p. 401	34. C, p. 398
7. B, p. 388	21. C, p. 404	35. B, p. 398
8. C, p. 390	22. C, p. 380	36. B, p. 387
9. A, p. 391	23. B, p. 380	37. C, p. 387
10. B, p. 391	24. A, p. 380	38. C, p. 388
11. B, p. 384	25. C, p. 380	39. D, p. 398
12. C, p. 392	26. D, p. 393	40. A, p. 398
13. D, p. 398	27. D, p. 381	
14. D, p. 398	28. B, p. 381	

TRUE/FALSE QUESTIONS

1. T, p. 381	8. T, p. 387	15. F, p. 405
2. F, p. 380	9. F, p. 383	16. F, p. 383
3. F, p. 398	10. T, p. 396	17. F, p. 392
4. T, p. 382	11. T, p. 396	18. T, p. 394
5. F, p. 384	12. T, p. 396	19. F, p. 393
6. F, p. 384	13. T, p. 397	20. F, p. 388
7. T, p. 387	14. T, p. 399	

FILL-IN-THE-BLANK QUESTIONS

1. chronological, p. 380; functional, p. 380
2. life expectancy, p. 381
3. aging, p. 380
4. gerontology, p. 382
5. age stratification, p. 386
6. ageism, p. 379
7. Elder abuse, p. 398
8. Medicare, p. 396
9. dying trajectory, p. 404
10. hospice, p. 405
11. Activity theory, p. 401
12. denial, p. 404
13. cohort, p. 381
14. Disengagement theory, p. 401
15. task-based, p. 404

13

THE ECONOMY AND WORK IN GLOBAL PERSPECTIVE

BRIEF CHAPTER OUTLINE

CHAPTER SUMMARY

While economists focus on how the limited resources and efforts of a society are allocated among competing ends, sociologists focus on interconnections among the economy, other social institutions, and the social organization of work, at both the microlevel and the macrolevel. The **economy** is the social institution that ensures the maintenance of society through the production, distribution, and consumption of goods and services. Preindustrial economies are characterized by **primary sector production** in which workers extract raw materials and natural resources from the environment and use them without much processing. Industrial economies engage in **secondary sector production**, which is based on the processing of raw materials (from the primary sector) into finished goods. The economic base in postindustrial economies shifts to **tertiary sector production**—the provision of services rather than goods. As an ideal type, **capitalism** has four distinct features: private ownership of the means of production, pursuit of personal profit, competition, and lack of government intervention. **Socialism** is characterized by public ownership of the means of production, the pursuit of collective goals, and centralized decision making. In a **mixed economy**, elements of a capitalist, market economy are combined with elements of a socialist economy. Mixed economies are often referred to as **democratic socialism**—an economic and political system that combines private ownership of some means of production, governmental distribution of some essential goods and services, and free elections. For many people, jobs and professions are key sources of identity. Functionalists focus on the impact of the economy on other parts of society, conflict theorists examine capitalist greed and the suppression of workers' wages, and symbolic interactionists study the factors that contribute to worker satisfaction and identity. Work is socially organized by type. **Occupations** are categories of jobs that involve similar activities at different work sites. **Professions** are high status, knowledge-based occupations characterized by abstract, specialized knowledge, autonomy, authority over clients and subordinate occupational groups, and a degree of altruism. Those in managerial occupations typically are responsible for workers, physical plants, equipment, and the finances of a bureaucratic organization. **Marginal jobs** are those that do not comply with the employment norms that job content should be legal, covered by government regulations, relatively permanent, and provide adequate hours and pay in order to make a living. **Contingent work** is part-time work, temporary work, and **subcontracted work** that offers advantages to employers but may be detrimental to workers. Unemployment remains a problem for many workers. The three types of unemployment are cyclical, seasonal, and structural. The **unemployment rate** is the percentage of unemployed persons in the labor force actively seeking jobs. U.S. **labor unions** have been credited with improving the work environment and gaining some measure of control over work-related activities. Many workers have joined unions in order to gain strength through collective actions. The number of persons in the U.S. having physical or mental disabilities is increasing. In 1990, the United States became the first nation to formally address the issue of equality for persons with a disability. Future trends of the U.S. economy and the global economy all indicate that the U.S. will remain a major player in the world economy and that transnational corporations will become even more significant. A global workplace is emerging in which telecommunications networks will link workers in distant locations. However, as nations become more dependent on each other, they also become more competitive in the economic sphere; therefore changes in the global economy may require people of all nations to make changes in their traditional ways of doing things.

LEARNING OBJECTIVES

After reading Chapter 13, you should be able to:

1. Describe the purpose of the economy and distinguish between sociological perspectives on the economy and the study of economics.

2. Trace the major historical changes that have occurred in economic systems and note the most prevalent form of production found in each.

3. Describe the four distinctive features of "ideal" capitalism and explain why pure capitalism does not exist.

4. Define socialism and describe its major characteristics.

5. Compare and contrast capitalism, socialism, and mixed economies.

6. Distinguish between the functionalist, conflict, and symbolic interactionist perspectives on the economy and work.

7. Distinguish between the primary and secondary labor markets and describe the types of jobs found in each.

8. Describe job satisfaction and alienation, and explain the impact of each on workers.

9. Discuss the major characteristics of professions and describe the process of deprofessionalization.

10. Compare scientific management (Taylorism) with mass production through automation (Fordism), noting the strengths and weaknesses of each.

11. Identify the occupational categories considered to be marginal jobs and explain why they are considered to be marginal.

12. Define contingent work and identify the role subcontracting often plays in contingent work.

13. Distinguish between the various types of unemployment.

14. Describe some of the means by which workers resist working conditions they consider to be oppressive.

15. Trace the historical development of labor unions.

16. Describe employment opportunities for persons with disabilities and laws that protect them.

KEY TERMS

(defined at page number shown and in glossary)

KEY PEOPLE

(identified at page number shown)

CHAPTER OUTLINE

I. COMPARING THE SOCIOLOGY OF ECONOMIC LIFE WITH ECONOMICS
   A. Economists attempt to explain how the limited resources and efforts of a society are allocated among competing ends.
      1. Economists focus on the complex workings of economic systems.
   B. Sociologists focus on interconnections between the economy, other social institutions, and the social organization of work.
      1. At the macrolevel, sociologists may study the impact of transnational corporations on industrial and low-income nations.
      2. At the microlevel, they may study people's satisfaction with their jobs.
II. ECONOMIC SYSTEMS IN GLOBAL PERSPECTIVE
   A. **The economy** is the social institution that ensures the maintenance of society through the production, distribution, and consumption of goods and services.
   B. In preindustrial societies, most workers engage in **primary sector production**—the extraction of raw materials and natural resources from the environment.

C. Industrialization brings sweeping changes to the economy as new forms of energy and machine technology proliferate as the primary means of producing goods and most workers engage in **secondary sector production**—processing raw materials into finished goods.

D. A postindustrial economy is based on **tertiary sector production**—the provision of services (such as food service, transportation, communication, education, and entertainment) rather than goods.

III. CONTEMPORARY WORLD ECONOMIC SYSTEMS

A. **Capitalism** is an economic system characterized by private ownership of the means of production, from which personal profits can be derived through market competition and without government intervention. "Ideal" capitalism is made up of the following four distinctive features:

1. Private ownership of the means of production, or the rights of individuals to own income-producing property, such as land and factories, and the right to "buy" people's labor.

   a. In 1869, workers created the first **labor unions**— groups of employees who join together to bargain with an employer or group of employers over wages, benefits, and working conditions.

   b. Under early monopoly capitalism (1890-1940), ownership quickly shifted to **corporations**—large-scale organizations that have legal powers, such as the ability to enter into contracts and buy and sell property, separate from business owners.

   c. In advanced monopoly capitalism (1940—present), ownership and control of major industrial and business sectors are increasingly concentrated, much of which in **transnational corporations**—large corporations that are headquartered in one country but sell and produce goods and services in many countries.

2. Pursuit of personal profit, or the belief that people are free to maximize their individual gains through personal profits

   a. Economic development is assumed to benefit capitalists, workers, and the entire society through public expenditures made possible from business tax revenues.

3. Competition, or producers vying with one another for customers, is supposed to act as a balance to excessive profits.

   a. In early monopoly capitalism, competition was diminished by increasing concentration within particular industries.

   b. An **oligopoly** exists when several companies overwhelmingly control an entire industry.

   c. A **shared monopoly** exists when four or fewer companies supply 50 percent or more of a particular market.

   d. Corporations with control within and across industries, formed by a series of mergers and acquisitions across industries, are referred to as **conglomerates**—combinations of businesses in different commercial areas, all of which are owned by one holding company.

246

  e. Competition is reduced over the long run by **interlocking corporate directorates**—members of the board of directors of one corporation who also sit on the board(s) of other corporations.

 4. Lack of government intervention—ideally, capitalism works best without government intervention in the workplace.

  a. Because the "ideal" of unregulated markets benefiting all citizens has seldom been realized, government regulations have been implemented in an effort to curb the excesses brought about by laissez-faire policies.

  b. Much government intervention has been in the form of aid to businesses, such as tax credits, subsidies, and loan guarantees.

B. **Socialism** is an economic system characterized by public ownership of the means of production, the pursuit of collective goals, and centralized decision making.

 1. Karl Marx predicted that socialism would arise as a temporary stage in the transition to *communism*—an economic system characterized by common ownership of all economic resources.

 2. "Ideal" socialism has three distinctive features, described below:

  a. Public ownership of the means of production—means of production are owned and controlled by a collectivity or the state, not private individuals or corporations; housing and medical care are considered rights of citizens.

  b. Pursuit of collective goals—the goal is to meet everyone's needs rather than personal profits for some individuals.

  c. Centralized decision making—a central government makes decisions in a hierarchical form.

C. A **mixed economy** combines elements of a market economy (capitalism) with elements of a command economy (socialism).

 1. **Democratic socialism** is an economic and political system that combines private ownership of some of the means of production, governmental distribution of some essential goods and services, and free elections.

 2. Examples are Sweden, Great Britain, and France.

IV. PERSPECTIVES ON ECONOMY AND WORK IN THE UNITED STATES

A. From a functionalist perspective, the economy is a vital social institution because it is the means by which goods and services are produced and distributed.

 1. When the economy runs smoothly, other parts of society function more effectively; however, if the system becomes unbalanced, maladjustment occurs.

 2. The *business cycle* is the rise and fall of economic activity relative to long-term growth in the economy.

B. From a conflict perspective, business cycles are the result of capitalist greed; in order to maximize profits, capitalists suppress the wages of workers.

 1. As the prices of the products increase, the workers are not able to purchase them in the quantities that have been produced.

 2. Consequently, surpluses occur that cause capitalists to reduce production, close factories, lay off workers, and thus contribute to the growth of the reserve army of the unemployed, which then helps to reduce the wages of the

remaining workers. In some situations, workers are replaced with machines or nonunionized workers.

C. Symbolic interactionists examine factors that contribute to worker satisfaction.

1. Work is an important source of self-identity for many people.
2. Work can help people feel positive about themselves or it can cause them to feel alienated.
3. *Job satisfaction* is highest when employees have some degree of control over their work, when they feel that they are part of the decision-making process, when they are not too closely supervised, and when they feel they play an important part in the outcome.
4. *Alienation* occurs when workers' needs for identity and meaning are not met, and when work is done strictly for material gain, not a sense of personal satisfaction.

V. THE SOCIAL ORGANIZATION OF WORK

A. **Occupations** are categories of jobs that involve similar activities at different work sites.

1. The **primary labor market** is comprised of high-paying jobs with good benefits that have some degree of security and the possibility for future advancement.
2. The **secondary labor market** is comprised of low-paying jobs with few benefits and very little job security or possibility for future advancement.

B. **Professions** are high-status, knowledge-based occupations.

1. Professions have five major characteristics: abstract, specialized knowledge, autonomy, self-regulation, authority, and altruism.
2. Parent education, race, and gender influence people's access to professions.
3. Some professions are undergoing *deprofessionalization*, in which some of the characteristics of a profession are eliminated.

C. The term *manager* often is used to refer to executives, managers, and administrators who typically have responsibility for workers, physical plants, equipment, and the financial aspects of a bureaucratic organization.

1. Scientific management (Taylorism) was developed by industrial engineer Frederick Winslow Taylor to increase productivity in factories by teaching workers to perform a task in a concise series of steps; this lead to the *deskilling* of work, and workers were paid by the number of units they produced, which increased estrangement between workers and managers.
2. Mass production through automation (Fordism) incorporated hierarchical authority structures and scientific management techniques into the manufacturing process. Assembly lines, machines, and robots became a means of technical control over the work process.

D. Lower Tier of the Service Sector and Marginal Jobs

1. Positions in the lower tier of the service sector are part of the **secondary labor market**, characterized by low wages, little job security, few chances for advancement, higher unemployment, and very limited (if any) unemployment benefits.

   a. Examples include janitors, waitresses, messengers, lower-level sales clerks, typists, file clerks, migrant laborers, and textile workers.

248

b. Many jobs in this sector are **marginal jobs** that differ from the employment norms of the society in which they are located. In the U.S., these norms are that a job is (a) legal; (b) covered by government work regulations; (c) relatively permanent; and (d) providing adequate pay with sufficient hours of work each week to make a living.

2. More than 11 million workers are employed in personal service industries, such as eating and drinking places, hotels, laundries, beauty shops, and household service, primarily maid service.

a. Service workers are often viewed by customers as subordinates or personal servants.

3. Private household service work—including launderers, cooks, maids, babysitters, etc.—has shifted from domestics who work full time for one employer to part-time work with several different employers.

4. Internationalization of marginal jobs—marginal jobs are more likely to be found in peripheral industries than in core, or essential, industries.

a. The use of workers in other countries to manufacture or assemble goods to be sold in higher-income nations is referred to as the *global assembly line*, in which corporations hire workers (often female) in low-income nations at very low wages, sometimes under hazardous conditions.

E. **Contingent work** is part-time work, temporary work, and subcontracted work that offers advantages to employers but often is detrimental to the welfare of workers.

1. Employers benefit by hiring workers on a part-time or temporary basis; they are able to cut costs, maximize profits, and have workers available only when they need them.

2. **Subcontracting**—a form of economic organization in which a larger corporation contracts with other (usually smaller) firms to provide specialized components, products, or services—is another form of contingent work that cuts employer costs, often at the expense of workers.

F. The *underground economy* is made up of a variety of activities through which people make money (sometimes through criminal behavior) that is not reported to the government.

VI. UNEMPLOYMENT

A. There are three major types of unemployment: (1) *cyclical unemployment* occurs as a result of lower rates of production during recessions in the business cycle; (2) *seasonal unemployment* results from shifts in the demand for workers based on weather or seasons, such as holidays and summer vacations; and (3) *structural unemployment* arises because the skills demanded by employers do not match the skills of the unemployed or because the unemployed do not live where jobs are located.

B. The **unemployment rate** is the percentage of unemployed persons in the labor force actively seeking jobs.

C. *Unemployment compensation* provides eligible unemployed workers with short-term income while they look for other jobs.

VII. WORKER RESISTANCE AND ACTIVISM

A. **Labor Unions**

1. U.S. labor unions have been credited with gaining an eight-hour workday, a five-day work week, health and retirement benefits, sick leave and unemployment insurance, and workplace health and safety standards for many employees through *collective bargaining*—negotiations between labor union leaders and employers on behalf of workers.
   2. Although the overall number of union members in the United States has increased since the 1960s, the proportion of all employees who are union members has declined.
   B. Absenteeism, Sabotage, and Resistance
   1. Absenteeism is one means by which workers resist working conditions they consider to be oppressive.
   2. Other workers use sabotage to bring about informal work stoppages, such as creating a halt in the movement of the assembly line.
   3. While most workers do not sabotage machinery, a significant number do resist oppressive conditions, which help them survive at work.
VIII. EMPLOYMENT OPPORTUNITIES FOR PERSONS WITH A DISABILITY
   A. The number of persons in the United States having physical or mental disabilities is increasing because:
   1. Advances in medical technology now make survival possible.
   2. As people live longer they are more likely to experience disabling diseases.
   3. Persons born with serious disabilities are more likely to survive infancy.
   B. In 1990, the United States passed the Americans with Disabilities Act.
   1. The ADA forbids discrimination on the basis of disability.
   2. Despite the law, about two-thirds of working-age persons with a disability in the U.S. are unemployed today.
IX. THE GLOBAL ECONOMY IN THE FUTURE
   A. The U.S. economy will experience dramatic changes in the next century as workers may find themselves fighting for a larger piece of an ever shrinking economic pie that includes trade deficit and a large national debt.
   B. Workers increasingly may be fragmented into two major divisions—those who work in the innovative, primary sector and those whose jobs are located in the growing secondary, marginal sector of the labor market.
   C. Most futurists predict that economic interdependence and competition will become even more significant in the global economy of the future, and multinational corporations will become even less aligned with the values of any one nation.

CRITICAL THINKING QUESTIONS

1) How does the sociology of economic life differ from economics? How do the two disciplines complement each other?
2) Compare the major economic systems. What are the strengths and weaknesses of each?
3) In what ways is the economic system in the U.S. consistent with "ideal" capitalism and in what ways is it different?

4) Which sociological perspective do you think is most useful for understanding the economic system in the U.S.? Why? What are the strengths and weaknesses of each?
5) In what ways has worker resistance and activism changed the nature of work? Which employment sectors and types of employment are less impacted by worker resistance and activism?
6) In what ways do you think work and the economy will change in the future?

PRACTICE TESTS

MULTIPLE CHOICE QUESTIONS

Select the response that best answers the question or completes the statement.

1. The _____ is the social institution that ensures the maintenance of society through the production, distribution, and consumption of goods and services.
    a. economy
    b. government
    c. state
    d. political arena

2. The owner of a large corporation has amassed several million dollars that she utilizes in expanding her enterprise. This wealth is referred to as:
    a. services
    b. labor
    c. goods
    d. capital

3. Most of sub-Saharan Africa is highly dependent on _____ sector production.
    a. primary
    b. secondary
    c. tertiary
    d. quartiary

4. Industrial economies are characterized by _____ sector production.
    a. primary
    b. secondary
    c. tertiary
    d. quartiary

5. _____ refers to the group of people who contribute their physical and intellectual services to the production process in return for wages.
    a. Goods
    b. Services
    c. Labor
    d. Economics

6. According to the text, all of the following are features of capitalism, **except**:
   a. private ownership of the means of production
   b. pursuit of personal profit
   c. competition
   d. governmental intervention in the marketplace

7. Huge corporations developed during the period of:
   a. early capitalism
   b. early monopoly capitalism
   c. advanced monopoly capitalism
   d. post capitalism

8. If Sony, Philips, Time Warner, and only a few other companies control the entire music industry, this arrangement would be a(n):
   a. oligopoly
   b. monopoly
   c. conglomerate
   d. amalgamation

9. _____ is an economic system characterized by public ownership of the means of production, the pursuit of collective goals, and centralized decision making.
   a. Capitalism
   b. Communism
   c. Socialism
   d. Communitarianism

10. Democratic socialism is characterized by
    a. extensive government action to provide support and services to its citizens.
    b. private ownership of the means of production, from which personal profits can be derived through market competition and without government intervention.
    c. public ownership of the means of production, the pursuit of collective goals, and centralized decision making.
    d. private ownership of some of the means of production, governmental distribution of some essential goods and services, and free elections.

11. According to the _____ perspective, when the economy runs smoothly, other parts of society function more effectively.
    a. conflict
    b. functionalist
    c. symbolic interactionist
    d. neo-Marxist

12. U.S. automobile manufacturers (referred to as the "Big Three") and cereal companies (three of which control 77 percent of the market) are examples of:
    a. shared monopolies
    b. conglomerates
    c. amalgamations
    d. interlocking corporate directorates

252

13. Studies have found that job satisfaction is highest when workers
    a. have some degree of control over their work.
    b. are removed from the decision making process.
    c. know that their supervisors closely watch all aspects of their work.
    d. do not feel the pressure of playing an important part in the outcome.

14. All of the following are characteristics of professions, **except**:
    a. broad based knowledge on a wide variety of topics
    b. authority
    c. concern for others, not just self-interest
    d. self regulation

15. Scientific management (Taylorism) is characterized by:
    a. increased need for skilled workers
    b. automation
    c. piece-rate system
    d. professionalization

16. According to George Ritzer, fast-food restaurants like McDonald's and Burger King illustrate:
    a. the industrial society
    b. the piece rate system
    c. Fordism
    d. robotics

17. _____ jobs differ from employment norms of the society in which they are located.
    a. Primary tier
    b. Marginal
    c. Peripheral
    d. Criminal

18. _____ occurs when workers' needs for self-identity and meaning are not mets and work is done strictly for material gain rather than personal satisfaction.
    a. Alienation
    b. Altruism
    c. Autonomy
    d. Authority

19. Which form of unemployment may be permanent?
    a. vertical
    b. seasonal
    c. cyclical
    d. structural

20. Despite the Americans with Disabilities Act, about _____ of working-age persons with a disability are unemployed today.
    a. one-fourth
    b. one-third
    c. two-thirds
    d. two-fifths

21. All of the following have been identified as characteristics central to the postindustrial economy, **except**:
    a. Information displaces property as the central preoccupation in the economy.
    b. Workplace culture shifts toward factories and away from increased diversification of work settings, the workday, and the employee.
    c. Conventional boundaries between work and home (public life and private life) are breached.
    d. New communication technologies make it possible for people to work around the clock from locations around the globe.

22. _____ includes a wide range of activities, such as fast-food service, transportation, communication, education, real estate, advertising, sports, and entertainment.
    a. Primary sector production
    b. Secondary sector production
    c. Tertiary sector production
    d. Quartiary sector production

23. Under early monopoly capitalism (1890-1940), most ownership rapidly shifted to:
    a. share holders
    b. individuals
    c. corporations
    d. bureaucracies

24. _____ are large-scale organizations that have legal powers, such as the ability to enter into contracts and buy and sell property, separate from their individual owners.
    a. Corporations
    b. Bureaucracies
    c. Conglomerates
    d. Interlocking directorates

25. _____ unemployment occurs as a result of lower rates of production during recessions in a business cycle.
    a. Seasonal
    b. Vertical
    c. Cyclical
    d. Structural

26. _____ are nonunion workers who try to take over the jobs of workers who are on strike.
    a. Scabs
    b. Bargainers
    c. Scums
    d. Free riders

27. _____ provides unemployed workers with short-term income while they look for other jobs.
    a. Compensatory income
    b. Unemployment income
    c. Unemployment compensation
    d. Income dispersion

28. Most gains made by labor unions occur through:
    a. strikes
    b. scabs
    c. collective bargaining
    d. sabotage

29. In 2005, the highest rates of union members in the United States were among _____.
    a. African American women
    b. white men
    c. Latinos/Latinas
    d. African American men

30. The _____ is made up of workers who are paid in cash without paying taxes and the selling and purchasing of goods that are not reported to the government.
    a. underground economy
    b. legitimate economy
    c. secondary economy
    d. tertiary economy

31. All of the following are ways employers reduce costs at the expense of workers **except**:
    a. subcontracting
    b. temporary workers
    c. part-time work
    d. unionization

32. _____ refers to the practice of U.S. companies moving certain operations outside of this country.
    a. Offshoring
    b. Onshoring
    c. Outshoring
    d. Farshoring

33. _____ are tangible objects that are necessary (such as food, clothing, and shelter) or desired (such as DVDs and electric toothbrushes).
   a. Products
   b. Materials
   c. Productions
   d. Goods

34. _____ are intangible activities for which people are willing to pay, such as dry cleaning, a movie, or medical care.
   a. Services
   b. Goods
   c. Products
   d. Info-tracs

35. Steel workers who process metal ore and autoworkers who then convert the ore into automobiles, trucks, and busses are engaged in _____ production.
   a. secondary sector
   b. tertiary sector
   c. essential sector
   d. primary sector

36. The ostentatious display of symbols of wealth, such as owning numerous mansions and expensive works of art and wearing extravagant jewelry and clothing, are examples of:
   a. conspicuous consumers
   b. conspicuous consumption
   c. conspicuous leisure
   d. conspicuous waste

37. According to sociologist George Ritzer, the number of lower-paying, second-tier service sector positions has increased, often at the expense of workers, which he referred to as the _____ of society.
   a. Marxism
   b. McDonaldization
   c. Fordism
   d. Taylorism

38. Large corporations that are headquartered in one country but sell and produce goods and services in many countries are referred to as:
   a. conglomerates
   b. transnational corporations
   c. oligopolies
   d. interlocking corporate directorates

39. Economist Adam Smith advocated a policy of _____, whereby the government does not intervene in the market.
   a. socialism
   b. oligopoly
   c. laissez-faire
   d. privatization

256

40. Competition is reduced over the long run by _____, where members of the board of directors of one corporation also sit on the board(s) of other corporations.
   a. oligopolies
   b. shared monopolies
   c. conglomerates
   d. interlocking corporate directorates

TRUE/FALSE QUESTIONS

1. The economy is the social institution responsible for the production, distribution, and consumption of goods and services.
   T   F

2. Preindustrial economies typically are based on the extraction of raw materials and natural resources from the environment.
   T   F

3. U.S. labor unions came into existence in the twentieth century when workers became tired of toiling for the benefit of capitalists.
   T   F

4. A conglomerate exists when four or fewer companies supply 50 percent or more of a particular market.
   T   F

5. According to Karl Marx, socialism and communism are virtually identical.
   T   F

6. According to a functionalist perspective, some problems in society are linked to peaks and troughs in the business cycle.
   T   F

7. Conflict theorists view capitalism as the solution to society's problems.
   T   F

8. The primary labor market consists of low-paying jobs with few benefits.
   T   F

9. Sociologists categorize most doctors, engineers, lawyers, professors, computer scientists, and certified public accounts as "professionals."
   T   F

10. Children whose parents are professionals are more likely to become professionals themselves.
   T   F

11. Scientific management increased worker satisfaction.
   T   F

12. Positions in the lower tier of the service sector are part of the secondary labor market, characterized by low wages, little job security, few chances for advancement, higher unemployment rates, and very limited unemployment benefits.
T   F

13. On average, union workers earn higher wages than nonunion workers.
T   F

14. Unemployment compensation is available for all workers who become unemployed.
T   F

15. The United States is the first nation to formally address the issue of equality for persons with a disability.
T   F

16. According to Box 13.1, workers' skills are usually upgraded when new technology is introduced in the workplace.
T   F

17. According to Box 13.2, McDonald's represents a form of U.S. cultural imperialism.
T   F

18. Most corporations in the U.S. have gained much more than they have lost as a result of government intervention in the economy.
T   F

19. The internet has changed, for the better, the lives of blind people, according to Box 13.3.
T   F

20. According to Box 13.4, the world is flat, which means that there is a less level global playing field in business in the twenty-first century.
T   F

FILL-IN-THE-BLANK QUESTIONS

1. The extraction of raw materials and natural resources is known as _____ production.

2. Postindustrial economics are characterized by _____ production.

3. A(n) _____ is a group of employees who join together to bargain with an employer over wages, benefits, and working conditions.

4. Large corporations that are headquartered in one country but sell and produce goods and services in many countries are known as _____.

5. A(n) _____ exists when several companies overwhelmingly control an entire industry.

258

6. _____ is an economic system characterized by private ownership of the means of production, from which personal profits can be derived through market competition and without government intervention.

7. Karl Marx described _____ as a temporary stage en route to an ideal communist society.

8. Frederick Taylor revolutionized management with a system called _____.

9. Fordism is named for _____, who used _____ technique in his manufacturing of automobiles.

10. A corporation that consists of a combination of businesses in different commercial areas that are all owned by one holding company is referred to as a _____.

11. _____ focus on the complex workings of economic systems, whereas _____ focus on interconnections among the economy, other social institutions, and the social organization of work.

12. _____ work is part-time, temporary, or subcontracted work.

13. _____ is a large-scale organization that has legal powers separate from its individual owners.

14. The _____ is the rise and fall of economic activity relative to long-term growth in the economy.

15. _____ are high-status, knowledge-based occupations.

SHORT ANSWER/ESSAY QUESTIONS

1. How far has the United States moved into postindustrialization? What are some of the fastest growing U.S. occupations?
2. What are the occupational categories considered to be marginal jobs in the United States? Why are these occupations considered to be marginal? What is it meant by internationalization of marginal jobs?
3. What are the three major contemporary economic systems? In what ways are they similar and different?
4. What are some similarities and differences in scientific management and mass production?
5. How do the functionalists, conflict theorists, and symbolic interactionists explain work in the United States?

STUDENT CLASS PROJECTS AND ACTIVITIES

1. This project calls for you to compare the U.S. budget deficit today to that for the years 1980-2009. During the years 1980-1992, the federal budget deficit shot up to between 3 percent and 7 percent of the gross domestic product. In the years of Presidents Kennedy, Johnson, and Nixon, the deficits generally hovered at a relatively harmless 1 percent or 2 percent of the GDP, except for a very brief increase to 3 percent at the end of the Johnson Administration to help pay for the Vietnam War. Research the reasons that federal spending outstripped revenues at such an astonishing rate of increase during the Reagan-Bush years. Analyze the effectiveness of "supply-side" economics during these years (the idea that the U.S. could cut income taxes and simultaneously pay for massive increases in defense and certain highly popular domestic programs). Depict the federal budget deficit by presentation of graphs (probably a bar graph would serve best here) from the years 1980 to 2009. Then compare the federal budget deficit of the mid 1990s to today's deficit. Present at least four well thought out solutions for cutting the deficit. Critique some of the present policies enacted in order to cut the deficit. Some research sources for this project would be in the government documents section of your library, or from the Office of Management and Budget. Summarize your findings in a paper and include a bibliography of references.

2. Conduct interviews with people who are employed in different parts of the labor market. Interview one person who is employed in the primary labor market, one in the secondary labor market, and one in the tertiary labor market. Provide the sex, race, and age of the workers, as well as the names of their employers. In your interviews, ask questions about (1) the worker's family and educational background; (2) training requirements for the job; (3) how and why they entered this particular segment of the labor market; (4) their typical day of work; (5) employment benefits, such as paid vacations, paid holidays, sick leave, health insurance, etc.; (6) job security and advancement possibilities; (7) worker satisfaction and any resistance strategies they use. In the writing up of this project, include a summary of the responses to each specific question and compare the responses based on their sector of the labor market.

3. Trace the history (including the origin and growth) of labor unions in the United States, for both male and female workers. Include in your research a statement of the present status of labor unions in the American marketplace.

4. Compare the economic system of the United States with the economic system of another Western industrialized society, such as Western European countries or Canada. Look for information on (1) the presence of corporations and their activities; (2) government interventions in the marketplace; (3) labor laws and regulations; (4) labor unions; (5) distribution of goods and services in the societies, such as income gaps, percent in poverty, health care distribution, college education distribution, and unemployment services or compensation; (7) other programs or services offered to citizens, such as day care subsidies, parental leave programs, etc. Summarize the similarities and differences between the U.S. economy and the economy of the country you chose, and evaluate both systems. Provide a bibliography of your references.

INTERNET ACTIVITIES

1. To read about one of the most powerful labor organizations in the United States, go to the **AFL-CIO** web site: **http://www.aflcio.org/**. Once there, click on unions affiliated with the AFL-CIO to see the large number of unions associated with this organization. Browse the web site to see what issues are important to labor organizations today.
2. Check out **Forbes** listing of the richest people in the world: **http://www.forbes.com/2005/03/09/bill05land.html**. Do a web search for some of these people and see what businesses they are associated with.
3. Read about current issues in employment, find out the current unemployment rate, look at earnings by different occupations and demographics, and much more at the **U.S. Bureau of Labor Statistics** web site: **http://www.bls.gov/**.

INFOTRAC COLLEGE EDITION EXERCISES

Visit the **InfoTrac College Edition** web site at: **http://www.wadsworthmedia.com/webtutor/infotrac.htm**. You will arrive at a screen that enables you to search topics.

1. Search for articles on both **seasonal** and **structural unemployment**. Bring to class a brief summary that compares and contrasts what we know about these two forms of unemployment.
2. **The Virtual Seminar in Global Political Economy**—The Virtual Seminar is a true international college of the Internet, with online classes for college credit exploring questions such as the Third World debt crisis and international social movements. Search InfoTrac for this topic and report your findings.
3. Look up articles on **economic interdependence**. Scan these articles and make a list of related issues to understand in order to get the full picture of this growing phenomenon.

SOLUTIONS

MULTIPLE CHOICE QUESTIONS

1. A, p. 410
2. D, p. 410
3. A, p. 411
4. B, p. 413
5. C, p. 410
6. D, p. 416
7. B, p. 417
8. A, p. 420
9. C, p. 422
10. D, p. 423
11. B, p. 423
12. A, p. 420
13. A, p. 426
14. A, p. 428

15. C, p. 429
16. C, p. 430
17. B, p. 431
18. A, p. 426
19. D, p. 435
20. C, p. 438
21. B, p. 414
22. C, p. 414
23. C, p. 417
24. A, p. 417
25. C, p. 434
26. A, p. 436
27. C, p. 435
28. C, p. 436

29. D, p. 436
30. A, p. 434
31. D, p. 433
32. A, p. 410
33. D, p. 410
34. A, p. 410
35. A, p. 413
36. B, p. 414
37. B, p. 415
38. B, p. 417
39. C, p. 421
40. D, p. 420

TRUE/FALSE QUESTIONS

1. T, p. 410
2. T, p. 411
3. T, p. 436
4. F, p. 420
5. F, p. 422
6. T, p. 423
7. F, p. 424-425

8. F, p. 427
9. T, p. 428
10. T, p. 428
11. F, p. 430
12. T, p. 427
13. T, p. 436
14. F, p. 435

15. T, p. 438
16. F, p. 412
17. T, p. 418
18. T, p. 421
19. T, p. 439
20. T, p. 441

FILL-IN-THE-BLANK QUESTIONS

1. primary sector, p. 411
2. tertiary sector, p. 413
3. labor union, p. 417
4. transnational corporation, p. 417
5. oligarchy, p. 420
6. Capitalism, p. 416
7. socialism, p. 422
8. scientific management, p. 429
9. Henry Ford, p. 430; scientific management, p. 430
10. conglomerate, p. 420
11. Economists, p. 410; sociologists, p. 410
12. Contingent work, p. 432
13. Corporation, p. 417
14. business cycle, p. 423
15. Professions, p. 428

14

POLITICS AND GOVERNMENT IN GLOBAL PERSPECTIVE

BRIEF CHAPTER OUTLINE

POLITICS, POWER, AND AUTHORITY
    Political Science and Political Sociology
    Power and Authority
    Ideal Types of Authority
POLITICAL SYSTEMS IN GLOBAL PERSPECTIVE
    Monarchy
    Authoritarianism
    Totalitarianism
    Democracy
PERSPECTIVES ON POWER AND POLITICAL SYSTEMS
    Functionalist Perspectives: The Pluralist Model
    Conflict Perspectives: Elite Models
    Critique of Pluralist and Elite Models
THE U.S POLITICAL SYSTEM
    Political Parties and Elections
    Political Participation and Voter Apathy
GOVERNMENTAL BUREAUCRACY
    Characteristics of the Federal Bureaucracy
    The Iron Triangle and the Military-Industrial Complex
THE MILITARY AND MILITARISM
    Explanations for Militarism
    Gender, Race, and the Military
TERRORISM AND WAR
    Types of Terrorism
    Terrorism in the United States
    War
POLITICS AND GOVERNMENT IN THE FUTURE

CHAPTER SUMMARY

The relationship between politics, power, and authority is a strong one in all countries. **Politics** is the social institution through which power is acquired and exercised by some people or groups. **Power**—the ability of persons or groups to carry out their will even when opposed by others—is a social relationship involving both leaders and followers. Most leaders seek to legitimate their power through **authority**—power that people accept as legitimate rather than coercive. **Government** is the formal organization that regulates relationships among members of a society and those outside its borders. Some social analysts refer to the government as the **state**—the political entity that possesses a legitimate monopoly of the use of force within its territory to achieve its goals. According to Max Weber, there are three ideal types of authority: **traditional**, **charismatic**, and **rational-legal** (bureaucratic). There are four main types of contemporary political systems: **monarchy**, **authoritarianism**, **totalitarianism**, and **democracy**. In a democracy, the people hold the ruling power either directly or through elected representatives. There are two key perspectives on how power is distributed in the U.S. Functionalists believe in a **pluralist model**, in which power is widely dispersed throughout many competing interest groups. Conflict theorists believe in an **elite model**, in which power is concentrated in a small group of elites, and the masses are relatively powerless. The **power elite** is comprised of influential business leaders, key government leaders, and the military. **Political parties** are organizations whose purpose is to gain and hold legitimate control of government. The Democrats and the Republicans have dominated the U.S. political system since the mid-nineteenth century, although party loyalties have been declining in recent years. People learn political attitudes, values, and behaviors through **political socialization**. The vast governmental bureaucracy is a major source of power and carries out the actual functioning of the government. One way that **special interest groups** exert powerful influence on the bureaucracy is through the iron triangle of power—a three-way arrangement in which a private interest group, a congressional committee, and bureaucratic agency make the final decision on a political issue. The military bureaucracy is so wide-ranging that it encompasses the **military-industrial complex**—the mutual interdependence of the military establishment and private military contractors. This complex is supported by **militarism**—a societal focus on military ideals and aggressive preparedness for war. **Terrorism**, the use of calculated, unlawful physical force or threats of violence against a government, organization, or individual to gain some political, religious, economic, or social objective, has recently created worldwide fear. Three types of terrorism—revolutionary, state-sponsored, and repressive—all extract a massive toll by producing rampant fear, widespread loss of human life, and extensive destruction of property. Recent acts of terrorism directed against the United States have been facts of life in some other countries for many years. **War**, an organized, armed conflict between nations or distinct political factions, has both direct and indirect effects. In examining politics and government in the future, one group of analysts views the U.S. as becoming less democratic. Other analysts focus more narrowly on specific questions of concern. Our response to these concerns will have a profound effect on people and governments all around the globe.

LEARNING OBJECTIVES

After reading Chapter 14, you should be able to:

1. Explain the relationship between politics, government, and the state, and note how political sociology differs from political science.

2. Distinguish between power and authority, and describe the three major types of authority.

3. Compare and contrast governments characterized by monarchy, authoritarianism, totalitarianism, and democracy.

4. State the major elements of the pluralist (functionalist) and elite (conflict) models of power and political systems and explain how they differ.

5. Explain the critiques of the functionalist and conflict models of power and political systems.

6. Describe the purpose of political parties.

7. Analyze how well U.S. political parties measure up to the ideal-type characteristics.

8. Explain the relationship between political socialization, political attitudes, and political participation.

9. Discuss the characteristics of the federal bureaucracy and explain what the term "permanent government" means.

10. Describe the military-industrial complex and explain why it is called an iron triangle.

11. Discuss militarism and explain why support for this ideology has been so strong in the United States.

12. Describe American cultural reactions to war and terrorism both here and abroad.

13. Examine the significance of current world events on the immediate and long-term future.

KEY TERMS

(defined at page number shown and in glossary)

authoritarianism, p. 453
authority, p. 449
charismatic authority, p. 450
democracy, p. 454
elite model, p. 457
government, p. 446
militarism, p. 469
military-industrial complex, p. 467

monarchy, p. 451
pluralist model, p. 454
political action committees, p. 456
political party, p. 460
political socialization, p. 463
political sociology, p. 447
politics, p. 446
power, p. 448

KEY PEOPLE

(identified at page number shown)

CHAPTER OUTLINE

I. POLITICS, POWER, AND AUTHORITY
    A. **Politics** is the social institution through which power is acquired and exercised by some people and groups.
    B. In contemporary societies, the primary political system is the **government**—the formal organization that has the legal and political authority to regulate the relationships among members of a society and between the society and those outside its borders.
    C. Sociologists often refer to the government as the **state**—the political entity that possesses a legitimate monopoly over the use of force within its territory to achieve its goals.
    D. *Political science* focuses on power and the distribution of power in different types of political systems, the branches of the federal government, operative political processes, and interrelationships between governments, multinational corporations, and international organizations. **Political sociology** examines the nature and consequences of power within or between societies, focusing on the social circumstances of politics and the interrelationships between politics and social structures.
    E. Power and Authority
        1. **Power** is the ability of persons or groups to achieve their goals despite opposition from others.
        2. **Authority** is power that people accept as legitimate rather than coercive.
    F. Ideal Types of Authority
        1. According to **Max Weber**, there are three ideal types of authority:
            a. **Traditional authority** is power that is legitimized on the basis of long-standing custom.
            b. **Charismatic authority** is power legitimized on the basis of a leader's exceptional personal qualities or the demonstration of extraordinary insight and accomplishment that inspire loyalty and obedience from followers. **Routinization of charisma** occurs when charismatic authority is

266

succeeded by a bureaucracy controlled by a rationally established authority or by a combination of traditional and bureaucratic authority.

    c. **Rational-legal authority** is power legitimized by law or written rules and regulations.

  2. These three types of authority demonstrate how different bases of legitimacy are tied to a society's economy.

II. POLITICAL SYSTEMS IN GLOBAL PERSPECTIVE

A. Emergence of Political Systems

  1. Political institutions first emerged in agrarian societies as they acquired surpluses and developed greater social inequality.

  2. Nation-states—political organizations that have recognizable national boundaries within which their citizens possess specific legal rights and obligations—developed first in Europe between the twelfth and fifteenth centuries.

B. **Monarchy** is a political system in which power resides in one person or family and is passed from generation to generation through lines of inheritance.

C. **Authoritarianism** is a political system controlled by rulers who deny popular participation in government.

D. **Totalitarianism** is a political system in which the state seeks to regulate all aspects of people's public and private lives.

E. **Democracy** is a political system in which the people hold the ruling power, either directly or through elected representatives.

III. PERSPECTIVES ON POWER AND POLITICAL SYSTEMS

A. Functionalist Perspectives

  1. The functionalist perspective assumes people share a consensus on central concerns (e.g., freedom and protection) and that the government serves important functions.

    a. Functions include: (1) maintaining law and order; (2) planning and directing society; (3) meeting social needs; and (4) handling international relations, including warfare.

  2. Associated with functionalism is the **pluralist model**, which states power in political systems is widely dispersed throughout many competing **special interest groups**—political coalitions comprised of individuals or groups that share a specific interest they wish to protect or advance with the help of the political system.

    a. Key elements of pluralism include: (1) decisions on behalf of the people by leaders who engage in bargaining, accommodation, and compromise; (2) competition among leadership groups makes power abuse by any one group more difficult; (3) people influence policy by forming or voting and participating in special interest groups; (4) power is widely dispersed in society; and (5) policy is not always based on majority preference but the balance between interest groups.

    b. Over the last two decades, special interest groups have become more involved in "single issue politics," such as abortion, gun control, gay and lesbian rights, or environmental concerns.

267

    c. **Political action committees (PACs)** are organizations of special interest groups that fund campaigns to help elect (or defeat) candidates based on their stances on specific issues.

B. Conflict Perspectives

  1. According to the **elite model**, power in political systems is concentrated in the hands of a small group of elites, and the masses are relatively powerless.

    a. Key elements are (1) decisions are made by the elite, who possess greater resources than the "masses" they govern; (2) consensus exists among the elite but not among most people on the basic values and goals of society; (3) power is highly concentrated at the top of the social hierarchy, and the powerful set policy for everyone; and (4) public policy reflects the values and preferences of the elite.

  2. According to **C. Wright Mills**, the **power elite** is comprised of leaders at the top of business, the executive branch of the federal government, and the military (especially the "top brass" at the Pentagon). The elites have similar class backgrounds and interests.

  3. **G. William Domhoff** referred to elites as the *ruling class*—a relatively fixed group of privileged people who wield sufficient power to constrain political processes and serve underlying capitalist interests. The ruling class consists of the corporate rich, who make up less than 1 percent of the U.S. population.

    a. The corporate rich influence the political process by financing campaigns and providing favors to political candidates, participating in special interest groups, and holding prestigious positions on government advisory committees or commissions.

  4. Class Conflict Perspectives

    a. Most contemporary elite models are based on the work of **Karl Marx**.

    b. *Instrumental Marxists* argue that the state acts to perpetuate the capitalist class; *structural Marxists* contend that the state is not only an instrument of the capitalist class (evidenced by social welfare programs) but it ultimately serves their interests.

  5. Critique of Pluralist and Elite Models

    a. Domhoff argues U.S. society appears to be pluralist when actually it is elitist.

    b. Mills suggests that the economic, political, and military elite are interrelated and may be a relatively cohesive group.

IV. THE U.S. POLITCIAL SYSTEM

A. Political Parties and Elections

  1. A **political party** is an organization whose purpose is to gain and hold legitimate control of government; they affect elections.

    a. Parties develop and articulate policy positions; educate voters about the issues and simplify the choices for them; recruit candidates who agree with those policies and help them win office; and hold elected officials responsible for implementing the party's policy positions.

    b. Since the Civil War, two political parties—the Democratic and the Republican—have dominated the political system in the United States and confronted two broad types of concerns:

  c. Social issues are those relating to moral judgments or civil rights.

  d. Economic issues involve the amount that should be spent on government programs and the extent to which these programs should encourage a redistribution of income and assets.

 B. Political Participation and Voter Apathy

  1. **Political socialization** is the process by which people learn political attitudes, values, and behavior. For young children, the family is the primary agent of political socialization.

  2. Socioeconomic status affects people's political attitudes, values, and beliefs.

  3. Political participation occurs at four levels:

   a. Voting

   b. Attending and taking part in political meetings

   c. Actively participating in political campaigns

   d. Running for or holding political office

  4. At most, about 10 percent of the voting age population in this country participates at a level higher than simply voting, and about 62 percent voted in the 2008 presidential election.

V. GOVERNMENTAL BUREAUCRACY

 A. Characteristics of the Federal Bureaucracy

  1. The size and scope of government has grown in recent decades, partially because of dramatic increases in technology and demands from the public that the government "do something" about various problems facing society.

  2. Much of the actual functioning of the government is carried on by its bureaucracy—the *permanent government* in Washington that is made up of top tier, civil service bureaucrats who have built a major power base.

  3. The governmental bureaucracy has been able to perpetuate itself and expand because it has many employees with highly specialized knowledge and skills who cannot easily be replaced by those from the "outside."

 B. The Iron Triangle and the Military-Industrial Complex

  1. The *iron triangle of power* is a three-way arrangement in which a private-interest group (usually a corporation), a congressional committee or subcommittee, and a bureaucratic agency make the final decision on a political issue that is to be decided by that agency.

  2. A classic example of the iron triangle is the **military-industrial complex**—the mutual interdependence of the military establishment and private military contractors that started during World War II and has continued to the present.

   a. Between 1945 and the early 1990s, the U.S. government spent an estimated $10.2 trillion for national defense (in constant 1987 dollars). In the 1990s, the U.S. spent more than $270 billion per year on the U.S. military budget.

   b. Several of the largest multinational corporations are among the largest defense contractors, and the defense divisions of these companies have a virtual monopoly over defense contracts. The military budgets for 2003 and 2004 were almost $400 billion per year.

   c. Some members of Congress have actively supported the military-industrial complex because it allowed them to provide economic assistance for their local voting constituencies in the form of funding for defense-related industries, military bases, and space centers in their home state. These activities are known as *pork* or *pork barrel projects.*

VI. THE MILITARY AND MILITARISM

    A. **Militarism** is a societal focus on military ideals and an aggressive preparedness for war; militarism is supported by core U.S. values such as patriotism, courage, reverence, loyalty, obedience, and faith in authority. Sociologists have proposed several reasons for militarization:

        1. The economic interests of capitalists, college and university faculty and administrators who are recipients of research grants, workers, labor union members, and others who depend on military spending.

        2. The role of the nation and its inclination toward coercion in response to perceived threats.

        3. The relationship between militarism and masculinity.

    B. Gender, Race, and the Military

        1. The introduction of the all-volunteer force in the early 1970s and the end of the draft shifted the focus of the military from training "good citizen soldiers" to an image of the "economic person"—one who enlists in the military in the same way that a person might take a job in the private sector.

        2. Considerable pressure has been placed on the military to recruit women; in 2009, women made up approximately 20 percent of the total uniformed force.

        3. Women must break through the "brass ceiling" to advance in the military, and many women experience sexual assault.

VII. TERRORISM and WAR

    A. **Terrorism** is the use of calculated, unlawful physical force or threats of violence against a government, organization, or individual to gain some political, religious, economic, or social objective.

    B. Types of Terrorism

        1. *Political terrorism* uses intimidation, coercion, threats of harm, and other violence that attempts to bring about a significant change in or overthrow an existing government. There are three types of political terrorism.

            a. *Revolutionary terrorism* refers to acts of violence against civilians that are carried out by enemies of the government who want political change.

            b. *State-sponsored terrorism* occurs when a government provides financial resources for terrorists who conduct their activities in other nations.

            c. *Repressive terrorism* is conducted by a government against its own citizens for the purpose of protecting an existing political order.

    C. Terrorism in the United States

        1. The bombing of the Federal Building in Oklahoma City in 1995 and the 2001 destruction of the World Trade Center stand out as the two worst terrorist attacks in the U.S.

    D. War

        1. **War**—organized, armed conflict between nations or distinct political factions—includes both declared and undeclared wars.

        2. Casualties in conventional warfare pale when compared to those in biological or chemical warfare.

VIII. Politics and Government in the Future
   A. Future issues in the U.S. include the future of political parties, global corporate interests, preventing future terrorist attacks, balancing national security and individual privacy, changing demographics of the population, immigration and employment policies, and media portrayals of government.

CRITICAL THINKING QUESTIONS

1. What are some examples of people who have power and authority? Based on Weber's classification of ideal types of authority, what types of authority do these people have?
2. Compare and contrast monarchy, authoritarianism, totalitarianism, and democracy.
3. Compare and contrast functionalist and conflict perspectives of power and political systems. Which theory do you think is most useful for understanding power and politics in the U.S. today? Why? What are the strengths and weaknesses of each?
4. What role do you think the media played in the 2008 presidential election?
5. Which political issues do you think will be most significant throughout the next decade? Why?

PRACTICE TESTS

MULTIPLE CHOICE QUESTIONS

Select the response that best answers the question or completes the statement.

1. The social institution through which power is acquired and exercised by some people and groups is known as:
   a. government
   b. politics
   c. the state
   d. the military

2. The state is
   a. the social institution through which power is acquired and exercised by some people and groups.
   b. the formal organization that has the legal and political authority to regulate the relationships among members of a society and between the society and those outside its borders.
   c. the political entity that possesses a legitimate monopoly over the use of force within its territory to achieve its goals.
   d. the political entity that seeks to regulate all aspects of people's public and private lives.

3. _____ is the power that people accept as legitimate rather than coercive.
   a. Influence
   b. Clout
   c. Authority
   d. Legitimation

4. Traditional authority is based on:
   a. a leader's exceptional personal qualities
   b. written rules and regulations of law
   c. documents such as the U.S. Constitution
   d. long-standing custom

5. Napoleon, Julius Caesar, Martin Luther King, Jr., Cesar Chavez, and Mother Teresa are examples of:
   a. charismatic authority
   b. traditional authority
   c. rational-legal authority
   d. nontraditional authority

6. _____ examines the nature and consequences of power within or between societies and the social and political conflicts that lead to change in the allocation of power.
   a. Political science
   b. Political sociology
   c. Political anthropology
   d. U.S. government

7. A unit of political organizations that has recognizable national boundaries where citizens possess legal rights and obligations is a:
   a. bureaucracy
   b. government
   c. nation-state
   d. city-state

8. A hereditary right to rule or a divine right to rule is most likely to be found in a(n):
   a. monarchy
   b. authoritarian regime
   c. totalitarian regime
   d. democracy

9. The National Socialist (Nazi) party in Germany during World War II is an example of a(n):
   a. monarchy
   b. authoritarian regime
   c. totalitarian regime
   d. democracy

10. Representative democracies
   a. require that citizens meet regularly to debate and decide the issues of the day.
   b. always use a winner-takes-all system in elections, such as that used in the U.S.
   c. assure that all citizens are equally represented.
   d. are systems in which elected representatives are supposed to convey the concerns of the people they represent.

11. According to the pluralist model, power in political systems is
   a. widely dispersed throughout many competing interest groups.
   b. concentrated in the hands of a small group of elites.
   c. comprised of leaders at the top of business, the executive branch of the federal government, and the military.
   d. controlled by members of the ruling class.

12. _____ are political coalitions made up of individuals or groups that share a specific interest they wish to protect or advance with the help of the political system.
   a. Power elites
   b. Elite groups
   c. Special interest groups
   d. PACs

13. Political action committees
   a. generally have been abolished by reforms in campaign finance laws.
   b. are comprised of people who volunteer their time but not money to political candidates and parties.
   c. are organizations of special interest groups that fund campaigns to help elect candidates based on their stances on specific issues.
   d. encourage widespread political participation by citizens at the grassroots level.

14. The power elite model was developed by:
   a. Emile Durkheim
   b. Karl Marx
   c. G. William Domhoff
   d. C. Wright Mills

15. The organizations responsible for developing and articulating policy positions, educating voters about issues, and recruiting candidates to run for office are known as:
   a. political parties
   b. political action committees
   c. interest groups
   d. federal election committees

16. _____ is the process by which people learn political attitudes, values, and behavior.
    a. Indoctrination
    b. Military training
    c. Resocialization
    d. Political socialization

17. According to the text, most of the actual functioning of the U.S. government is carried on by:
    a. political action committees
    b. the federal bureaucracy
    c. the criminal justice system
    d. political parties

18. The three-way arrangement in which a private interest group, a congressional committee, and a bureaucratic agency make the final decision on a political issue that is to be decided by that agency is known as:
    a. political subversion
    b. the iron law of oligarchy
    c. the iron triangle of power
    d. the power elite

19. _____ is a societal focus on military ideals and an aggressive preparedness for war.
    a. Militarism
    b. Authoritarianism
    c. Totalitarianism
    d. Warmongerism

20. Women in the U.S. military
    a. have largely eliminated sexual harassment through greater representation in higher ranks.
    b. have disturbed the existing world of male militarism.
    c. must break through the "brass ceiling" if they hope to advance in rank.
    d. currently outnumber men.

21. _____ argue that divergent viewpoints lead to a system of political pluralism in which the government serves as a mediator between competing interests and viewpoints.
    a. Functionalists
    b. Conflict theorists
    c. Symbolic interactionists
    d. Postmodernists

22. Sociologist G. William Domhoff argues that the U.S. is governed by:
    a. a pluralistic model
    b. political action committees
    c. a ruling class made up of the corporate rich
    d. a monarchy

23. Sociologist Max Weber suggested that most leaders do not want to base their power on force alone; they seek to legitimize their power by turning it into _____, which is the power that people accept as legitimate rather than coercive.
    a. authority
    b. control
    c. dominance
    d. democracy

24. _____ is a political system controlled by rulers who deny popular participation in government.
    a. Utilitarianism
    b. Authoritarianism
    c. Totalitarianism
    d. Monarchy

25. _____ occurs when charismatic authority is succeeded by a bureaucracy controlled by a rationally established authority or by a combination of traditional and bureaucratic authority.
    a. Charismatic authority
    b. Rational-legal authority
    c. Bureaucratic charisma
    d. Routinization of charisma

26. _____ argue that the state exists only to support the interests of the dominant class.
    a. Pluralists
    b. Structural Marxists
    c. Instrumental Marxists
    d. Postmodernists

27. Although the U.S. Constitution grants _____ authority to the office of the presidency, a president who fails to uphold the public trust may be removed from office.
    a. coercive
    b. charismatic
    c. traditional
    d. rational-legal

28. A _____ is a unit of political organization that has recognizable national boundaries and whose citizens possess specific legal rights and obligations.
    a. national-state
    b. state
    c. bureaucracy
    d. city-state

29. In dictatorships, power is gained and held by a single individual. Pure dictatorships are rare; all rulers need the support of the military and the backing of business elites to maintain their position. Dictatorships occur in _____ political systems.
    a. democratic
    b. authoritarian
    c. monarchy
    d. totalitarian

30. In the United States, people have a voice in the government through _____, whereby citizens elect representatives to serve as bridges between themselves and the government.
    a. representative democracy
    b. monarchy
    c. direct democracy
    d. independent involvement democracy

31. A key element of pluralism is
    a. the inheritance of power from family members.
    b. veto groups are strictly prohibited.
    c. public policy reflects a balance among competing interest groups.
    d. power is concentrated among a few elites in society.

32. The _____ perspective argues that capitalists control the government through special interest groups, lobbying, campaign financing, and other types of "influence peddling" to get legislatures and the courts to make decisions favorable to their class, and the state exists only to support the interests of the dominant class.
    a. instrumental Marxist
    b. structural Marxist
    c. restrictive Marxist
    d. normative Marxist

33. A political party does all of the following, **except**:
    a. develops and articulates policy positions
    b. educates political candidates about critical social issues
    c. recruits political candidates and helps them win office
    d. holds the political candidates responsible for the party's policy positions

34. People with a _____ perspective tend to focus on equality of opportunity, the need for government regulation, and social safety nets.
    a. liberal
    b. socialist
    c. conservative
    d. libertarian

35. People with a _____ perspective are more likely to emphasize economic liberty and freedom from government interference.
    a. liberal
    b. socialist
    c. conservative
    d. radical

36. According to the text's discussion of the federal bureaucracy,
    a. during the nineteenth century, the government held a central role in everyday life.
    b. the government bureaucracy has been able to expand more in recent decades.
    c. the federal bureaucracy employs less than 1 million people.
    d. the public expects much less from the government.

37. Acts of violence against civilians that are carried out by enemies of the government in an attempt to bring about political change are referred to as:
    a. revolutionary terrorism
    b. state-sponsored terrorism
    c. repressive terrorism
    d. egalitarian terrorism

38. _____ activities constitute a long-standing political practice—projects designed to bring jobs and public monies to the home state of members of Congress, for which they can take credit.
    a. Cracker barrel
    b. Grease men
    c. Pork barrel
    d. Bread and butter

39. In Libya, Colonel Muammar Gaddafi has provided money and training for terrorist groups such as the Arab National Youth Organization, which was responsible for skyjacking a Lufthansa airplane over Turkey and forcing the German government to free the surviving members of Black September. This is an example of:
    a. repressive terrorism
    b. state-sponsored terrorism
    c. national terrorism
    d. jihad terrorism

40. _____ terrorism is conducted by a government against its own citizens for the purpose of protecting an existing political order.
    a. Repressive
    b. State-sponsored
    c. National
    d. Jihad

TRUE/FALSE QUESTIONS

1. More people voted in the 2008 presidential election than any other election in U.S. history.
   T   F

2. Power is a social relationship that involves both leaders and followers.
   T   F

3. Charismatic authority tends to be temporary and relatively unstable.
   T   F

4. Globally, today everyone is born, lives, and dies under the auspices of a nation-state.
   T   F

5. In a representative democracy, elected representatives are expected to keep the "big picture" in mind, and they do not necessarily need to convey the concerns of those they represent in all situations.
   T   F

6. Functionalists suggest that divergent viewpoints lead to a system of political pluralism in which the government functions as an arbiter between competing interests and viewpoints.
   T   F

7. Over the past two decades, special interest groups have become more involved in multiple-issue politics because so many political issues are intertwined with other concerns.
   T   F

8. Sociologist G. William Domhoff argues that the ruling class wields sufficient power to constrain political processes in the United States.
   T   F

9. Structural Marxists contend that the state is simply a passive instrument of the capitalist class.
   T   F

10. Children do not typically identify with the political party of their parents until they reach college age.
    T   F

11. People in the upper classes tend to vote based on a philosophy of *noblesse oblige*.
    T   F

12. The United States has one of the highest percentages of voter turnout of all Western nations.
    T   F

13. The U.S. governmental bureaucracy has been able to perpetuate itself and expand because many of its employees have highly specialized knowledge and skills and cannot be replaced easily by "outsiders."
   T   F

14. Pork barrel projects are a recent political practice in the United States.
   T   F

15. The most significant terrorist attacks on U.S. soil took place during World War II.
   T   F

16. Historically, the development of manhood and male superiority has been linked to militarism and combat.
   T   F

17. According to Box 14.1, television coverage of political scandals first began in the mid-1990s.
   T   F

18. Some analysts, as depicted in Box 14.2, suggest that the European Union has produced more than fifty years of peace in Europe.
   T   F

19. According to Box 14.3, California Governor Arnold Schwarzenegger has not been able to use media framing to develop favorable responses to his political initiatives.
   T   F

20. According to Box 14.4, recently all twenty-five of the top news stories were adequately covered by the U.S. media.
   T   F

FILL-IN-THE-BLANK QUESTIONS

1. Some social analysts refer to the government as the _____, which possesses a legitimate monopoly over the use of force.

2. The main types of political systems are _____, _____, _____, and _____.

3. _____ is the ability of persons or groups to achieve their goals, despite the opposition of others.

4. Power legitimized by law or written rules and regulations is _____.

5. According to C. Wright Mills, the _____ is comprised of political, economic, and military elite.

6. The mutual interdependence of the military establishment and private military contractors is known as the _____.

7. The three types of political terrorism are: _____, _____, and _____.

8. _____ is a political system in which people hold the ruling power either directly or through elected representatives.

9. According to sociologist _____, the U.S. actually has a ruling class that wields power to constrain political processes.

10. The two worst terrorist attacks that ever occurred in the U.S. were in the years _____ and _____.

11. According to _____, the three ideal types of authority are traditional, charismatic, and rational-legal.

12. When cities first developed, the _____ became the center of political power.

13. The people who are paid to influence legislation on behalf of special clients are referred to as _____.

14. A three-way arrangement in which a political decision is made by a private interest group, a congressional committee, and a bureaucratic agency is referred to as _____.

15. According to sociologist _____, one purpose of government is to socialize people to be good citizens.

SHORT ANSWER/ESSAY QUESTIONS

1. What is the role of the individual in politics, considering such issues as political socialization, political action, lobbying activities, voter apathy, and political elites?
2. What is the importance of political parties? What do they do? Are they good for democracies? Why? Why not?
3. What are the major political systems around the world?
4. What are the three ideal types of authority? In what ways are some more effective than others in achieving certain goals?
5. How do the pluralist and the elite models explain power in the United States? Which do you think best represents the U.S. political system? Why?

STUDENT CLASS PROJECTS AND ACTIVITIES

1. The text notes the low rate of voter participation in our country. Try to discover some explanations for this condition. Conduct an informal survey of a cross section of twenty people of different age, ethnicity, social class, race, and sex. On your survey forms, you must include the basic background information of your respondents (race, sex, address, age, ethnicity, and occupation or student status). Obtain responses to the following questions: (1) When was the last time you voted? Why? (2) Did you vote for the President in the last election? (3) Do you usually vote in all elections? Why? Why not? (4) Do you typically vote along political party lines, such as for Democrats or Republicans? If so, which party? Why? (5) Do other members of your family vote? (6) Do you know at least one elected official? Who? (7) Do you support the present position of the President of the United States on most issues?

Explain. Formulate any other questions you may want to ask. Provide a conclusion, an analysis, and an evaluation of the data in the writing up of this project.

2. Every year (usually in January), *Parade* magazine reports on the "The World's Worst Ten Dictators." Locate the latest issue and record each of the ten, summarizing the basic information of each, noting the country that each dictates, the length of the rule of each, the reason for being on this dubious list, and any other pertinent information provided in the article. Then, select two people from the list and look up additional information about them. Based on the information you are able to find, assess the validity of the information provided in the magazine's list.

3. Compare and contrast college students' perceptions of the two main political parties in the U.S. Try to find and interview ten students who identify themselves as Democrats and ten who identify as Republicans. Ask each student the following questions: (1) Which political party do you identify with most? (2) Why do you identify with that party? (3) What do you think are the core ideas of the political party you identify with? (4) Which of these core ideas do you agree with? (5) Are there any core ideas held by your political party that you disagree with? (6) What do you think are the core ideas of the other major political party? (7) Do you agree with any of these core ideas? (8) Which of the main ideas of the other political party do you disagree with? Why? Write a paper that compares and contrasts the views expressed by students who affiliated with each of the parties. In particular, compare the Republican students' perceptions of the core ideas of the Republican party with the Democrat students' perceptions of the Republican party. Then compare the Democrat students' perceptions of the core ideas of the Democratic party to the Republican students' perceptions of the Democratic party. To what extent does each group view the parties similarly and to what extent do they view them differently?

INTERNET ACTIVITIES

1. Sociologists are frequently interested in using surveys and opinion polls to determine how members of a society feel about a particular issue. One organization that is well known for conducting such research is the **Gallup** organization. Search the site for polling data related to this chapter: **http://www.gallup.com/**

2. In addition to the Gallup organization, the **Roper Center for Public Opinion Research**, **http://www.ropercenter.uconn.edu/**, also conducts research on public opinion. Investigate the Public Opinion Matters section of the web site to find data and articles on several topics that bear relevance to this chapter.

3. Most of the major political parties now have an online presence. After visiting these sites, compare the major similarities or differences among the political parties. The Democratic Party can be found at **http://democrats.org/**. The Republican National Committee can be found at **http://rnc.org/**. The Libertarian Party can be found at **http://www.lp.org/**. The Socialist Party is located at **http://sp-usa.org/**.

4. As discussed in the text, **lobbying groups** are a powerful force in politics in the United States. Each of these organizations provides information about their legislative concerns and activities. For instance, the NRA keeps track of bills and proposed laws that might impact gun ownership. Overall, how do the agendas of these organizations compare? Some of the best-known lobbying groups are: The

National Rifle Association, **http://www.nra.org/**, (NRA); The American Association of Retired Persons (AARP), **http://www.aarp.org/**; and The National Education Association (NEA), **http://www.nea.org/**.

5. **The Center for Responsive Politics**, http://www.opensecrets.org/, maintains an interesting site with valuable information on both PACs (political action committees) and lobbyist groups. For example, they maintain an online lobbyist report called "Influence, Inc." This provides background data on an issue, analysis of industry spending, and a relevant database. They also profile every political action committee registered in the Federal Election Commission.

INFOTRAC COLLEGE EDITION EXERCISES

Visit the **InfoTrac College Edition** web site at: **http://www.wadsworthmedia.com/webtutor/infotrac.htm**. You will arrive at a screen that enables you to search topics.

1. Search for articles and news stories about **voter apathy**. The United Sates has a low voting turnout. Investigate research on this phenomenon.

2. Opponents of the **death penalty** contend that it is cruel and unusual punishment, and that our legal system makes mistakes and wrongfully executes innocent people. Research this issue to find out how often this happens, and what states are doing about it.

3. Hundreds of **interest groups** seek to influence the outcome of elections and specific legislation. Investigate the range of interest groups, or focus on one group in particular.

4. **Campaign reform** is an important current issue, as the amount of money candidates can raise dictates the quality of their campaign, or even if they will be able to run at all. Look at news stories about this issue. See if you can collect information to present a pro and con class discussion.

5. There are an enormous number of articles under the topic **terrorism** on InfoTrac. Be prepared to make an oral presentation, create posters, or write a brief report.

SOLUTIONS

MULTIPLE CHOICE QUESTIONS

1. B, p. 446
2. C, p. 447
3. C, p. 449
4. D, p. 449
5. A, p. 450
6. B, p. 447
7. C, p. 451
8. A, p. 451
9. C, p. 453
10. D, p. 454
11. A, p. 454
12. C, p. 456
13. C, p. 456
14. D, p. 458

15. A, p. 460
16. D, p. 463
17. B, p. 465
18. C, p. 467
19. A, p. 469
20. C, p. 469
21. A, p. 454
22. C, p. 458
23. A, p. 449
24. B, p. 453
25. D, p. 450
26. C, p. 459
27. D, p. 450
28. A, p. 451

29. B, p. 453
30. A, p. 454
31. C, p. 454
32. A, p. 459
33. B, p. 460
34. A, p. 461
35. C, p. 461
36. B, p. 465
37. A, p. 470
38. C, p. 468
39. B, p. 471
40. A, p. 471

TRUE/FALSE QUESTIONS

1. T, p. 464
2. T, p. 448
3. T, p. 450
4. T, p. 451
5. F, p. 454
6. T, p. 484
7. F, p. 456

8. T, p. 458
9. F, p. 459
10. F, p. 463
11. T, p. 463
12. F, p. 464
13. T, p. 465
14. F, p. 468

15. F, p. 470
16. T, p. 469
17. F, p. 448
18. T, p. 452
19. F, p. 455
20. F, p. 473

FILL-IN-THE-BLANK QUESTIONS

1. state, p. 454
2. monarchy, p. 459; authoritarian systems, p. 461; totalitarian systems, p. 461; democratic systems, p. 463
3. Power, p. 455
4. rational-legal authority, p. 458
5. power elite, p. 465
6. military-industrial complex, p. 475
7. revolutionary, p. 478; state-sponsored, p. 478; repressive, p. 479
8. Democracy, p.
9. G. William Domhoff, p. 466
10. 1995, p. 479; 2001, p. 479
11. Max Weber, p. 457
12. city-state, p. 459
13. lobbyists, p. 464
14. iron triangle of power, p.
15. Emile Durkheim, p.463

15

FAMILIES AND INTIMATE RELATIONSHIPS

BRIEF CHAPTER OUTLINE

FAMILIES IN GLOBAL PERSPECTIVE
    Family Structure and Characteristics
    Marriage Patterns
    Patterns of Descent and Inheritance
    Power and Authority in Families
    Residential Patterns
THEORETICAL PERSPECTIVES ON FAMILIES
    Functionalist Perspectives
    Conflict and Feminist Perspectives
    Symbolic Interactionist Perspectives
    Postmodernist Perspectives
DEVELOPING INTIMATE RELATIONSHIPS AND ESTABLISHING FAMILIES
    Love and Intimacy
    Cohabitation and Domestic Partnerships
    Marriage
    Housework and Child-care Responsibilities
CHILD-RELATED FAMILY ISSUES AND PARENTING
    Deciding to Have Children
    Adoption
    Teenage Pregnancies
    Single-parent Households
    Two-parent Households
TRANSITIONS AND PROBLEMS IN FAMILIES
    Family Violence
    Children in Foster Care
    Divorce
    Remarriage
DIVERSITY IN FAMILIES
    Diversity among Singles
    African American Families
    Latina/o Families
    Asian American Families
    Native American Families
    Biracial Families
FAMILY ISSUES IN THE FUTURE

CHAPTER SUMMARY

Families are relationships in which people live together with commitment, form an economic unit, care for any young, and consider their identity to be significantly attached to the group. While the **family of orientation** is the family into which a person is born and in which early socialization usually takes place, the **family of procreation** is the family a person forms by having or adopting children. Sociologists investigate marriage patterns (such as **monogamy** and **polygamy**), descent and inheritance patterns (such as **patrilineal**, **matrilineal**, and **bilateral descent**), familial power and authority (such as **patriarchal**, **matriarchal**, and **egalitarian families**), residential patterns (such as **patrilocal**, **matrilocal**, and **neolocal residence**), and in-group or out-group marriage patterns (i.e. **endogamy** and **exogamy**). Functionalists emphasize that families fulfill important societal functions, including sexual regulation, socialization of children, economic and psychological support, and the provision of social status. By contrast, conflict and feminist perspectives view the family as a source of social inequality and focus primarily on the problems inherent in relationships of dominance and subordination. Symbolic interactionists focus on family communication patterns and subjective meanings that members assign to everyday events. Postmodern perspectives view the family as *permeable*—one that is subject to modification or change. The **nuclear family**, therefore, is one of many family forms. Families have changed dramatically in the United States, where there have been significant increases in **cohabitation**, **domestic partnerships**, **dual-earner marriages**, single-parent families, and rates of divorce and remarriage. Housework and child-care responsibilities remain key issues for the family, as many women who work outside the home work a **second shift** at home. Violence in the family is more common than violence among other groups of people. Problems in the family contribute to the large numbers of children who are in *foster care*, where adults other than a child's own parents or biological relatives serve as caregivers. Divorce has contributed to greater diversity in family relationships, including stepfamilies or *blended families*. Most people who divorce get remarried. While some never-married singles choose to remain single, others do so out of necessity. Support systems and extended family networks are important in African American, Latino/a, Asian American, and Native American families; however, factors such as age and class may reduce such family ties. Biracial families have greatly increased; today, skin color and place of birth are not reliable indicators of a person's identity and origin. In the future, people's perceptions of family will continue to change, as will the major issues facing the family.

LEARNING OBJECTIVES

After reading Chapter 15, you should be able to:

1. Define the social institution of the family.

2. Explain why it has become increasingly difficult to develop a concise definition of family.

3. Describe kinship ties and distinguish between families of orientation and families of procreation.

4. Compare and contrast extended and nuclear families.

5. Describe the different forms of marriage found across cultures.

6. Discuss the system of descent and inheritance, and explain why such systems are important in societies.

7. Distinguish between patriarchal, matriarchal, and egalitarian families.

8. Explain the differences in residential patterns and note why most people practice endogamy.

9. Describe functionalist, conflict and feminist, symbolic interactionist, and postmodernist perspectives on families.

10. Describe cohabitation and domestic partnerships, and note key social and legal issues associated with each.

11. Discuss issues related to housework and parenting, including problems faced by people in dual-earner marriages and the second shift.

12. Discuss the major issues associated with adoption, teenage pregnancies, single-parent households, and two-parent households.

13. Discuss the issue of domestic violence and its impact on families.

14. Explain the major causes and consequences of divorce and remarriage in the United States.

15. Describe the diversity found in contemporary U.S. families.

16. Describe the changing definition of the term "interracial marriage" and family.

KEY TERMS

(defined at page number shown and in glossary)

bilateral descent, p. 482
cohabitation, p. 484
domestic partnerships, p. 490
dual-earner marriages, p. 492
egalitarian family, p. 483
endogamy, p. 484
exogamy, p. 484
extended family, p. 480
family, p. 478
family of orientation, p. 479
family of procreation, p. 479
homogamy, p. 484
kinship, p.478
marriage, p. 481

matriarchal family, p. 483
matrilineal descent, p. 482
matrilocal residence, p. 484
monogamy, p. 481
neolocal residence, p. 484
nuclear family, p. 481
patriarchal family, p. 482
patrilineal descent, p. 482
patrilocal residence, p. 483
polyandry, p. 481
polygamy, p. 481
polygyny, p. 481
second shift, p. 492
sociology of family , p. 484

KEY PEOPLE

(identified at page number shown)

Jean Baudrillard, p. 487
Peter Berger and Hansfried Kellner,
p. 486
Jessie Bernard, p. 486
Emile Durkheim , p. 484

Arlie Hochschild, p. 492
Talcott Parsons, p. 484
Lillian Rubin, p. 507

CHAPTER OUTLINE

I. FAMILIES IN GLOBAL PERSPECTIVE
    A. **Families** are relationships in which people live together with commitment, form an economic unit, care for any young, and consider their identity significantly attached to the group.
    B. Family Structure and Characteristics
        1. In preindustrial societies, the primary social organization is through **kinship**—a social network of people based on common ancestry, marriage, or adoption.
        2. In industrialized societies, other social institutions fulfill some functions previously fulfilled by kinship ties; families are responsible primarily for regulating sexual activity, socializing children, and providing affection and companionship for family members.

3. Many of us will be members of two types of families: a **family of orientation**—the family into which we are born or adopted and in which early socialization takes place, and a **family of procreation**—the family we form by having or adopting children.
4. Extended and nuclear families
   a. An **extended family** is composed of relatives (e.g., grandparents, uncles, and aunts) in addition to parents and children who live in the same household.
   b. A **nuclear family** is composed of one or two parents and their dependent children, all of whom live apart from other relatives.

C. Marriage Patterns
1. **Marriage** is a legally recognized and/or socially approved arrangement between two or more individuals that carries certain rights and obligations and usually involves sexual activity.
2. In the U.S., **monogamy**—a marriage between two partners, usually a woman and a man—is the only form of marriage sanctioned by law.
3. **Polygamy** is the concurrent marriage of a person of one sex with two or more members of the opposite sex.
   a. The most prevalent form of polygamy is **polygyny**—the concurrent marriage of one man with two or more women.
   b. **Polyandry**—the marriage of one woman with two or more men—is very rare.

D. Patterns of Descent and Inheritance
1. In preindustrial societies, the most common pattern is **patrilineal descent**—tracing descent through the father's side of the family—whereby a son inherits his father's property and sometimes his position upon the father's death.
2. **Matrilineal descent** traces descent through the mother's side of the family; however, inheritance of property and position usually is traced from the maternal uncle (mother's brother) to his nephew (mother's son).
3. In industrial societies such as the U.S., **bilateral descent**—a system of tracing descent through both the mother's and father's sides of the family—is more common.

E. Power and Authority in Families:
1. **Patriarchal family**: a family structure in which authority is held by the eldest male (usually the father), who acts as head of household and holds power over the women and children.
2. **Matriarchal family**: a family structure in which authority is held by the eldest female (usually the mother), who acts as head of household.
3. **Egalitarian family**: a family structure in which both partners share power and authority equally.

F. Residential Patterns:
1. **Patrilocal residence**: the custom of a married couple living in the same household (or community) with the husband's family.
2. **Matrilocal residence**: the custom of a married couple living in the same household (or community) with the wife's parents.

288

3. In industrialized nations, most couples hope to live in a **neolocal residence**: the custom of a married couple living in their own residence apart from both the husband's and the wife's parents.

G. Endogamy and Exogamy:
   1. Most people in the U.S. practice **endogamy**—marrying within one's own group.
   2. **Exogamy** is the practice of marrying outside one's social group or category. The family, church, and state are the most important sources of positive or negative sanction for practicing exogamy.
   3. Most people engage in **homogamy**—the pattern of individuals marrying those who have similar characteristics, such as race/ethnicity, religious background, age, education, or social class.

II. THEORETICAL PERSPECTIVES ON FAMILIES

A. The **sociology of family** is the subdiscipline of sociology that attempts to describe and explain patterns of family life and variations in family structure.

B. Functionalist Perspectives
   1. The family is important in maintaining the stability of society and the well being of individuals.
   2. According to **Emile Durkheim**, both marriage and society involve a mental and moral fusion of individuals; division of labor contributes to greater efficiency in all areas of life.
   3. **Talcott Parsons** further defined the division of labor in families: the husband/father fulfills the *instrumental role* (meeting the family's economic needs, making important decisions, and providing leadership), while the wife/mother fulfills the *expressive role* (doing housework, caring for children, and meeting the emotional needs of family members).
   4. Four key functions of families in advanced industrial societies are:
      a. Sexual regulation
      b. Socialization
      c. Economic and psychological support for members
      d. Provision of social status and reputation

C. Conflict and Feminist Perspectives
   1. Families are a source of conflict over values, goals, and access to resources and power.
   2. According to some conflict theorists, families in capitalist economies are similar to workers in a factory: women are dominated at home by men the same way workers are dominated by capitalists in factories; reproduction of children and care for family members at home reinforce the subordination of women through unpaid (and devalued) labor.
   3. Some feminist perspectives focus on patriarchy rather than class because men's domination over women existed long before private ownership of property; contemporary subordination is rooted in men's control over women's labor power.

D. Symbolic Interactionist Perspectives
   1. Interactionists examine the roles of husbands, wives, and children as they act out their own parts and react to the actions of others.

2. According to **Peter Berger** and **Hansfried Kellner**, interaction between marital partners contributes to a shared reality: newlyweds bring separate identities to a marriage but gradually construct a shared reality as a couple.
3. According to **Jessie Bernard**, women and men experience marriage differently: there is "his" marriage and "her" marriage.

E. Postmodern Perspectives
   1. In the information age, the postmodern family is *permeable*—capable of being diffused or invaded in such a way that an entity's original purpose is modified or changed.
      a. The nuclear family is only one of many family forms.
      b. Maternal love is transformed into shared parenting.
      c. The individual values autonomy of the individual more than the family unit.
   2. Urbanity is another characteristic of the postmodern family.
      a. The boundaries between the workplace and the home become more open and flexible.
      b. New communications technologies integrate and control labor.
   3. Some paint a bleak future for families in the age of the "integrated circuit."
      a. There is a growing "digital divide" and a type of "cyber class warfare."
      b. Many families are left out of Internet access.

III. DEVELOPING INTIMATE RELATIONSHIPS AND ESTABLISHING FAMILIES
   A. Love and Intimacy
      1. Although the ideal culture emphasizes romantic love, men and women may not share the same perceptions about love: women tend to express their feelings verbally, while men tend to express their love through nonverbal actions.
      2. Scholars suggest that love and intimacy are closely intertwined.
         a. Perspectives about sexual activities vary according to culture and time period.
   B. Cohabitation and Domestic Partnerships
      1. **Cohabitation** refers to a couple who lives together without being legally married.
      2. Characteristics of persons most likely to cohabit are as follows: under age 45, have been married before, or are older individuals who do not want to lose financial benefits (such as retirement benefits) that are contingent upon not marrying.
      3. Many lesbian and gay couples cohabit because they cannot enter into a legally recognized marriage; some have sought recognition of **domestic partnerships**—household partnerships in which an unmarried couple lives together in a committed, sexually intimate relationship and is granted some of the same benefits as those accorded to married heterosexual couples.
   C. Marriage
      1. Couples marry for reasons such as being "in love," desiring companionship and sex, wanting to have children, feeling social pressure, attempting to escape from their parents' home, or believing they will have greater resources if they get married.

290

2. Communication and support are crucial to the success of marriages; problems that cause the most concern are lack of emotional intimacy, poor communication, and lack of companionship.

D. Housework and Child-care Responsibilities
1. Over 50 percent of all U.S. marriages are **dual-earner marriages**—marriages in which both spouses are in the labor force. Over half of all employed women hold full-time, year-round jobs.
2. Many married women also have a **second shift**—the domestic work that employed women perform at home after they complete their workday on the job; this amounts to an extra month of work for women each year.
3. In families where the wife's earnings are essential to family finances, more husbands have attempted to share some of the household and child-care responsibilities.
   a. Women and men perform different household tasks.
   b. Recurring tasks tend to be women's responsibilities, while men are more likely to do periodic tasks.

IV. CHILD-RELATED FAMILY ISSUES AND PARENTING
A. Deciding to have children
1. Couples deciding not to have children may consider themselves "child-free," while those who do not produce children through no choice of their own may consider themselves "childless."
   a. Advances in birth control techniques now make it possible to determine choice of parenthood, number of children, and the spacing of their births.
   b. Some couples experience *involuntary infertility*; a leading cause is sexually transmitted diseases.
   c. Women who are involuntarily childless engage in "information management" to combat the social stigma associated with childlessness.

B. Adoption
1. *Adoption* is a legal process through which rights and duties of parenting are transferred from a child's biological and/or legal parents to new legal parents.
2. About 6.4 million women become pregnant each year in the United States; of that number, about 44 percent of pregnancies are intended, while 56 percent are unintended.

C. The U.S. has the highest teen pregnancy rate in the Western industrialized world. Reasons include:
1. Lack of information about and use of contraceptives
2. Beliefs among males that females should be responsible for contraception and that pregnancy is a sign of male status
3. Lack of educational and employment opportunities to deter teen pregnancy
4. Religious and political opposition to addressing reproduction openly
5. Media portrayals of sexuality without consequences

D. Recently, single- or one-parent households have increased significantly due to divorce and births outside of marriage.

E. Two-parent households
1. The percent of children living in two-parent households has declined in recent decades.
2. Parenthood in the United States is idealized, especially for women.

291

V. TRANSITIONS IN FAMILIES AND PROBLEMS IN FAMILIES
  A. Family Violence
    1. *Spouse abuse* refers to any intentional act or series of acts—whether physical, emotional, or sexual—that causes injury to a female or male spouse. For sociologists, this includes cohabiting couples and separated or former spouses.
    2. Women are about five times more likely to be victims of domestic violence than men.
    3. Children are especially affected by household violence.
  B. Children in Foster Care
    1. *Foster care* refers to institutional settings or residences where adults other than a child's parents or biological relatives are caregivers.
    2. Many foster children have been in violent or dysfunctional homes.
  C. Divorce
    1. *Divorce* is the legal process of dissolving a marriage that allows former spouses to remarry if they so choose.
    2. Most divorces are based on *irreconcilable differences* (there has been a breakdown of the marital relationship for which neither partner is specifically blamed).
    3. Under *no-fault divorce laws*, proof of the blame is generally no longer necessary.
    4. Studies show that 43 percent of first marriages end in separation or divorce within fifteen years; one in three first marriages ends within ten years, where one in five ends within five years.
    5. Causes of Divorce
       a. At the macrolevel, societal factors contributing to higher rates of divorce include changes in social institutions such as religion and family.
       b. At the microlevel, characteristics that appear to contribute to divorce are:
          i. Marriage at an early age
          ii. A short acquaintanceship before marriage
          iii. Disapproval of the marriage by relatives and friends
          iv. Limited economic resources
          v. Having a high-school education or less
          vi. Parents who are divorced or have unhappy marriages
          vii. The presence of children (depending on gender and age) at the beginning of marriage
    6. Consequences of Divorce
       a. An estimated 60 percent of divorcing couples have one or more children.
       b. By age 16, about one in every three white and two in every three African American children will experience divorce within their families. Some children experience more than one divorce during their childhood because one or both of their parents may remarry and subsequently divorce again.
  D. Remarriage
    1. Most people who divorce get remarried: more than 40 percent of all marriages take place between previously married brides and/or grooms, and about half of all persons who divorce before age 35 will remarry within three years.

292

2. As a result of divorce and remarriage, some people become part of *blended families*, which consist of a husband and wife, children from previous marriages, and children (if any) from the new marriage.

VI. DIVERSITY IN FAMILIES

A. Diversity among Singles
1. Some never-married singles choose to remain single because of opportunities for a career (especially for women), the availability of sexual partners without marriage, a belief that the single lifestyle is full of excitement, and a desire to be self-sufficient and have the freedom to change and experiment.
2. Other never-married singles remain single out of necessity; they cannot afford to marry and set up their own household.
3. Among persons age 15 and over, over 40.8 percent of African Americans have never married, as compared with almost 30.1 percent of Latino/as, 26.8 percent of Asian and Pacific Islander Americans, and 22.6 percent of whites.
4. Lower marriage rates among African American women result from
   a. more African American women than men in the population and higher education among women.
   b. limited opportunities for African American men, which lead to military or criminal activity.
   c. higher rates of homosexuality among African American men than women.
   d. More African American men marrying women of other racial-ethnic groups.
   e. African American daughters being encouraged to choose education over marriage.

B. African American Families
1. A higher proportion of African Americans than whites live in extended family households that may provide emotional and financial support not otherwise available.
2. Among middle and upper-middle class African American families, nuclear families are more prevalent than extended family ties.

C. Latino/a Families
1. Family support systems found in many Latino/a families—*la familia*—cover a wide array of relatives, including parents, aunts, uncles, cousins, brothers and sisters, and their children.
2. Some sociologists question the extent that *familialism* exists across social classes. Norma Williams found extended family networks are disappearing, especially among advantaged urban Latino/as.

D. Asian American Families
1. While many Asian Americans live in nuclear families, others (especially those residing in Chinatowns) have extended family networks. Some are referred to as semi-extended families because other relatives live in close proximity but not necessarily in the same household.
2. Extended family networks of some Vietnamese Americans are limited because family members died in the war in Vietnam, and others did not migrate to the United States.

E. Native American Families
   1. Family ties remain strong among many Native Americans, who have always been known for these strong ties.
   2. Extended family patterns are common among lower-income Native Americans living on reservations; most others live in nuclear families.
F. Biracial Families
   1. Since the 1970s, there has been a 300 percent increase in marriages between people of different races; when these couples produce offspring, their children are considered to be biracial.
   2. Interracial marriage was illegal in sixteen states until 1967.
   3. At the beginning of the twenty-first century, skin color and place of birth have ceased to be reliable indicators of a person's identity or origin; the term *interracial marriage* has taken on a much broader interpretation.
VII. FAMILY ISSUES IN THE FUTURE
   A. Some people believe that the family as we know it is doomed; others believe that a return to traditional family values will save this social institution and create greater stability in society.
   B. Sociologist Lillian Rubin suggests that clinging to a traditional image of families is hypocritical in light of our society's failure to support them. Some laws have the effect of hurting children whose families do not meet the traditional model. For example, cutting down on government programs that provide food and medical care for pregnant women and infants will result in seriously ill children rather than model families.
   C. People's perceptions about what constitutes a family will continue to change in the next century: the family may become those persons on whom one can depend for emotional support, who are available in crises and emergencies, or who provide continuing affection, concern, and companionship.

CRITICAL THINKING QUESTIONS

1. Why have definitions of the family changed over time? What do you think would be the best definition?
2. What is the relationship between patterns of descent, inheritance, and power and authority in families, and the broader social structure and institutions?
3. Which theoretical perspective do you think is most useful for understanding families in contemporary U.S. society? What are the strengths and weaknesses of each?
4. What do you think are the top three most significant issues in families today? What do you think should be done to address these issues?
5. In what ways do you think families will change in the next century?

PRACTICE TESTS

MULTIPLE CHOICE QUESTIONS

Select the response that best answers the question or completes the statement.

1. According to the text, traditional definitions of the family
   a. are still highly applicable to today's families.
   b. include all persons in a relationship who wish to consider themselves a family.
   c. need to be expanded to provide a more encompassing perspective on what constitutes a family.
   d. are indistinguishable from contemporary definitions of the family.

2. A social network of people based on common ancestry, marriage, or adoption is known as:
   a. kinship
   b. a family
   c. a clan
   d. subculture

3. The family of procreation is defined as the family
   a. into which a person is born.
   b. composed of one or two parents and their dependent children, all of whom live apart from other relatives.
   c. that is composed of relatives in addition to parents and children who live in the same household.
   d. a person forms by having or adopting children.

4. Families that include grandparents, uncles, aunts, or other relatives who live in close proximity to the parents and children are known as a(n):
   a. clan
   b. extended family
   c. nuclear family
   d. family of procreation

5. _____ is the concurrent marriage of one man with two or more women, while _____ is the concurrent marriage of one woman with two or more men.
   a. Monogamy; polygamy
   b. Polygamy; monogamy
   c. Polygyny; polyandry
   d. Polyandry; polygyny

6. The most prevalent pattern of power and authority in families is:
   a. matriarchy
   b. monarchy
   c. oligarchy
   d. patriarchy

7. All of the following statements are correct regarding power and authority in families, **except**:
   a. Recently, there has been a trend toward more egalitarian family relationships in a number of countries.
   b. Some degree of economic independency makes it possible for women to delay marriage.
   c. The egalitarian family places a heavier burden on women around the globe than does patriarchal.
   d. Scholars have found no historical evidence to indicate that true matriarchies ever existed.

8. The custom of a married couple living in their own residence apart from both the husband's and the wife's parents is known as:
   a. isolated residence
   b. neolocal residence
   c. neutral local residence
   d. exogamous residence

9. When a person marries someone who comes from the same social class, racial-ethnic group, and religious affiliation, sociologists refer to this marital pattern as:
   a. endogamy
   b. exogamy
   c. inbreeding
   d. intraclass reproduction

10. Sociologist Jessie Bernard's research showed that
    a. interaction between marital partners contributes to a shared reality.
    b. husbands and wives often experience marriage differently.
    c. gender socialization contributes to the roles of battered and batterer in domestic violence situations.
    d. the postmodern family is permeable.

11. According to functionalists, family problems
    a. are related to changes in social institutions such as religion, economy, and government.
    b. are related to cyberspace, consumerism, and hyperreality.
    c. reflect social patterns of dominance and subordination.
    d. depend on individuals' interpretations of family interactions.

12. According to _____ theorists, interaction between marital partners contributes to a shared reality.
    a. feminist
    b. conflict
    c. functionalist
    d. symbolic interactionist

13. Sociologist _____ argued that husbands/fathers fulfill the instrumental roles in families, whereas wives/mothers fulfill the expressive role.
    a. Emile Durkheim
    b. Jessie Bernard
    c. Jean Baudrillard
    d. Talcott Parsons

14. At what point in history did people in the United States come to view home and work as separate spheres?
    a. during colonial times
    b. during the Industrial Revolution
    c. at the end of World War II
    d. in the 1980s, when more middle-class white women entered the paid workforce

15. Which of the following statements regarding cohabitation is true?
    a. Attitudes about cohabitation have not changed very much in the past two decades.
    b. The Bureau of the Census recently developed a more inclusive definition of cohabitation.
    c. People most likely to cohabit are those who are under age 30 and who have not been married before.
    d. In the United States, many lesbian and gay couples cohabit because they cannot enter into a legally recognized marital relationship.

16. Sociologist _____ coined the term "second shift" to refer to the domestic work that employed women perform at home after they complete their workday on the job.
    a. Arlie Hochschild
    b. Talcott Parsons
    c. Kath Weston
    d. Francesca Cancian

17. _____ argue that family life may be negatively affected by the decreasing distinction between work and family time.
    a. Structural functionalists
    b. Symbolic interactionists
    c. Postmodernists
    d. Conflict theorists

18. Each of the following are primary reasons for high rates of teenage pregnancy in the U.S., **except**:
    a. lack of contraceptive use among teenagers
    b. lack of educational and employment opportunities in some inner-city and rural areas
    c. recent increases in sexual activity among teenagers
    d. lack of accurate information regarding contraceptives

19. All of the following are cited in the text as primary social characteristics of those most likely to get divorced, **except**:
    a. marriage at a later age and being set in one's ways
    b. a short acquaintanceship before marriage
    c. disapproval of marriage by relatives and friends
    d. parents who are divorced or have unhappy marriages

20. In discussing diversity in families, the text points out that
    a. a higher percentage of whites live in extended family households than do African Americans.
    b. there is no such thing as "the" African American, Latino/a, Asian American, or Native American family.
    c. working-class African American women often are encouraged by their families to choose marriage over education.
    d. extended family patterns are no longer common among lower-income Native Americans living on reservations.

21. Sociologist Lillian Rubin suggests that clinging to a traditional image of families
    a. is needed in today's society.
    b. is hypocritical in light of our society's failure to support families.
    c. maintains our faith in family values.
    d. would lower the rate of divorce.

22. Many Asian Americans live in nuclear families; however, others (especially those residing in Chinatowns) have extended family networks. In a(n) _____ family, other relatives live in close proximity but not necessarily in the same household.
    a. extended
    b. partially extended
    c. non-extended
    d. semi-extended

23. Today, extended family patterns are common among lower-income Native Americans who
    a. live near the reservation.
    b. live on the reservation.
    c. live in large, urban areas.
    d. live in small, rural areas.

24. _____ argue that women are dominated by men in the home the same way that workers are dominated by capitalists in factories.
    a. Symbolic interactionists
    b. Conflict theorists
    c. Postmodernists
    d. Structural functionalists

298

25. Interracial marriage—which is often thought of as marriage between whites and blacks—was illegal in sixteen states until the U.S. Supreme Court overturned _____ in 1967.
    a. miscegenation laws
    b. segregation laws
    c. racial rules
    d. endogamy laws

26. A comparison of Census Bureau data from 1970 to 2000 shows that there has been _____ in the percentage of U.S. households comprising a married couple with their own children under 18 years of age.
    a. a significant decline
    b. a significant increase
    c. no change
    d. a doubling

27. The societal assumption that having children is the norm and those who choose not to have children believe they must justify their decision to others is referred to as:
    a. compulsory childhood
    b. pronatal bias
    c. nuclear family
    d. procreative assumption

28. _____ is a legally recognized and/or socially approved arrangement between two or more individuals that carries certain rights and obligations, and usually involves sexual activity.
    a. Cohabitation
    b. Love
    c. Marriage
    d. Monogamous habitation

29. In the United States, the only legally sanctioned form of marriage is _____, which refers to a marriage between two partners, usually a woman and a man.
    a. polyandry
    b. homogamy
    c. polygamy
    d. monogamy

30. The _____ refers to family units that are composed of relatives in addition to parents and children who live in the same household.
    a. extended family
    b. nuclear family
    c. conventional family
    d. blended family

31. In horticultural and agricultural societies, _____ are extremely important; having a large number of family members participate in food production may be essential for survival.
    a. families of procreation
    b. extended families
    c. blended families
    d. nuclear families

32. With the advent of industrialization and urbanization, maintaining the _____ family pattern becomes more difficult in societies. Increasingly, young people move from rural to urban areas in search of employment in the industrializing sector of the economy. At that time, the _____ family typically becomes the predominant family pattern in the society.
    a. nuclear; extended
    b. extended; blended
    c. extended; nuclear
    d. nuclear; blended

33. _____ refers to family units composed of one or two parents and their dependent children, all of whom live apart from other relatives.
    a. Married family
    b. Parent-child family
    c. Conventional family
    d. Nuclear family

34. According to the sociologist Norma Williams, extended family networks among Latino/as have been
    a. disappearing.
    b. growing stronger.
    c. strongest in urban areas.
    d. strongest in large cities.

35. In regard to housework, sociologist Arlie Hochschild notes that
    a. domestic work is shared equally by both husbands and wives.
    b. women and men perform different household tasks, but they spend equal amounts of time on these activities.
    c. even when husbands share some of the household responsibilities, they typically spend much less time on these activities than do their wives.
    d. husbands now cook as often as their wives.

36. Couples with more _____ ideas about women's and men's roles tend to share more equally in food preparation, housework, and child care.
    a. patriarchal
    b. matriarchal
    c. equal-partner
    d. egalitarian

300

37. In 2000, Latinas (Hispanic women) had a total fertility rate of 2.6, which was 50 percent above that of white (non-Hispanic) women. Among Latina women, the highest fertility rate is found among _____ women.
    a. Puerto Rican
    b. Cuban American
    c. Mexican American
    d. Latin American

38. _____ families consist of a husband and wife, children from previous marriages, and children (if any) from the new marriage.
    a. Blended
    b. Nuclear
    c. Extended
    d. Kinship

39. All of the following are primary factors that contribute to lower marriage rates of African American women, **except**:
    a. high mortality rates among young African American men
    b. discriminatory practices that limit economic opportunities of African American men
    c. more African American men marrying members of other racial/ethnic groups
    d. higher rates of homosexuality among African American women than other groups

40. Some single fathers who do not have custody of their children remain actively involved in their children's lives. Others may become _____, who take their children to recreational activities and buy them presents for special occasions but have a very small part in the children's day-to-day lives.
    a. phantom fathers
    b. Disneyland daddies
    c. big day dads
    d. just for fun fathers

TRUE/FALSE QUESTIONS

1. Sociologists currently define family as a group of people related by blood, marriage, or adoption who live together, form an economic unit, and bear and raise children.
   T   F

2. In industrialized societies, other social institutions fulfill some of the functions previously taken care of by the kinship network.
   T   F

3. In the United States, the only legally sanctioned form of marriage is monogamy.
   T   F

4. Today, there are no known societies that practice polyandry.
   T   F

5. In industrial societies, kinship is usually traced through patrilineal descent.
   T   F

6. Functionalists suggest that the erosion of family values may occur when religion becomes less important in everyday life.
   T   F

7. According to Parsons, the husband/father fulfills the expressive role in the family.
   T   F

8. According to some feminist scholars, hostility and violence against women and children may be attributed to patriarchal attitudes, economic hardship, and rigid gender roles in society.
   T   F

9. Postmodernist perspectives describe the contemporary family as permeable.
   T   F

10. A domestic partnership is a household partnership in which an unmarried couple lives together in a committed, sexually intimate relationship that is granted some of the same rights and benefits as those accorded to married heterosexual couples.
    T   F

11. Over 50 percent of all marriages in the United States are dual-earner marriages.
    T   F

12. The U.S. has the highest rate of teenage pregnancy in the Western industrialized world.
    T   F

13. Women who are childless are often thought of as "selfish."
    T   F

14. An estimated 60 percent of divorcing couples have one or more children.
    T   F

15. Polyandry is more common worldwide than polygyny.
    T   F

16. According to Box 15.1, in today's society, people in the U.S. are more inclined to get married than at any other time in history.
    T   F

17. The number of children living with grandparents and with no parent in the home increased by more than 50 percent between 1990 and 2000, according to Box 15.1
    T   F

18. According to Box 15.2, most couples who hire Indian surrogates are too lazy or selfish to carry out a pregnancy on their own.
    T   F

19. According to Box 15.3, in 1996, Congress passed the Defense of Marriage Act, which provides that every U.S. state is required to recognize a same-sex marriage under the laws of another state.
    T   F

20. Since its commencement, according to Box 15.4, Hope Meadows has been largely successful in helping children get adopted.
    T   F

FILL-IN-THE-BLANK QUESTIONS

1. The family of _____ is the one into which a person is born.

2. The most prevalent form of polygamy is _____.

3. A system of tracing descent through the mother's side is known as _____ descent.

4. The practice of marrying those with similar characteristics is known as _____ and is somewhat similar to the practice of _____.

5. _____ is a legally recognized arrangement between two or more individuals that carries certain rights and obligations, and usually involves sexual activity.

6. _____ refers to two people who live together and think of themselves as a couple without being married.

7. Household partnerships wherein an unmarried couple lives together in a committed, sexually intimate relationship and is granted some of the same benefits as those accorded to married heterosexual couples are defined as _____.

8. According to sociologist Arlie Hoschild, the _____ is the domestic work that employed women perform at home after they complete their work day on the job.

9. _____ theorists focus on family dynamics, including communication patterns and subjective meanings.

10. In _____ families, both partners share power and authority equally.

11. The U.S. culture emphasizes _____ love, which refers to a deep emotion, the satisfaction of significant needs, and a caring for and acceptance of the person we love.

12. The sociologist _____ pointed out that women and men experience marriage differently, with a "her" and "his" marriage.

13. Violence between men and women in the home is often called domestic violence or _____.

303

14. In the U.S. today, most divorces are granted on the grounds of _____.

15. Under _____ laws, proof of spouse blame is generally no longer necessary.

SHORT ANSWER/ESSAY QUESTIONS

1. How do sociologists define family? Explain why this is the preferred definition among sociologists.
2. What are the functionalist, symbolic interactionist, conflict, feminist, and postmodernist perspectives on families?
3. What are some of the problems of the family in the United States today? What would you recommend as solutions to these problems?
4. How could the family be described as both a "haven in a heartless world" and a "cradle of violence?"
5. What are some of the causes and consequences of divorce?

STUDENT CLASS PROJECTS AND ACTIVITIES

1. Select and then research a topic of unconventional reproduction, such as in vitro fertilization, (also known as test-tube fertilization), artificial insemination, or surrogacy. After selecting the topic, (1) trace the history of the development of that specific type of reproduction; (2) provide a description and definition of the reproduction procedure; (3) describe the widespread use of the procedure; (4) discuss any complications, moral, legal, or medical, that have surfaced related to the practice; and (5) examine and report any court cases that relate to your specific topic. In the writing of your paper, ensure that you provide a conclusion and a bibliography.
2. Write a paper on the American family in the century ahead. The paper should center on the topic: "The American Family in the Twenty-First Century." Issues to be included are the following: (1) What type of family form will be dominant? (2) As families change, will the type of community change as well? (3) What is the status of commuter marriage? (4) Describe dating practices. Will the dating game be a "dangerous sport" because of the health risk? (5) What about life expectancy? Has the life span increased? What impact does this have on families? (6) Are there new categories of families, such as gay and lesbian married couples with and without children, single women having babies by donor insemination, etc.? (7) What type of child care is available? (8) What is the size of the family? Include any other information that pertains to this topic.
3. Construct a "sociological" family tree of your relatives of at least three generations (your grandparents, parents, aunts, uncles, and your own generation). Record names, dates, and information as you gather this information. Examine your genealogy for some of the following norms of marriage: (1) Who married, who did not? (2) Who did your relatives marry? (3) At what age were they married? (4) Did the norms of homogamy operate? (5) Did they marry "up" or "down"?; (6) How many children did they have? (7) How many divorces were there? (8) What religion were they?. You may need to interview several family members to gather the information.

After recording the information, examine how closely your family's marital choices follow the patterns of the typical American explained in the text and why variations might have occurred in your family tree.

INTERNET ACTIVITIES

1. This **family educational page**, http://www.familyeducation.com/home/, contains gateways to a number of sites with information relating to studies of the family, with emphasis on children. Explore these sites and write up a critique or summary about the information each contains.
2. Go to this site on the **economic and legal benefits of marriage**: http://www.religioustolerance.org/mar_bene.htm. Search the site and present a debate on the pros and cons of cohabitation.
3. Here, **http://www.ndvh.org/**, you will find resources and information related to **domestic violence**. Read about current issues and news stories related to domestic violence.
4. Use these sites to produce a report on the **state of children** both in the United States and globally. Describe the ways that the welfare of children is directly connected to the state of families: **http://www.unicef.org/**, **http://www.futureofchildren.org/**, and **http://www.researchforum.org/**.

INFOTRAC COLLEGE EDITION EXERCISES

Visit the **InfoTrac College Edition** web site at: **http://www.wadsworthmedia.com/webtutor/infotrac.htm**. You will arrive at a screen that enables you to search topics.

1. Investigate **domestic partnerships** (legal relationships between gay/lesbian couples that grant them some of the same rights as married couples) from a global perspective, and from a state by state pattern in the United States. Which country has the most restrictive policy on these partnerships? Which has the most nonrestrictive policy?
2. Search the category **polygamy**. Compare research about cross-cultural marriage practices. Apply the functionalist perspective to these kinds of practices.
3. Divide up the subdivisions on **teenage pregnancy** among class members. Construct a collective report on this phenomenon using a contribution from each class member.
4. Search for the keywords **dual-career families**. Many students in the class will probably fall into this category. Search through these reports and compose a list of good advice that they might put to use in their own future.
5. The number of reported incidents of **child abuse** in this country is very high. Analyze the incidence and types of child abuse, as well as risk factors.

SOLUTIONS

MULTIPLE CHOICE QUESTIONS

1. C, p. 478	15. D, p. 489	29. D, p. 481
2. A, p. 478	16. A, p. 492	30. A, p. 480
3. D, p. 479	17. C, p. 486	31. B, p. 480
4. B, p. 479	18. C, p. 495	32. C, p. 480
5. C, p. 481	19. A, p. 501	33. D, p. 481
6. D, p. 482	20. B, p. 504	34. A, p. 505
7. C, p. 483	21. B, p. 507	35. C, p. 492
8. B, p. 484	22. D, p. 505	36. D, p. 483
9. A, p. 484	23. B, p. 505	37. C, p. 493
10. B, p. 486	24. B, p. 485	38. A, p. 502
11. A, p. 484	25. A, p. 506	39. D, pp. 503-504
12. D, p. 486	26. A, p. 481	40. B, p. 498
13. D, p. 484	27. B, p. 493	
14. B, p. 488	28. C, p. 481	

TRUE/FALSE QUESTIONS

1. F, p. 478	8. T, p. 485	15. F, p. 481
2. T, p. 478	9. T, p. 486	16. F, p. 480
3. T, p. 481	10. T, p. 490	17. T, p. 480
4. F, p. 481	11. T, p. 492	18. F, p. 491
5. F, p. 482	12. T, p. 495	19. F, p. 496
6. T, p. 484	13. T, p. 493	20. T, p. 501
7. F, p. 484	14. T, p. 502	

FILL-IN-THE-BLANK QUESTIONS

1. orientation, p. 479
2. polygyny, p. 481
3. matrilineal, p. 482
4. homogamy, p. 484; endogamy, p. 484
5. Marriage, p. 481
6. Cohabitation, p. 484
7. domestic partnerships, p. 490
8. sociology of family, p. 484
9. Symbolic interactionist, p. 486
10. egalitarian, p. 483
11. romantic love, p. 488
12. Jessie Bernard, p. 486
13. spouse abuse, p. 486
14. irreconcilable differences, p. 499
15. no-fault divorce, p. 499

16

EDUCATION

BRIEF CHAPTER OUTLINE

AN OVERVIEW OF EDUCATION
EDUCATION IN HISTORICAL-GLOBAL PERSPECTIVE
    Informal Education in Preliterate Societies
    Formal Education in Preindustrial, Industrial, and Postindustrial Societies.
    Contemporary Education in Other Nations
SOCIOLOGICAL PERSPECTIVES ON EDUCATION
    Functionalist Perspectives
    Conflict Perspectives
    Symbolic Interactionist Perspectives
    Postmodernist Perspectives
INEQUALITY AMONG ELEMENTARY AND SECONDARY SCHOOLS
    Inequality in Public vs. Private Schools
    Unequal Funding of Public Schools
    Inequality within Public School Systems
    Racial Segregation and Resegregation of Schools
PROBLEMS WITHIN ELEMENTARY AND SECONDARY SCHOOLS
    School Discipline and Teaching Styles
    Bullying and Sexual Harassment
    Dropping Out
SCHOOL SAFETY AND SCHOOL VIOLENCE
OPPORTUNITIES AND CHALLENGES IN COLLEGES AND UNIVERSITIES
    Opportunities and Challenges in Community Colleges
    Opportunities and Challenges in Four-Year Colleges and Universities
    The Soaring Cost of a College Education
    Racial and Ethnic Differences in Enrollment
    The Lack of Faculty Diversity
    The Continuing Debate over Affirmative Action
FUTURE ISSUES AND TRENDS IN EDUCATION
    Academic Standards and Functional Illiteracy
    The Debate over Bilingual Education
    Equalizing Opportunities for Students with Disabilities
    School Vouchers
    Charter Schools and "For-Profit" Schools
    Home Schooling
    Education in the Future

CHAPTER SUMMARY

Education is the social institution responsible for the systematic transmission of knowledge, skills, and cultural values within a formally organized structure. Education is responsible for **cultural transmission**, or the process by which children and recent immigrants become acquainted with the dominant cultural beliefs, values, norms, and accumulated knowledge of a society. People in preliterate societies acquire knowledge and skills through **informal education**; in preindustrial and industrial societies, people acquire specific knowledge, skills, and thinking processes through **formal education**. **Mass education** refers to providing free public schooling for wide segments of a nation's population. Japan and Germany both have systems that place children into different academic tracks in middle school that have significant consequences for their future education and job opportunities. Functionalists suggest that education performs a number of essential functions for society. Conflict theorists emphasize that education perpetuates class, racial-ethnic, and gender inequalities; students from diverse backgrounds enter school with different amounts of **cultural capital**, or assets such as values, beliefs, attitudes and competencies in language and culture, which provide them unequal opportunities in schools. Unequal opportunities occur through **tracking**—the practice of assigning students to specific curriculum groups and courses on the basis of their test scores, previous grades, or other criteria—and the **hidden curriculum**—the transmission of cultural values and attitudes, such as conformity and obedience to authority, through implied demands found in the rules, routines, and regulations of schools. Over time, students whose values and attitudes do not conform to those in the school system may find themselves disqualified from gaining educational credentials in a society that emphasizes **credentialism**—a process of social selection in which class advantage and social status are linked to academic qualifications. Symbolic interactionists point out that education may create a *self-fulfilling prophecy* for students who perform up—or down—to the expectations held for them by teachers. Postmodernists highlight the permeability of both lower and higher education today, with urbanity, autonomy, and consumption as major characteristics. The U. S. public schools today are a microcosm of many of the problems facing the country: inequality in public vs. private schools, unequal funding of public schools, inequality within public school systems, and racial and ethnic segregation and resegregation. Problems within elementary and secondary schools include school discipline and teaching styles, bullying and sexual harassment, and students dropping out. School safety and school violence are major concerns in education. Opportunities and challenges in community colleges and in four-year colleges and universities include the soaring cost of a college education, racial and ethnic differences in enrollment, lack of faculty diversity, and the debate over affirmative action. Some analysts believe academic standards are not high enough in U.S. schools; suggestions for reducing illiteracy and **functional illiteracy** focus on school reforms. Other concerns include *bilingual education* and providing better educational opportunities for students with disabilities. Some controversial solutions include school vouchers, which give parents the choice of what school their child will attend, charter schools, which are public schools that operate under charter contracts, and home schooling. In 2001, Congress passed the No Child Left Behind Act, intended to challenge the state's public school systems to meet certain education goals. In the U.S., education in the future must accommodate the increasing diversity of the population at all levels of education.

LEARNING OBJECTIVES

After reading Chapter 16, you should be able to:

1. Trace the history of education in preliterate, preindustrial, and industrial societies to the present.

2. Compare and contrast contemporary education in other nations, using Japan and Germany as the models.

3. Explain the functionalist perspective on education and describe the manifest and latent functions fulfilled by the institution of education.

4. Describe conflict perspectives on education and discuss tracking and the hidden curriculum.

5. Differentiate symbolic interactionist perspectives on education from other paradigms.

6. Describe the significance of the self-fulfilling prophecy and labeling on educational achievement.

7. Discuss the postmodernist perspective on education and the significance of the "students as consumers" model.

8. Discuss inequalities among elementary and secondary schools in the realms of differences between public and private schools, unequal funding of public schools, inequalities within public school systems, and racial segregation of schools.

9. List and discuss some of the major problems in U. S. elementary and secondary education, including school discipline and teaching styles, bullying and sexual harassment, and dropping out.

10. Describe opportunities and challenges in colleges and universities, including rising costs of education and lack of diversity among students and faculty.

11. Trace the history of the debate over affirmative action in higher education, citing the two U. S. Supreme Court decisions and the influence of those decisions.

12. Distinguish between illiteracy and functional illiteracy and discuss these problems as they impact U.S. adults.

13. Discuss the issues of bilingual education and educating students with disabilities in historical context, and summarize current controversies.

14. Describe alternative forms of education, including school vouchers, charter schools, "for profit" schools, and home schooling, and discuss the strengths and weaknesses of each.

15. Use trends in education to formulate ideas about the future of this institution.

KEY TERMS

(defined at page number shown and in glossary)

KEY PEOPLE

(identified at page number shown)

CHAPTER OUTLINE

I. AN OVERVIEW OF EDUCATION:
   A. **Education** is the social institution responsible for the systematic transmission of knowledge, skills, and cultural values within a formally organized structure.
      1. Education is a socializing institution that imparts values, beliefs, and knowledge considered essential to the social reproduction of individual personalities and entire cultures.
      2. The *sociology of education* is devoted to the study of education as an important social institution.
II. EDUCATION IN HISTORICAL-GLOBAL PERSPECTIVE
   A. **Cultural transmission**—the process by which children and recent immigrants become acquainted with the dominant cultural beliefs, values, norms, and accumulated knowledge of a society—occurs through formal and informal education.
   B. Informal Education in Preliterate Societies
      1. *Preliterate societies* existed before the invention of reading and writing.
      2. Education in preliterate societies consisted of **informal education**, which is learning that occurs in a spontaneous, unplanned way.
   C. Formal Education in Preindustrial, Industrial, and Postindustrial Societies
      1. **Formal education**—learning that takes place within an academic setting such as a school, which has a planned instructional process and teachers who convey specific knowledge, skills, and thinking process to students—is found in preindustrial and industrial societies.

310

2. As societies industrialize, the need for formal education for the masses increases. **Mass education** refers to providing free public schooling for wide segments of a nation's population.
3. **Horace Mann** started the free public school movement in the U.S., and stated that education should be the "great equalizer."

D. Contemporary Education in Other Nations
1. In Japan, when the country began to industrialize during the Meiji Period (1868-1912), public education became mandatory for children.
   a. Education was linked with economic and national development.
   b. Today, many toddlers are sent to cram schools (*jukus*) to ensure enrollment in good preschools.
   c. Middle school academic achievement determines the high school; to enter colleges and universities, students must score well on a variety of college entrance exams prepared by each college or university.
   d. Fewer educational opportunities are available for women, resulting in an absence of female students and professors at the college and university level.
2. In Germany, belief in compulsory education can be traced back to the sixteenth century.
   a. All children attend elementary school for 4-6 years, where children are not tracked into ability groups and often have the same class and teacher for 4 years.
   b. The type of middle and high school a student attends is based on admissions tests, grade point average, teacher recommendations, and parents' wishes.
   c. Universities in Germany are free, but students must pay for books or housing; students may apply for financial aid.

III. SOCIOLOGICAL PERSPECTIVES ON EDUCATION
A. Functionalists view education as one of the most important components of society.
1. **Emile Durkheim** believed education was important for creating social solidarity and stability and advocated *moral education*, which he believed was the foundation of a cohesive social order.
2. Education serves five major **manifest functions**—open, stated, and intended goals or consequences of activities within an organization or institution:
   a. Socialization
   b. Transmission of culture
   c. Social control
   d. Social placement
   e. Change and innovation.
3. Education has at least three **latent functions**—hidden, unstated, and sometimes unintended consequences of activities within an organization or institution:
   a. Restricting some activities
   b. Matchmaking and production of social networks
   c. Creation of a generation gap.

311

B. According to conflict theorists, schools perpetuate class, race-ethnic, and gender inequalities as some groups seek to maintain their privileged position at the expense of others.
1. Cultural Capital and Class Reproduction: education is a vehicle for reproducing existing class relationships.
a. According to **Pierre Bourdieu**, children have less chance of academic success when they lack **cultural capital**—social assets that include values, beliefs, attitudes, and competencies in language and culture.
b. Children from middle- and upper-income families are endowed with more cultural capital than children from working-class and poverty-level families.
2. Tracking and Social Inequality
a. Class reproduction also occurs through standardized tests, ability grouping, and **tracking**—the assignment of students to specific courses and educational programs based on their test scores, previous grades, or other criteria.
3. The Hidden Curriculum
a. The **hidden curriculum** is the transmission of cultural values and attitudes, such as conformity and obedience to authority, through implied demands found in rules, routines, and regulations of schools.
b. Lower-class students may be disqualified from higher education and the credentials needed in a society that emphasizes **credentialism**—a process of social selection in which class advantage and social status are linked to the possession of academic qualification.
c. Credentialism is closely related to *meritocracy*—a social system in which status is assumed to be acquired through individual ability and effort.
4. Gender Bias and the Hidden Curriculum
a. Studies show that gender bias pervades the overall academic environment and cuts across lines of race and class.
b. Although some improvement has been made in girls' education in recent years, girls are still underrepresented in higher-level science and computer science classes.
5. Ethnicity, Language, and Hidden Curriculum
a. When teaching English as a second language, instructors frequently find that they not only teach children English, but social skills as well.
b. The children are exposed to a wide range of beliefs, values, attitudes and behavior expectations that are not directly related to their subject matter.
c. Conflict theorists argue that inequality is structurally produced and reproduced by formal and informal socialization processes in schools and other educational settings.
C. Symbolic Interactionists focus on classroom communication patterns and educational practices.
1. Labeling and the Self-Fulfilling Prophecy
a. For some students, *labeling* in schools may become a *self-fulfilling prophecy*—an unsubstantiated belief or prediction that results in behavior which makes the originally false belief come true.
2. Using Labeling Theory to Examine the IQ Debate

312

a. An experiment showed that teacher beliefs about students' IQs can impact student performance.
b. **Richard J. Herrnstein** and **Charles Murray** argued that people of different races are inherently "smarter" than others based on average IQ scores. Their findings have been disputed by many scholars.
c. IQ testing has resulted in the labeling of students (e.g., African American and Mexican American children have been placed in special education classes on the basis of IQ scores when they could not understand the tests).

D. Postmodernist Perspectives
   1. Postmodernists often highlight differences and inequality in society; therefore education is characterized by its permeability.
   2. Urbanity is reflected in multicultural and anti-bias curriculum introduced in early childhood education.
   3. Autonomy is evidenced in policies such as voucher systems.
   4. In higher education, the permeability is exemplified by "educational consumption," the offering of "high-tech" or "wired" campuses, virtual classrooms, and student centers offering extravagant activities and services.
   5. According to Ritzer, "McUniversity" can be thought of as a means of educational consumption that allows students to consume educational services and eventually obtain "goods" such as degrees and credentials.

IV. INEQUALITY AMONG ELEMENTARY AND SECONDARY SCHOOLS
   A. Inequality in Public vs. Private Schools
      1. Almost 90 percent of U. S. elementary and secondary students are educated in public schools; about 9.5 percent of all students are educated in low-tuition private schools, primarily Catholic; 1.5 percent of all students attend high-tuition private school.
      2. Private schools are perceived as better than public schools but, according to some social analysts, there is little to substantiate this claim.
   B. Unequal Funding of Public Schools
      1. Most educational funds are derived from local property taxes and state legislative appropriations.
      2. Children living in affluent suburbs often attend relatively new schools and have access to the latest equipment, which students in central city schools and poverty-ridden rural areas lack.
   C. Inequality within Public School Systems
      1. Using a case study by John Devine of New York City public schools as an example, research shows that schools are stratified into higher-, middle-, and lower-tiered schools.
      2. Higher-tier schools are highly specialized. Admission is limited based upon competitive entrance examinations; dropout rates are low. A large number of graduates attend prestigious universities.
      3. Middle-tier schools, located in fairly well-integrated neighborhoods, are often referred to as "ed op"(educational option) schools; they offer training for specific careers. Some serve as *magnet schools*, offering a specialized curriculum and enrolling high-achieving students from other neighborhoods.

4. Lower-tier high schools are very large, overcrowded, and located in highly segregated, deteriorating, and violent neighborhoods. Students have the highest dropout rate, the lowest graduate rates, the worst scores on standardized tests, the poorest attendance patterns, and the worst statistics on assaults and possession of weapons.

D. Racial Segregation and Resegregation of Schools
   1. In many areas of the U.S., schools remain segregated or have become resegregated after earlier attempts at integration failed.
      a. Five decades after the 1954 U.S. Supreme Court ruled in *Brown v. The Board of Education of Topeka, Kansas*, that segregated schools were unconstitutional, racial segregation remains.
      b. Racial segregation is increasing in many school districts, and efforts to bring about *desegregation* or *integration* have failed in many school districts.
   2. Racially segregated schools often have low retention rates, students with below-grade level reading skills, high teacher-student ratios, less qualified teachers, and low teacher expectations.
      a. Even in supposedly integrated schools, tracking and ability grouping may produce resegregation at the classroom level.
      b. African American and white achievement differences increase with every year of schooling; thus, schools may reinforce rather than eliminate the disadvantages of race and class.

V. PROBLEMS WITHIN ELEMENTARY AND SECONDARY SCHOOLS
   A. School Discipline and Teaching Styles
      1. Leading discipline problems include violence, drug abuse, suicide, robbery, assault, and other forms of aggressive behavior.
      2. Most states now prohibit the use of corporal punishment, and some school districts prohibit various forms of disciplinary actions.
      3. Jean Anyon suggests that structural inequalities need to be addressed in order to achieve better discipline in schools.
   B. Bullying and Sexual Harassment
      1. Sexual harassment is defined as "unwanted and unwelcome sexual behavior that interferes with your life" and does not include behaviors that are enjoyed or wanted (such as kissing, touching, flirting).
      2. Both boys and girls report that they have experienced harassment, bullying, and teasing from classmates; many believe their academic experience has been diminished by other students, even as early at elementary school.
      3. Thirty percent of students reported moderate or frequent bullying.
   C. Dropping Out
      1. There has been a slight decrease in school dropout rates in the 2000s; however, each year almost one-third of public high school students fail to graduate from high school.
      2. Males are more likely to dropout than females.
      3. Latino/a students have the highest dropout rate, followed by African Americans, then non-Hispanic whites.

4. Romo and Falbo found many students were taking the GED instead of graduating from high school; the most common reasons given for dropping out included: lack of interest in school, serious personal problems, serious family problems, poor grades, and alcohol and/or drug problems.

VI. SCHOOL SAFETY AND SCHOOL VIOLENCE

A. Problems such as bullying, harassment, and dropout rates are a concern because schools should provide a safe and supportive environment so students and teachers can feel secure.
1. Some schools are creating anger management and peer mediation programs, and partnerships with law enforcement agencies and social service organizations.
2. Studies now show that U.S. schools are among the safest places for young people.
3. Many schools use weapons-scanning metal detectors, security guards, and magnetic door locks.

B. Violence and fear of violence continue to be pressing problems in schools throughout the U.S.
1. Numerous school, college, and university students, faculty, and other victims have been subjected to acts of violence and multiple deaths.
2. Many students must learn in an academic environment that is similar to a maximum security prison.
3. Gun control advocates are calling for greater control over the licensing and ownership of firearms, among several other suggestions.

VII. OPPORTUNITIES AND CHALLENGES IN COLLEGES AND UNIVERSITIES

A. Opportunities and Challenges in Community Colleges
1. Community colleges are one of the fastest-growing areas of U.S. higher education today; they are more affordable and offer more flexible class schedules.
2. Community colleges offer significant educational opportunities to students across lines of income, gender, and race/ethnicy.
3. Limited resources are one of the major problems facing community colleges today.

B. Opportunities and Challenges in Four-Year Colleges and Universities
1. They offer degrees ranging from bachelor's to doctorate and professional degrees; most provide a liberal education in a general education curriculum.
2. Challenges include the cost of higher education, problems with students completing a degree program, racial and ethnic differences in enrollment, and lack of faculty diversity.

C. The Soaring Cost of a College Education
1. The cost of attending private and public institutions has increased dramatically over the past decade, more than the overall rate of inflation.
2. This high cost of college reproduces the class system: students who lack money may be denied access to higher education, and those who are able to attend college tend to receive different types of education based on their ability to pay.

315

D. Racial and Ethnic Differences in Enrollment
   1. People of color are underrepresented in higher education; however, some increases in minority enrollment have occurred over the past three decades.
   2. Gender differences are evident: women accounted for more than 65 percent of all African American college students in 2006.
   3. Despite increasing enrollments, students of color continue to experience prejudice and discrimination at predominantly white colleges and universities.
E. Lack of Faculty Diversity
   1. African Americans, Latino/as, Asian Americans, and other people of color account for only about 12 percent of all faculty members.
   2. Minority faculty members are less likely to be full professors with tenure.
   3. Women are underrepresented at the level of full professor and overrepresented at the assistant professor and instructor levels.
F. The Continuing Debate over Affirmative Action
   1. *Affirmative action* refers to policies intended to promote equal opportunity for categories of people deemed to have been previously excluded from equality in education, employment, and other fields.
   2. In 2003, the Supreme Court ruled in *Grutter v. Bollinger* (involving admissions policies of the University of Michigan's law school) and in *Gratz v. Bollinger* (involving the undergraduate admissions policies of the same university) that race can be a factor for universities in shaping their admissions programs, but only within carefully defined limits.
VIII. CONTINUING ISSUES AND FUTURE TRENDS IN EDUCATION
   A. Academic Standards and Functional Illiteracy
      1. Some believe academic standards are not high enough in the U.S., as evidenced by the rate of **functional illiteracy**, which is the inability to read and/or write at the skill level necessary for carrying out everyday tasks.
      2. Estimates are 56 percent of adult Latino/as, 44 percent of adult African Americans, and 16 percent of adult (non-Latino/a) whites are functionally illiterate in English.
      3. Recommendations focus on school reforms, such as more testing of students and teachers, and increasing high school graduation requirements and the number of school days per year.
   B. The Debate over Bilingual Education
      1. The issue of bilingual education is not a new one.
         a. Congress mandated in 1968 that public schools must provide non-English speaking students with *bilingual education*—instruction in both a non-English language and in English.
         b. The U. S. Supreme Court decision, *Lau v. Nichols* (1974) affirmed the responsibility of the states and local districts for providing appropriate education for minority-language students.
         c. Critics of TBE (transitional bilingual education) believe that TBE programs are less effective than classes that are conducted primarily in English.
         d. Supporters of TBE believe that these programs are effective in helping children build competency both in their native language and in English.

316

C. Equalizing Opportunities for Students with Disabilities
1. *Disability* is regarded as any physical and/or mental condition that limits students' access to, or full involvement, in school life.
2. The Americans with Disabilities Act of 1990 requires schools and government agencies to make their facilities, services, activities, and programs accessible to people with disabilities.
3. The Individuals with Disabilities Education Act (IDEA) mandates that students with disabilities receive free and appropriate education.
4. Many schools have attempted to *mainstream* children with disabilities by *inclusion programs* under which the special education curriculum is integrated with the regular education program, and with *individualized education plans* that provide annual education goals for each child
D. School Vouchers
1. School vouchers are public funds provided to families to pay for their choice of school, which can be public or private.
2. Parents who receive vouchers are generally "very satisfied" with their children's education; however, critics believe they undermine the public school system by using public funds for private education.
E. Charter Schools and "For-Profit" Schools
1. Charter schools are public schools that provide more autonomy for students and teachers and are less bureaucratic than regular public schools.
2. Critics argue charter schools take money from conventional public schools; proponents argue charter schools provide many minority students a higher quality education than regular public schools.
3. "Contracting out," or hiring for-profit companies to operate public schools, is a more recent approach. Critics believe privatizing education undermines the public school system.
F. Home Schooling
1. Some parents choose to home school to avoid the problems of public school while providing a quality education for their children.
2. Primary reasons for home schooling include concern about the school environment, religious or moral instruction, and dissatisfaction with academic instruction available.
3. Proponents believe that their children are receiving a better education at home; critics question the knowledge and competence of the typical parent to educate their children at home.
IX. EDUCATION IN THE FUTURE
1. The No Child Left Behind Act of 2001 is intended to close the achievement gap between rich and poor students by holding schools accountable for demonstrating student learning.
2. Ultimately, the institution of education must be maintained and enhanced in various ways. Spending larger sums of money on education does not guarantee that many of the problems of education will be resolved.

CRITICAL THINKING QUESTIONS

1. What is the purpose of education in U.S. society? How effective is the U.S. educational system?
2. Which theoretical perspective do you think is most useful for analyzing the U.S. educational system? What are the strengths and weaknesses of each perspective?
3. In what ways have your race, class, and gender shaped your educational experiences? How do you think your educational experiences would be different if you were of a different race, class, and/or gender?
4. What are some ways to solve issues of inequality among elementary and secondary schools?
5. What are some ways to solve issues facing colleges and universities today?

6. In what ways do you think education will change in the U.S. in the future?

PRACTICE TESTS

MULTIPLE CHOICE QUESTIONS

Select the response that best answers the question or completes the statement:

1. _____ is the social institution responsible for the systematic transmission of knowledge, skills, and cultural values within a formally organized structure.
   a. Religion
   b. Mass media
   c. The government
   d. Education

2. _____ refers to providing free public schooling for wide segments of a nation's population.
   a. Mass education
   b. Public education
   c. Contemporary education
   d. Federal education

3. Learning that takes place in an academic setting that has a planned instructional process and teachers who convey knowledge and skills to students is called:
   a. informal education
   b. preliterate education
   c. inclusive education
   d. formal education

4. In Japan, students' futures are based on the academic track they are placed in during:
    a. preschool
    b. elementary school
    c. middle school
    d. high school

5. The restriction of activities, matchmaking and producing social networks, and creating a generation gap are some _____ of education.
    a. latent functions
    b. manifest functions
    c. dysfunctions
    d. self-fulfilling prophecies

6. According to sociologist _____, education is crucial for promoting solidarity and stability in society.
    a. Emile Durkheim
    b. Pierre Bourdieu
    c. George Ritzer
    d. Max Weber

7. Which of the following is a manifest function of education?
    a. creation of a generation gap
    b. restricting some activities
    c. matchmaking and production of social networks
    d. social control

8. _____ functions are hidden, unstated, and sometimes unintended consequences of activities within an organization or institution.
    a. Manifest
    b. Dormant
    c. Latent
    d. Covert

9. Sociologist _____ has suggested that students come to school with differing amounts of cultural capital.
    a. Emile Durkheim
    b. Pierre Bourdieu
    c. George Ritzer
    d. Max Weber

10. The assignment of students to specific courses and educational programs based on their test scores, previous grades, or both is known as:
    a. tracking
    b. the hidden curriculum
    c. equitable assessment
    d. class reproduction

319

11. According to the conflict perspective, the _____ includes the unarticulated and unacknowledged things students are taught in school, which tend to enhance the self-esteem of members of dominant groups and have negative effects on the self-esteem of subordinate groups.
    a. labeling effect
    b. dysfunction of education
    c. tracking
    d. hidden curriculum

12. The _____ perspective focuses on classroom communication patterns and educational practices such as labeling that affect students' self-concept and aspirations.
    a. conflict
    b. functionalist
    c. symbolic interactionist
    d. credentialist

13. _____ argue that schools perpetuate class, racial-ethnic, and gender inequalities as some groups seek to maintain privilege at the expense of others.
    a. Structural functionalists
    b. Symbolic interactionists
    c. Conflict theorists
    d. Postmodernists

14. About _____ of U.S. elementary and secondary students are educated in public schools.
    a. 54 percent
    b. 65 percent
    c. 80 percent
    d. 90 percent

15. In an experiment, researchers told teachers that a few randomly selected students had extremely high IQ scores. Teachers treated these students differently and, in return, the students performed better than their peers. This experiment demonstrates:
    a. the labeling effect
    b. hidden curriculum
    c. cultural capital
    d. tracking

16. In public schools in the 1940s, a leading discipline problem was:
    a. violence
    b. suicide
    c. talking without permission during class
    d. drug abuse

17. The National Literacy Act of 1991 defines _____ as an individual's ability to read, write, and speak English, and compute and solve problems at levels of proficiency necessary to function on the job and in society.
    a. functional illiteracy
    b. minimum literacy
    c. literacy
    d. functional literacy

18. In _____, the Supreme Court affirmed the responsibility of the states and local districts for providing appropriate education for minority-language students.
    a. *Brown v. Board of Education*
    b. *Lau v. Nichols*
    c. *Hopwood v. State of Texas*
    d. *Bakke v. The University of California at Davis*

19. In their controversial book *The Bell Curve*, Herrnstein and Murray argued that _____ groups differ in average IQ and are likely to differ in "intelligence genes" as well.
    a. gender
    b. racial-ethnic
    c. age
    d. subcultural

20. The _____ perspective emphasizes the irregularity and permeability of education.
    a. structural functionalist
    b. conflict
    c. symbolic interactionist
    d. postmodernist

21. _____ is the process by which children and recent immigrants become acquainted with the dominant cultural beliefs, values, norms, and accumulated knowledge of a society.
    a. Cultural transmission
    b. Education
    c. Socialization
    d. Cultural diffusion

22. _____ societies existed before the invention of reading and writing. These societies have no written language and are characterized by very basic technology and a simple division of labor.
    a. Modern
    b. Preindustrial
    c. Preliterate
    d. Industrial

23. _____ is learning that occurs in a spontaneous, unplanned way.
    a. Spontaneous education
    b. Informal education
    c. Traditional education
    d. Latent education

24. The _____ perspective emphasizes educational consumption, in which students consume educational services and obtain "goods" such as degrees and credentials.
    a. functionalist
    b. interactionist
    c. conflict
    d. postmodernist

25. In the _____ court case, the Supreme Court ruled that segregated schools are unconstitutional because they are inherently unequal.
    a. *Lau v. Nichols*
    b. *Brown v. Board of Education*
    c. *Grutter v. Bollinger*
    d. *Hopwood v. Texas*

26. Many schools have attempted to _____ children with disabilities by providing inclusion programs under which the special education curriculum is integrated with the regular education program.
    a. mainstream
    b. include
    c. individualize
    d. exclude

27. About half of all undergraduates in the U.S. are educated in:
    a. private universities
    b. public universities
    c. community colleges
    d. bilingual programs

28. Children with disabilities may interact with their regular education teacher, the special education teacher, a speech therapist, an occupational therapist, a physical therapist, and a resource teacher, depending on the child's individual needs. This is a(n) example of:
    a. mainstreaming
    b. individualized education
    c. inclusion
    d. exclusion

29. Regarding race in higher education,
    a. minority students have disproportionately high enrollment in colleges and universities due to scholarship money available to them.
    b. minority faculty members now outnumber white faculty members at most colleges and universities.
    c. minority students experience prejudice and discrimination at predominantly white colleges and universities.
    d. Native Americans have the highest proportion of student enrollment of all minority groups.

322

30. A program in which public money is used to pay the tuition of a public or private school of the family's choice is called:
    a. school vouchers
    b. magnet schools
    c. charter schools
    d. home schooling

31. In 2003, the U.S. Supreme Court ruled in _____ and _____ that race can be a factor for universities in shaping their admission programs within carefully defined limits.
    a. *Brown v. Board of Education; Lau v. Nicholas*
    b. *Hopwood v. State of Texas; Lau v. Nichols*
    c. *Bukke v. U.C. Board; Hopwood v. Texas*
    d. *Grutter v. Bollinger; Gratz v. Bollinger*

32. Along with other provisions, the _____ requires schools and government agencies to make their facilities, services, activities, and programs accessible to people with disabilities.
    a. Individuals with Disabilities Education Act (IDEA)
    b. Americans with Disabilities Act (ADA) of 1990
    c. Handicapped Americans Recovery Program (HARP)
    d. Road to Rehabilitation Reform Law (RRR-L)

33. The law that specifically covers the treatment of children with disabilities is the _____, which mandates that students with disabilities must receive free and appropriate education.
    a. Americans with Disabilities Act (ADA)
    b. Individuals with Disabilities Education Act (IDEA)
    c. Americans with Physical Handicaps Act (APHA)
    d. Individuals with Special Needs Act (ISNA)

34. In the country of _____, national achievement tests are administered by the government, and students must pass them in order to advance to the next level of education.
    a. Japan
    b. the United States
    c. Bosnia
    d. England

35. The No Child Left Behind Act of 2001 primarily focuses on
    a. holding schools accountable for students' learning.
    b. desegregating public schools.
    c. equalizing the funding public schools receive.
    d. creating more charter and magnet schools.

36. Among parents whose children attend private secondary schools, an important factor for a majority (51 percent) is the emphasis on:
    a. religion
    b. academics
    c. money
    d. ethical standards

37. According to the text's discussion of unequal funding as a source of inequality in education,
    a. of the small proportion of school funding that comes from the federal government, most funds are earmarked for special programs that specifically target disadvantaged students or students with disabilities.
    b. most educational funds are derived from state and federal income taxes.
    c. the property tax base for central city schools has continued to grow in most regions.
    d. recent redistribution of funds has made many schools' resources more equitable that in the past.

38. In the United States, the free public school movement was started in 1848 by _____, who stated that education should be the "great equalizer."
    a. Noah Webster
    b. Horace Mann
    c. Thomas Jefferson
    d. Theodore Roosevelt

39. With the rapid growth of capitalism and factories during the _____, it became necessary for workers to have basic skills in reading, writing, and arithmetic.
    a. Industrial Revolution
    b. Enlightenment
    c. Renaissance
    d. Middle Ages

40. During the _____, the focus of education shifted to the importance of developing well-rounded and liberally educated people.
    a. Industrial Revolution
    b. Enlightenment
    c. Renaissance
    d. Middle Ages

TRUE/FALSE QUESTIONS

1. Only girls report that they experience harassment at school.
   T   F

2. Each year, almost one-third of public high school students fail to graduate from high school.
   T   F

324

3. Public schools are among the most dangerous places in society today.
   T   F

4. Manifest functions in education include teaching specific subjects, such as science, history, and reading.
   T   F

5. Meritocracy is the process of social selection in which class advantage and social status are linked to the possession of academic qualifications.
   T   F

6. Labeling students as "gifted" can result in a self-fulfilling prophecy.
   T   F

7. Many students who drop out of high school believe a high school diploma will not increase their job opportunities.
   T   F

8. Minority students in racially segregated schools typically perform better academically than minority students in integrated schools.
   T   F

9. Today, most schools in the U.S. are racially integrated.
   T   F

10. The total cost of attending public colleges and universities has increased dramatically over the past decade.
    T   F

11. Most public educational funds come from the federal government.
    T   F

12. Increases in available scholarship funds have facilitated an increase in the number of low-income students attending college since the 1980s.
    T   F

13. Most states now prohibit the use of corporal punishment on students in their public schools.
    T   F

14. The *Lau v. Nichols* case affirmed the responsibility of states and local districts for providing bilingual education to non-English speaking students.
    T   F

15. A critique of the No Child Left Behind Act is that it does not address the unequal distribution of money and resources to public schools.
    T   F

16. Public colleges and universities have received increased public funding in recent years.
    T   F

17. According to Box 16.1, in the U.S., core classes such as science and math were always taught in English before the 1960s civil rights movement.
T   F

18. College students tutoring students in local public schools appears to be, according to Box 16.3, a lose-lose situation.
T   F

19. The original idea of school vouchers, according to Box 16.4, was to prevent a student from entering a private school.
T   F

20. According to Box 16.4, all states in the U.S. have now accepted school voucher programs.
T   F

FILL-IN-THE-BLANK QUESTIONS

1. _____ refers to providing free public schooling for wide segments of a nation's population.

2. Intended goals or consequences of education, such as socialization, transmission of culture, social control, social placement, and change and innovation, are _____.

3. _____ is the assignment of students to specific courses and educational programs based on their test scores, previous grades, or both.

4. _____ is a process of social selection in which class advantage and social status are linked to the possession of academic qualifications.

5. _____ is defined as unwanted and unwelcome sexual behavior that interferes with one's life.

6. _____ offer specialized curriculum that focuses on areas such as science, music, or art, and enroll high-achieving students from other neighborhoods.

7. In the United States, the free public school movement was started in 1848 by _____.

8. Policies and procedures that are intended to promote equal opportunities for categories of people previously excluded from equality in education and employment are referred to as _____.

9. The inability to read and/or write at the skill level necessary for carrying out everyday tasks is referred to as _____.

10. In 2001, President George W. Bush passed the _____ Act, representing the most far-reaching federal education reform since compulsory education laws were passed in the early twentieth century.

11. Instruction for non-English speaking students in both a non-English language and in English is called _____.

12. The term _____ refers to the transmission of cultural values and attitudes, such as conformity and obedience to authority, through implied demands found in the rules, routines, and regulations of a school.

13. About one-half of the nation's undergraduates are educated in _____ colleges.

14. A(n) _____ prophecy may result from the labeling of students by teachers and administrators.

15. Many schools have attempted to _____ children with disabilities by providing inclusion programs with the regular education program.

SHORT ANSWER/ESSAY QUESTIONS

1. Is education a vehicle for decreasing social inequality or does education reproduce existing class relationships? Explain your answer.
2. Compare and contrast the U.S. public schools with public schools in Japan and Germany. Discuss the strengths and weaknesses of each educational model.
3. Discuss four major problems in public schools and/or higher education today. Describe and evaluate proposed solutions to these problems, or propose your own solutions.
4. What do you think are the most significant changes in education over the past twenty years? What changes would you predict for the future?
5. Based on the postmodernist perspectives on education, what educational services and goods will students in both early childhood and higher education consume in the future? What will be the major attractions in the future educational setting?

STUDENT CLASS PROJECTS AND ACTIVITIES

1. Research the topic of inequalities between public schools and school funding. Search for articles that discuss the conditions of public schools in inner-city and rural areas and compare them to the issues discussed regarding surburban schools. Write a paper on the issue that includes the following information: (1) a discussion of the major issues facing inner-city schools; (2) a discussion of the major issues facing rural schools; (3) a discussion of the issues facing suburban schools; (4) a critical comparison of the issues facing the three types of schools; (5) your own analysis and discussion of inequalities in schools; and (6) a proposal for solutions regarding inequalities in public schools.
2. Some solutions that have been implemented to resolve issues in public schools include school vouchers, charter schools, magnet schools, and "for-profit" schools. Investigate these options further. Begin by searching for examples in your own community. If none are available in your community, try to find the nearest community that has some of these options. Write a paper that includes the following: (1) the availability of alternative forms of public education; (2) a description of each of the programs available; (3) a description of the problem(s) each program is

designed to combat, and how it addresses them; (4) a comparison of each of the available programs; (5) a summary of what you see as the strengths and weaknesses of each; and (6) any alternative solutions you can think of or ways to improve the current solutions available.

3. Research the topic of public education in several high-income countries, such as France, Japan, New Zealand, the Netherlands, Sweden, England, Norway, Canada, or Germany. In the writing up of your paper, include the following information: (1) a description of their educational system; (2) a description of their specific or unique feature; (3) a description of their curriculum; (4) a description of some of the strengths and weaknesses of their program; (5) a critique of their programs (Which do you think work well? Which do not?); (6) any other information that will enhance our understanding of their education; and (7) a bibliography of your references.

4. The U.S. Congress passed the No Child Left Behind Act in 2001. Research this act further, and its implementation. Write a paper that includes the following information: (1) a list of the major provisions of the act and an explanation of each; (2) a summary of the ways five states (of your choice) have responded to and/or implemented the act; (3) a summary of the impact of NCLB on each of the states you chose; and (4) a critical reaction to the legislation, including your opinion of its effectiveness in meeting its goals.

INTERNET ACTIVITIES

1. The **National Education Association** (NEA), **http://www.nea.org/**, is currently the largest teachers' organization in the United States. Its purpose is to advance the interests of the teaching profession, push for teaching reform, and press for quality education in the United States. Visit their site to see their core concerns.

2. Read about the NEA's proposals for improving public education and their response to the No Child Left Behind Act at:
http://www.nea.org/home/NoChildLeftBehindAct.html.

3. The **American Federation of Teachers**, **http://www.aft.org/index.html**, (affiliated with the labor union the AFL-CIO) is one of the largest teachers' unions in the United States. Like the NEA, the AFT tries to represent teacher interests and further the quality of education. The AFT sponsors research and reports on education. Investigate the current concerns of this teachers' group.

4. The **National Center for Education Statistics** (NCES) is the primary federal entity that collects and analyzes educational data for the United States and other countries. This site gives you fast facts on topics like the transition from high school to college, various statistics on educational systems, earnings of college graduates, and much more. Search this site: **http://nces.ed.gov/index.asp**.

5. Visit the following sites to examine available information on **home schooling**. These sites provide information about the home school movement and resources for homeschoolers: **http://www.home-school.com**, **http://www.homeschool.com**, and **http://www.hslda.org/**.

INFOTRAC COLLEGE EDITION EXERCISES

Visit the **InfoTrac College Edition** web site at: **http://www.wadsworthmedia.com/webtutor/infotrac.htm**. You will arrive at a screen that enables you to search topics.

1. More and more states are turning to **standardized tests** to assess proficiency as a measure of school accountability and to determine whether students should pass to the next grade. What are some the costs and benefits of this approach?
2. **Home schooling** has become increasingly popular. What kinds of research questions are sociologists and educators asking? Compose a list of five research questions from articles listed on InfoTrac.
3. In the aftermath of Columbine and other school shootings, parents, students, and teachers are increasingly fearful of **school violence**. Investigate this issue with respect to its incidence and social policy implications.
4. Many states allow parents vouchers that permit them to send their child to a school outside their district. Investigate the **school voucher** system and examine the costs and benefits.
5. **Charter schools** operate in accordance with a "bottom line" that is specified by their charter. In exchange for meeting their educational goals, they are exempt from meeting traditional state rules and restrictions. Research the trend towards charter schools. What makes one school succeed and another fail?
6. There is currently an explosion of **online education** at the college level. Online courses offer convenience to the student, but do not offer the same type of classroom interaction, even if the professor and students interact in a "chat room." Assess this trend and its significance for American education.

SOLUTIONS

MULTIPLE CHOICE QUESTIONS

1. D, p. 512
2. A, p. 514
3. D, p. 513
4. C, p. 515
5. A, pp. 519-520
6. A, p. 518
7. D, p. 519
8. C, p. 519
9. B, p. 522
10. A, p. 522
11. D, p. 524
12. C, p. 526
13. C, p. 522
14. D, p. 529

15. A, p. 527
16. C, p. 534
17. C, p. 541
18. B, p. 542
19. B, p. 527
20. D, p. 528
21. A, p. 512
22. D, p. 513
23. B, p. 513
24. D, p. 528
25. B, p. 532
26. A, p. 543
27. C, p. 538
28. C, p. 543

29. C, p. 539
30. A, p. 543
31. D, p. 541
32. B, p. 543
33. B, p. 543
34. D, p. 529
35. A, p. 545
36. B, p. 529
37. A, p. 530
38. B, p. 514
39. A, p. 514
40. C, p. 514

TRUE/FALSE QUESTIONS

1. F, p. 534
2. T, p. 535
3. F, p. 536
4. T, p. 518
5. F, p. 525
6. T, p. 527
7. T, p. 535

8. F, p. 532
9. F, p. 532
10. T, p. 539
11. F, p. 530
12. F, p. 539
13. T, p. 534
14. T, p. 542

15. T, p. 545
16. F, p. 539
17. F, p. 514
18. F, p. 544
19. F, p. 546
20. F, p. 546

FILL-IN-THE-BLANK QUESTIONS

1. Mass education, p. 514
2. manifest functions, pp. 518-519
3. Tracking, p. 522
4. Credentialism, p. 525
5. Sexual harassment, p. 534
6. Magnet school, p. 531
7. Horace Mann, p. 514
8. affirmative action, p. 541

9. functional illiteracy, p. 541
10. No Child Left Behind, p. 545
11. bilingual education, p. 542
12. hidden curriculum, p. 524
13. community, p. 538
14. self-fulfilling, p. 527
15. mainstream, p. 543

17

RELIGION

BRIEF CHAPTER OUTLINE

THE SOCIOLOGICAL STUDY OF RELIGION
    Religion and the Meaning of Life
    Religion and Scientific Explanations
SOCIOLOGICAL PERSPECTIVES ON RELIGION
    Functionalist Perspectives on Religion
    Conflict Perspectives on Religion
    Symbolic Interactionist Perspectives on Religion
WORLD RELIGIONS
    Hinduism
    Buddhism
    Confucianism
    Judaism
    Islam
    Christianity
TYPES OF RELIGIOUS ORGANIZATION
    Ecclesia
    The Church-Sect Typology
    Cults
TRENDS IN RELIGION IN THE UNITED STATES
    Religion and Social Inequality
    Race and Class in Central-City and Suburban Churches
    Secularization and the Rise of Religious Fundamentalism
RELIGION IN THE FUTURE

CHAPTER SUMMARY

Religion is one of the most significant social institutions in society. It can be a highly controversial topic; one group's deeply held religious practices may be a source of irritation to another. It is a source of both stability and conflict not only in the U.S. but throughout the world. **Religion** is a system of beliefs, symbols, and rituals, based on some sacred or supernatural realm that guides human behavior, gives meaning to life, and unites believers into a moral community. According to Durkheim, **sacred** refers to the extraordinary or supernatural aspects of life, while those things that people do not set apart as sacred are referred to as **profane**. Religions have been classified into four main categories based on their dominant belief, including **simple supernaturalism**, **animism**, **theism** (including **monotheism** and **polytheism**), and **nontheism**. Although religion seeks to answer important questions such as why we exist, why people suffer and die, and what happens when we die, increases in scientific knowledge have contributed to **secularization**—the process by which religious beliefs, practices, and institutions lose their significance in sectors of society and culture. According to functionalists, religion provides meaning and purpose to life, promotes social cohesion and a sense of belonging, and provides social control and support for the government. The conflict perspective argues that religion can be used as a tool of domination and retard social change, or be a catalyst for social change. Symbolic interactionists examine the meanings people give to religion and the meanings they attach to religious symbols in their everyday life. Religions vary around the world. *Hinduism*, one of the world's oldest religions, is complex and diverse, comprising numerous varieties of beliefs and practices. *Buddhism* first emerged in India about 2,500 years ago. It is based on the teachings of Buddha, who perceived that life is suffering and pain; the answer to world's problems is achieved through mediation, which leads to personal transformation and a spiritual existence. *Confucianism*, based upon the teachings of Confucius, was the official religion of China until the Communist takeover and establishment of the People's Republic of China. Order in human relationships and a strict code of moral conduct are some of the basic concepts. Despite centuries of religious hatred and discrimination, *Judaism* persists as one of the world's most influential religions. The three key components of Judaism are: God (the deity), Torah (God's teachings), and Israel (the community or holy nation). *Islam* is based on the teachings of Muhammad, a prophet, not a deity; followers must adhere to the Five Pillars of Islam. Almost one-third of the world's population is *Christian*, the dominant world religion today. The Christian deity is a sacred Trinity comprised of God the Father, Jesus the Son, and the Holy Spirit. Contemporary religious organizations may be categorized as **ecclesia**, **churches**, **denominations**, **sects**, and **cults**. Religion in the United States is diverse; pluralism and religious freedom are among the cultural values most widely espoused. Social science research continues to identify significant patterns between social class and religious denominations. With the increase of secularization, there has been a resurgence of religious **fundamentalism**, as well as debates over religious beliefs and values. New religious forms, such as **liberation theology**, the *Goddess movement*, and *religious nationalism* have recently emerged. Maintaining an appropriate balance between religion and other aspects of social life will be an important challenge for the United States and other nations of the world in the future.

LEARNING OBJECTIVES

After reading Chapter 17, you should be able to:

1. Define and discuss the four main categories of religion, and link them to the types of societies in which they tend to occur.

2. List the six major world religions and state their basic beliefs.

3. Define the concept of religion and indicate how the sociological investigation of religion is different than a theological investigation.

4. Summarize the controversy about teaching intelligent design in the classroom.

5. Describe the ways that a group may move from one type of religious organization to another over time.

6. Discuss the relationship between religion and questions about the meaning of life.

7. Describe the functionalist perspective on religion, including its major functions in societies.

8. Describe symbolic interactionist perspectives on religion and explain how women and men may view religion differently.

9. Discuss the effects of race and class on central-city and suburban churches in the U.S.

10. Explain Durkheim's ideas about religion focusing on the sacred and the profane.

11. Describe conflict perspectives on religion and distinguish between the approaches of Karl Marx and Max Weber.

12. Describe the relationship between religion and social inequality.

13. Distinguish between the different types of religious organization.

14. Compare and contrast civil religion with other forms of religion.

15. Explain the relationship between the growth of secularization and the rise of religious fundamentalism.

16. Describe the future of religion in the U.S. and in other nations of the world.

17. Appraise the major arguments for and against the theory of secularization.

KEY TERMS

(defined at page number shown and in glossary)

animism, p. 556
church, p. 573
civil religion, p. 560
cult, p. 575
denomination, p. 573
ecclesia, p. 572
faith, p. 553
fundamentalism, p. 577
liberation theology, p. 577
monotheism, p. 556

nontheism, p. 556
polytheism, p. 556
profane, p. 554
religion, p. 552
rituals, p. 554
sacred, p. 553
sect, p. 573
secularization, p. 556
simple supernaturalism, p. 556
theism, p. 556

KEY PEOPLE

(identified at page number shown)

CHAPTER OUTLINE

I. THE SOCIOLOGICAL STUDY OF RELIGION
   A. **Religion** is a system of beliefs, symbols, and rituals, based on some sacred or supernatural realm, that guides human behavior, gives meaning to life, and unites believers into a single moral community.
   B. Religion and the Meaning of Life
      1. Religion seeks to answer important questions, such as why we exist, why people suffer and die, and what happens when we die.
         a. According to **Peter Berger**, religion is a *sacred canopy*, a sheltering fabric hanging over people, providing security and answers to questions of life.
         b. This sacred canopy requires **faith**, unquestioning belief that does not require proof or scientific evidence.
      2. Sacred and Profane
         a. According to **Emile Durkheim**, **sacred** refers to those aspects of life that are extraordinary or supernatural; those things that are set apart as "holy."
         b. Those things people do not set apart as sacred are referred to as **profane**—the everyday, secular, or "worldly" aspects of life.
      3. In addition to beliefs, religion also is comprised of symbols and **rituals**—symbolic actions that represent religious meanings and are regularly repeated—that range from songs and prayers to offerings and sacrifices.
      4. Religions have been classified into four main categories based on their dominant belief:
         a. **Simple supernaturalism** is the belief that supernatural forces affect people's lives either positively or negatively.
         b. **Animism** is the belief that plants, animals, or other elements of the natural world are endowed with spirits or life forces having an impact on events in society.
         c. **Theism** is a belief in a god or gods.
            i. **Monotheism** is a belief in a single, supreme being or god who is responsible for significant events, such as the creation of the world. Christianity, Judaism, and Islam are monotheistic.
            ii. **Polytheism** is a belief in more than one god. Examples include Hinduism and Shinto.
         d. **Nontheism** is based on a belief in divine spiritual forces, such as sacred principles of thought and conduct, rather than a god or gods.

C. Religion and Scientific Explanation
   1. During the Industrial Revolution, rapid growth in scientific and technological knowledge gave rise to the idea that science ultimately would answer questions that previously had been in the realm of religion.
   2. Many scholars believed that scientific knowledge would result in **secularization**—the process by which religious beliefs, practices, and institutions lose their significance in sectors of society and culture—but others point out a resurgence of religious beliefs and an unprecedented development of alternative religions in recent years.

II. SOCIOLOGICAL PERSPECTIVES ON RELIGION
   A. Functionalist Perspectives on Religion
      1. According to Emile Durkheim, all religions share in common three elements: (a) beliefs held by adherents; (b) practices (rituals) engaged in collectively by believers; and (c) a moral community that results from the group's shared beliefs and practices pertaining to the sacred.
      2. Religion has three important functions in any society:
         a. Providing meaning and purpose to life
         b. Promoting social cohesion and a sense of belonging
         c. Providing social control and support for the government
      3. The informal relationship between religion and the state is referred to as **civil religion**—the set of beliefs, rituals, and symbols that make sacred the values of the society and place the nation in the context of the ultimate system of meaning.
   B. Conflict Perspectives on Religion
      1. According to **Karl Marx**, the capitalist class uses religious *ideology* as a tool of domination to mislead the workers about their true interests; thus, religion is the "opiate of the masses."
      2. By contrast, **Max Weber** argued that religion could be a catalyst to produce social change.
         a. In *The Protestant Ethic and the Spirit of Capitalism*, Weber linked the teachings of John Calvin with the rise of capitalism.
         b. John Calvin emphasized the doctrine of *predestination*—the belief that all people are divided into two groups, the saved and the damned, and only God knows who will go to heaven (the elect) and who will go to hell, even before they are born.
         c. Because people cannot know whether they will be saved, they look for signs that they are among the elect. As a result, people work hard, save their money, and do not spend it on worldly frivolity; instead, they reinvest it in their land, equipment, and labor.
         d. As people worked ever harder to prove their religious piety, structural conditions in Europe led to the Industrial Revolution, free markets, and the commercialization of the economy, which worked hand in hand with Calvinist religious teachings.
   C. Symbolic Interactionist Perspectives on Religion
      1. For many people, religion serves as a reference group to help them define themselves. Religious symbols, for example, have a meaning for large bodies

335

of people (e.g., the Star of David for Jews, the crescent moon and star for Muslims, and the cross for Christians).

2. *Her* Religion and *His* Religion: Women and men may belong to the same religion, but their individual religion will not necessarily be a carbon copy of the group's entire system of beliefs.

3. Religious symbolism and language typically create a social definition of the roles of men and women.

III. WORLD RELIGIONS

A. Hinduism

1. Hinduism is one of the world's oldest current religions, originating in Pakistan between 3,500 and 4,500 years ago.

2. Hinduism does not have a specific founder or worship of a single god; it is considered by many to be polytheistic.

3. This religion does not have a sacred book inspired by a god or gods and is not based on the teachings of any one person. Some religion scholars refer to it as an *ethical religion*—a system of beliefs that calls upon adherents to follow an ideal way of life, which is achieved by adhering to the duties of one's caste.

4. Individuals pass through cycles of life, death, and rebirth until the soul earns liberation; the soul's acquisition of each new body is tied to the law of *karma*. The ultimate goal is to enter the state of *nirvana* —becoming liberated from the world by uniting the individual soul with the universal soul.

5. Of the more than 700 million Hindus, 95 percent reside in India. Over 80 percent of India's population is Hindu.

B. Buddhism

1. Buddhism first emerged in India about 2,500 years ago, but expanded into other nations in various forms.

2. The founder, Siddhartha Gautama, spent his life teaching others how to obtain enlightenment and how to reach nirvana.

3. The Four Noble Truths and the Eightfold Path to Nirvana are incorporated in some form in the three major branches of Buddhism today.

4. Buddhism is one of the fastest-growing Eastern religions in the United States.

C. Confucianism

1. Confucianism, which means the "family of scholars," existed before its eventual leader, Confucius, was born. He emerged as a teacher about the same time that the Buddha became a significant figure in India.

2. Important teachings are: order in human relationships, a strict code of moral conduct, respect for others, and benevolence and reciprocity. Humans are by nature good and learn from good role models.

3. Confucius established the social hierarchy of Five Constant Relationships: ruler-subject, husband-wife, elder brother-younger brother, elder friend-junior friend, and father-son.

4. Confucianism was the official religion of China until the Communist takeover and the establishment of the People's Republic of China.

D. Judaism
1. About 18 million Jews reside in about 134 countries; the majority reside in the U.S. or in Israel.
2. Jewish belief is monotheism; a single god, called Yahweh, made a covenant with his chosen people following a critical element in Judaism, the Exodus from Egypt in 1200-1300 B.C.E.
3. Believers of Judaism find the source of their beliefs in the Ten Commandments, and in the moral, ceremonial, and cultural laws contained in four books of the Torah: Exodus, Leviticus, Numbers, and Deuteronomy. The Torah, also called the Pentateuch, is the sacred book of contemporary Judaism.
4. Jews worship in synagogues on the Sabbath, which is observed from sunset Friday to sunset Saturday. They celebrate a set of holidays that are distinct from U. S. dominant cultural religious celebrations, and believe that one day the Messiah will come to earth ushering in an age of peace and justice for all.
5. Throughout their history, Jews have been the object of prejudice and discrimination. The Holocaust that took place in Nazi Germany between 1933 and 1945 took the lives of six million Jews.
6. The three branches of Judaism—Orthodox, Reformed, and Conservative—provide for a variety of cultural practices and teachings.
E. Islam
1. Islam, whose followers are known as Muslims, is based on the life and teachings of Muhammad, who was born about 570 C.E. in the city of Mecca, located today in Saudi Arabia.
2. Followers must adhere to the Five Pillars of Islam.
3. Allah is the one god of Islam; the Qur' an is their holy book and was revealed to the Prophet Muhammad through the Angel Gabriel at the command of God.
4. A core belief is the idea of *jihad*, meaning struggle; the Greater Jihad is the internal struggle against sin; the Lesser Jihad is the external struggle that takes place in the world, including violence and war. The term *jihad* is typically associated with religious fundamentalism.
5. Some social analysts believe that Islamic fundamentalism is uniquely linked to the armed struggles of some groups.
6. Because of a recent wave of migration and a relatively high rate of conversion, Islam is one of the fastest-growing religions in the United States.
F. Christianity
1. As with Judaism and Islam, Christianity follows the Abrahamic tradition; although Jews and Christians share some common scriptures in what Christians call the "Old Testament," they interpret them differently.
   a. God offers a new covenant to the followers of Jesus Christ.
   b. The central themes in the teachings of Jesus are the Kingdom of God and standards of personal conduct.
   c. Central teachings of Christianity are: the death and resurrection of Jesus—establishing that he is the son of God—and the ascension of

337

Jesus into heaven forty days after his resurrection, with an eventual "second coming" marking the end of the world as it is now known.

2. The Christian deity is a sacred Trinity, comprised of God the Father, Jesus the Son, and the Holy Spirit.

3. Almost one-third of the world's population refers to themselves as Christian; it is the world's dominant religion.

4. Today, about 160 million people in the U.S. are associated with Christian churches, of whom about 67 million consider themselves Catholics.

IV. TYPES OF RELIGIOUS ORGANIZATION

A. Some countries have an official or state religion known as an **ecclesia**—a religious organization that is so integrated into the dominant culture that it claims as its membership all members of a society. Examples include: the Anglican Church (the official Church of England), the Lutheran Church in Sweden and Denmark, the Catholic Church in Italy and Spain, and Islam in Iran and Pakistan.

B. The Church-Sect Typology.

1. A **church** is a large, bureaucratically organized, religious organization that tends to seek accommodation with the larger society in order to maintain some degree of control over it.

2. Midway between the church and the sect is a **denomination**—a large, organized religion characterized by accommodation to society, but frequently lacking in ability or intention to dominate society.

3. A **sect** is a relatively small religious group that has broken away from another religious organization to renew what it views as the original version of the faith.

C. A **cult** is a religious group with practices and teachings outside the dominant cultural and religious traditions of a society

1. Some major religions (including Judaism, Islam, and Christianity) and some denominations (such as the Mormons) started as cults.

2. Cult leadership is based on charismatic characteristics of the individual, including an unusual ability to form attachments with others.

3. Over time, some cults disappear; others undergo transformation into sects or denominations.

V. TRENDS IN RELIGION IN THE UNITED STATES

A. Religion and Social Inequality

1. Religion in the United Sates is diverse; pluralism and religious freedom are among the cultural values.

2. Social scientists identify patterns between social class and religious denominations.

   a. Denominations with affluent members and high social status are the mainline liberal churches; the earliest members were immigrants from Britain who helped establish the U. S. government and capitalist economy.

   b. Denominations that adhere to conservative religious beliefs are often in the middle to lower socioeconomic status and tend to discourage women from working outside the home.

338

B. Race and Class in Central-City and Suburban Churches
   1. As more middle- and upper-income individuals and families moved to the suburbs, some churches relocated to the suburbs.
   2. The former building site is often replaced by a minority congregation: working class, older members living on fixed income, or recent immigrant groups.
   3. Racial and cultural minorities who lack economic resources may be drawn to churches that help them establish a sense of dignity.
C. Secularization and the Rise of Religious Fundamentalism
   1. **Secularization**—the decline in the significance of the sacred in daily life—has occurred, causing an increase of religious fundamentalism among some groups.
   2. **Fundamentalism** is a traditional religious doctrine that is conservative, typically opposed to modernity, and rejects worldly pleasures in favor of otherworldly spirituality.
      a. In the United States, it originally appealed to people from low-income, rural, southern backgrounds.
      b. The "new" fundamentalist movement has a wider appeal to people from all socioeconomic levels, geographical areas, and occupations.
      c. Some political leaders vow to bring the Christian religion "back" into schools and public life.
VI. RELIGION IN THE FUTURE
   A. New religious forms are being created, while traditional forms of religious life are experiencing dramatic revitalization; liberation theology is one example.
      1. **Liberation theology** is the Christian movement that advocates freedom from political subjugation within a traditional perspective and advocates the need for social transformation to benefit the poor and powerless.
      2. Some feminist movements have turned to pagan religions and witchcraft to counter the patriarchal structure and content of the world's religions.
         a. The *Goddess movement* acknowledges the legitimacy of female power and encompasses a variety of counterculture beliefs based on paganism and feminism.
      3. In other nations, the rise of *religious nationalism* has led to the blending of strongly held religious and political beliefs; it is strong in the Middle East, where Islamic nationalism has spread, greatly affecting women and children.
      4. In the U. S., the influence of religion will be evident in ongoing battles over school prayer, abortion, gay rights, and women's issues.
   B. Maintaining an appropriate balance between religion and other aspects of social life will be an important challenge in the future—the debate continues over what religion it is and what it should do.

CRITICAL THINKING QUESTIONS

1. What do you think is the appropriate role of religion in U.S. society?
2. In what ways are religious and scientific explanations similar and different?
3. What theoretical perspective do you think is most useful for understanding religion in society? What are the strengths and weaknesses of each?
4. What are the commonalities among the various world religions? What are some of their main differences?
5. In what ways do you think the institution of religion will change and/or stay the same throughout the next century?

PRACTICE TESTS

MULTIPLE CHOICE QUESTIONS

Select the response that best answers the question or completes the statement:

1. _____ is a system of beliefs and practices based on some sacred or supernatural realm that guides human behavior, gives meaning to life, and unites believers into a single moral community.
   a. Secularism
   b. Religion
   c. Denomination
   d. Ecclesia

2. _____ study specific religious doctrines or belief systems, including answers to questions such as the nature of a god or gods, and the relationships between the sacred and human beings.
   a. Sociologists
   b. Prophets
   c. Theologians
   d. Millennialists

3. Sociologist _____ referred to religion as a sacred canopy.
   a. Emile Durkheim
   b. Clifford Geertz
   c. Peter Berger
   d. Robert Bellah

4. According to Emile Durkheim, _____ refers to those aspects of life that are extraordinary or supernatural.
   a. religion
   b. sacred
   c. profane
   d. superhuman

5. The belief that land, animals, or other elements of the natural world are endowed with spirits or life forces that have an impact on events in society is known as:
    a. polytheism
    b. animism
    c. theism
    d. monotheistic religion

6. Three of the major world religions—Christianity, Judaism, and Islam—are characterized as:
    a. simple supernaturalism
    b. animism
    c. polytheism
    d. monotheism

7. Which of the following is **not** one of the elements Durkheim believed to be common to all religions?
    a. a liberating theology
    b. a system of beliefs
    c. a moral community
    d. practices or rituals

8. According to the functionalist perspective, religion offers meaning for the human experience by
    a. providing an explanation for events that create a profound sense of loss on both an individual and a group basis.
    b. offering people a reference group to help them define themselves.
    c. reinforcing existing social arrangements, especially the stratification system.
    d. encouraging the process of secularization.

9. Celebrations on Memorial Day and the Fourth of July are examples of:
    a. religious tolerance
    b. civil religion
    c. patriotic ethnocentrism
    d. separation of church and state

10. According to _____, the capitalist class uses religious ideology as a tool of domination.
    a. Emile Durkheim
    b. C. Wright Mills
    c. Karl Marx
    d. Max Weber

11. In regard to religion, Max Weber asserted that
    a. church and state should be separated.
    b. religion could be a catalyst to produce social change.
    c. religion retards social change.
    d. the religious teachings of the Catholic church were directly related to the rise of capitalism.

12. "Her religion" and "his religion" have been examined from a(n) _____ perspective.
    a. functionalist
    b. postmodernist
    c. conflict
    d. interactionist

13. The ideas of dharma, karma, and nirvana are associated with:
    a. animism
    b. Buddhism
    c. Confucianism
    d. Hinduism

14. The "Four Noble Truths" and the practice of the "Eightfold Path" are today recognized as teachings of:
    a. Hinduism
    b. Buddhism
    c. Confucianism
    d. Islam

15. All of the following are one of the "Five Constant Relationships of Confucianism," **except**:
    a. mother-daughter
    b. father-son
    c. husband-wife
    d. elder brother-younger brother

16. Which one of the following is **not** associated with Judaism?
    a. People are responsible for making ethical choices.
    b. The Messiah will come to earth one day.
    c. God rested on the seventh day after He created the world.
    d. The Messiah has already come to earth.

17. Which of the following is not associated with Islam?
    a. Muslim
    b. the Qur' an
    c. The Greater Jihad
    d. Four Noble Truths

18. The Anglican Church in England and the Lutheran Church in Sweden are examples of a(n):
    a. ecclesia
    b. sect
    c. denomination
    d. secular institution

19. In a _____, membership is largely based on birth, and children of members typically are baptized as infants.
    a. church
    b. sect
    c. denomination
    d. cult

20. According to the text, religious nationalism—the blending of strongly held religious and political beliefs—is especially strong today in:
    a. the United States
    b. Middle Eastern nations
    c. Japan
    d. Brazil

21. An unquestioning belief that does not require proof or scientific evidence is referred to as:
    a. faith
    b. sacred
    c. profane
    d. ritual

22. Activities such as bowing toward Mecca and participating in the celebration of communion are examples of:
    a. faith
    b. theism
    c. sacred
    d. ritual

23. Some analysts believe that the global village is a _____, in which religious institutions and traditions compete for adherents, and worshippers shop for a religion in much the same way that consumers decide which goods and services they will purchase.
    a. sacred canopy
    b. religious marketplace
    c. simple supernaturalism
    d. secularization

24. A _____ religion is based on a belief in divine spiritual forces, such as sacred principles of thought and conduct, rather than a god or gods.
    a. theistic
    b. polytheistic
    c. nontheistic
    d. monotheistic

25. During the Industrial Revolution, many scholars believed increases in scientific knowledge would result in _____, whereby religion loses some significance in society.
    a. liberation theology
    b. monotheism
    c. secularization
    d. profanity

26. _____ assert that religion is used by the dominant classes to impose their own control over society and its resources.
    a. Functionalists
    b. Conflict theorists
    c. Symbolic interactionists
    d. Postmodernists

27. _____ asserted that the religious teachings of John Calvin were directly related to the rise of capitalism.
    a. Max Weber
    b. Karl Marx
    c. Emile Durkheim
    d. Peter Berger

28. _____ analyze the role of religion in people's self-definitions.
    a. Functionalists
    b. Conflict theorists
    c. Symbolic interactionists
    d. Postmodernists

29. The religion referred to as an ethical religion because it calls upon adherents to follow an ideal way of life but is not based on scriptures inspired by a god or gods is:
    a. Confucianism
    b. Christianity
    c. Islam
    d. Hinduism

30. _____ started as a school of thought or a tradition of learning before its eventual leader was born.
    a. Confucianism
    b. Hinduism
    c. Judaism
    d. Islam

31. _____ teaches that individuals can achieve a perfect existence by accomplishing a way of life that avoids indulgence.
    a. Hinduism
    b. Buddhism
    c. Islam
    d. Confucianism

32. Central to contemporary _____ belief is the idea of a single god, called Yahweh.
    a. Christian
    b. Jewish
    c. Hindu
    d. Buddhist

33. Which of the following is **not** one of the Five Pillars of Islam?
    a. participation in five periods of prayer daily
    b. fasting during the daylight hours in the month of Ramadan
    c. believing in multiple gods
    d. paying taxes to help support the needy

34. A _____ is a large, organized religion characterized by accommodation to society but frequently lacking in ability or intention to dominate society.
    a. church
    b. sect
    c. denomination
    d. cult

35. A _____ is a loosely organized religious group with practices and teachings outside the dominant cultural and religious traditions of a society.
    a. church
    b. sect
    c. denomination
    d. cult

36. _____ constitute the largest religious body in the U.S.
    a. Protestants
    b. Roman Catholics
    c. Jews
    d. Muslims

37. _____ is a traditional religious doctrine that is conservative, is typically opposed to modernity, and rejects "worldly pleasures" in favor of otherworldly spirituality.
    a. Secularization
    b. Fundamentalism
    c. Ecclesia
    d. Liberation theology

38. _____, which advocates freedom from political subjugation and social transformation to benefit the poor and other oppressed groups, is becoming more popular among African Americans and feminists.
    a. Secularization
    b. Fundamentalism
    c. Ecclesia
    d. Liberation theology

39. Durkheim refers to the everyday, secular aspects of life as:
    a. sacred
    b. secular
    c. profane
    d. atheistic

40. The dominant world religion today is:
    a. Christianity
    b. Islam
    c. Judaism
    d. Buddhism

TRUE/FALSE QUESTIONS

1. According to Peter Berger, the sacred canopy is the institution of the family.
   T   F

2. Across cultures and in different eras, a wide variety of things have been considered sacred.
   T   F

3. Secularization is the process by which religious beliefs, practices, and institutions lose their significance in sectors of society and culture.
   T   F

4. Marx used the phrase, the "opiate of the masses" to refer to the organized activity of religious people seeking a change in society.
   T   F

5. Karl Marx wrote *The Protestant Ethic and the Spirit of Capitalism* to explain how religion may be used by the powerful to oppress the powerless.
   T   F

6. Ninety-five percent of all Hindus in the world today reside in Pakistan.
   T   F

7. The Dalai Lama is associated with Hinduism.
   T   F

8. Today, Confucianism is the official religion of the People's Republic of China.
   T   F

9. Because it has fewer adherents worldwide than some other major religions, the influence of Judaism is weak in Western culture today.
   T   F

10. Sociologists who study religion are primarily interested in people's beliefs.
    T   F

11. More people in the world today refer to themselves as Christians than any other religion.
    T   F

12. Even though Christianity spread outward from Europe to other cultures, it changed very little, utilizing one highly integrated body of religious beliefs.
    T F

13. Denominations tend to be more tolerant and less likely than churches to expel or excommunicate members.
    T F

14. "New" fundamentalists have encouraged secular humanism in U.S. schools.
    T F

15. Liberation theology initially emerged in Latin America.
    T F

16. In the United States, traditional fundamentalism primarily appealed to people from lower-income, rural, and southern backgrounds.
    T F

17. According to Box 17.1, the U.S. Constitution originally specified that religion should be taught in the public schools.
    T F

18. According to Box 17.2, world-renowned physicist Albert Einstein suggested that "true religion" and scientific knowledge are incompatible.
    T F

19. According to Box 17.3, the First Amendment to the Constitution mandates the separation of church and state.
    T F

20. According to Box 17.4, Islam, Christianity, and Judaism believe in one god, are rooted in the same part of the world, and share some holy sites.
    T F

FILL-IN-THE-BLANK QUESTIONS

1. Durkheim referred to those things that are considered secular or "worldly" as
   _____.

2. _____ is the belief that supernatural forces affect people's lives either positively or negatively.

3. The belief in one or more gods is known as _____.

4. The process by which religious beliefs lose their significance in sectors of society is known as _____.

5. According to anthropologist _____, religion is a set of cultural symbols that establishes powerful motivation.

6. According to _____, religion is the opiate of the masses.

7. _____ wrote in his classical book *The Protestant Ethic and the Spirit of Capitalism* that religion could be a catalyst to produce social change.

8. In the United States today, one of the fastest-growing Eastern religions is _____.

9. A _____ is a relatively small religious group that has broken away from another religious organization to renew what it views as the original version of the faith.

10. The informal relationship between religion and the state is referred to as _____, the set of beliefs, rituals, and symbols that make sacred the values of the society and place the nation in the context of the ultimate system of meaning.

11. _____ is an unquestioning belief that does not require proof or scientific evidence.

12. A religious organization that claims as its membership all members of a society is known as a(n) _____.

13. The _____ encompasses a variety of countercultural beliefs based on paganism and feminism, and acknowledgement of the legitimacy of female power.

14. The rise of _____ has led to the blending of strongly held religious and political beliefs, especially in the Middle East.

15. The Christian movement _____ advocates freedom from political subjugation within a traditional perspective and seeks to benefit the poor and downtrodden.

SHORT ANSWER/ESSAY QUESTIONS

1. What are some of the central beliefs of the six largest world religions?
2. What is religion and how do three major sociological perspectives—functionalist, conflict, and symbolic interactionist—differ in their perspectives on religion?
3. What are some of the relationships among race, class, gender, and religion?
4. What is meant by civil religion? In what ways are people affected in their everyday lives by civil religion?
5. What are some of the current trends and changes in worldwide religion? Do you think there will be a continuing tendency toward secularization, or toward fundamentalism? Will religion continue as a major social institution?

STUDENT CLASS PROJECTS AND ACTIVITIES

1. Compare and contrast the religious books of law of at least two different religions. The Torah, the Qur' an, or the New Testament are some suggestions. Answer the following questions: (1) What are some similarities of the two books? (2) What are some of the differences? (3) How does each book treat the issue of gender and gender roles? (4) What are the expectations of a believer dealing with a

348

nonbeliever? (5) What are some of the ethical laws, such as: (a) How do the believers relate to the poor, or the rich? (b) How do the believers relate to those who have disobeyed the rules? (c) What are the expected rites and rituals of the believer? (d) What are the "sacreds" (as defined by Durkheim) of the religion? (6) What are the major beliefs? and (7) What happens to believers after their deaths? Provide a summary of your findings and a bibliography.

2. Research a particular cult found in the United States or elsewhere. You should provide in your paper: (1) a definition of the cult; (2) the purpose of the cult; (3) a general description of the cult which should include: (a) its origin; (b) its membership and method of acquiring members; (4) the nature of its group life, including advantages and disadvantages of membership in the cult. After providing this basic information, include a two-page paper on a "day in the life" of the cult. Pretend to be a member and take the membership point of view in the writing up of the paper. This day could be a very special day, such as your wedding day, or a typical school, work, or ceremonial day.

3. Conduct an informal survey of any twenty people in order to obtain information on their religious affiliation and their religious beliefs. Try to interview people of different religious faiths and perhaps people who do not identify with any organized religion. Ask the following questions in your survey: (1) What religion do you identify with? (2) What are the main beliefs of your religion? (3) What are the main rituals? (4) What kind of worship do you engage in and how often? (5) What do you like best about your religion? (6) Are there any things you do not like or disagree with that your religion teaches? (7) Were you always the same religion or did you change at some time? Based on your survey results, write a paper that compares the religious beliefs of the individuals you surveyed. Look for similarities and differences in the different religions represented in your sample.

INTERNET ACTIVITIES

1. Peggy Wehmeyer, a reporter for abcnews.com, contends that the Internet is having much the same effect on **transforming religion** as printing the Bible did. Explore a comparative religious community that provides information on the major world religions, morality and culture, family, and prayer circles. **http://www.beliefnet.com/**

2. To illustrate the extent to which the Internet is shaping religious experience, go to **http://www.hollywoodjesus.com/** and explore the uniquely American interplay between **religion and entertainment**.

3. The **American Atheists** have a number of articles about prayer in public school. Go to **http://www.atheists.org/** and search "school prayer." Use information from this site for in-class discussions about the interconnections between religion and education.

4. Explore information on the **Islamic faith**, news, and social organizations: **http://islamworld.net/** and **http://www.islamicity.com/**

5. Compare and contrast Internet sites of insiders and outsiders on the subject of **Islam and terrorism**: **http://www.submission.org/terrorism.html**, **http://www.islam101.com/terror/**, **http://www.islam-guide.com/ch3-11.htm**, **http://groups.colgate.edu/aarislam/response.htm**, and **http://answering-islam.org/Terrorism/**

INFOTRAC COLLEGE EDITION EXERCISES

Visit the **InfoTrac College Edition** web site at:
http://www.wadsworthmedia.com/webtutor/infotrac.htm. You will arrive at a screen that enables you to search topics.

1. The Southern Baptist Convention, the largest Protestant denomination in the U.S., is deeply divided over many issues. One issue is that of the status of women serving in the ministry. Research the issue of the role of **women in religious life**.
2. To what extent can you find evidence of how religious Americans are? Use the keyword **religiosity** (the sociological concept for religious belief and behavior).
3. Many churches and religious groups explicitly condemn **homosexuality**. Frame the debate among the various major world religions. Where do the various religious groups stand on this issue?
4. Fundamentalist religious organizations continue to grow, while other religious organizations have experienced declining memberships. Research **religious fundamentalism** and why it appeals to new adherents.
5. Many Islamic groups advocate a "holy war" against westernization. What does this term really mean to them? Does a holy war necessarily mean resorting to terrorist tactics? Look up the words: **jihad**, **holy war**, **religious conflict**, and **terrorism**. Don't limit your search to stories from the Middle East and Islam. Investigate the religious war that has been going on in Northern Ireland for decades. Look up articles on Bosnia-Herzegovina as well.
6. The issue of prayer in schools continues to be important as some groups contend that **school prayer** would improve the moral fiber of children, while others contend that it violates the premise of separation of church and state.

SOLUTIONS

MULTIPLE CHOICE QUESTIONS

1. B, p. 552
2. C, p. 553
3. C, p. 553
4. B, p. 553
5. B, p. 556
6. D, p. 556
7. A, p. 557
8. A, p. 557
9. B, p. 560
10. C, p. 560
11. B, p. 561
12. D, p. 563
13. D, p. 564
14. B, p. 566

15. A, p. 567
16. D, p. 568
17. D, p. 570
18. A, p. 572
19. A, p. 573
20. B, p. 578
21. A, p. 553
22. D, p. 554
23. B, p. 555
24. C, p. 556
25. C, p. 556
26. B, p. 560
27. A, p. 553
28. C, p. 563

29. D, p. 564
30. A, p. 567
31. B, p. 566
32. B, p. 569
33. C, p. 570
34. C, p. 573
35. D, p. 575
36. A, p. 575
37. B, p. 577
38. D, p. 577
39. C, p. 554
40. A, p. 571

TRUE/FALSE QUESTIONS

1. F, p. 553
2. T, p. 553
3. T, p. 556
4. F, p. 560
5. F, p. 561
6. F, p. 565
7. F, p. 567

8. F, p. 567
9. F, p. 568
10. F, p. 552
11. T, p. 572
12. F, p. 572
13. T, p. 573
14. F, p. 577

15. T, p. 577
16. T, p. 577
17. F, p. 554
18. F, p. 558
19. T, p. 562
20. T, p. 578

FILL-IN-THE-BLANK QUESTIONS

1. profane, p. 554
2. Simple supernaturalism, p. 556
3. theism, p. 556
4. secularization, p. 556
5. Clifford Geertz, p. 554
6. Karl Marx, p. 560
7. Max Weber, p. 561
8. Buddhism, p. 566
9. sect, p. 573
10. civil religion, p. 560
11. Faith, p. 553
12. ecclesia, p. 572
13. Goddess movement, p. 577
14. religious nationalism, p. 578
15. liberation theology, p. 577

18

HEALTH, HEALTH CARE, AND DISABILITY

BRIEF CHAPTER OUTLINE

CHAPTER SUMMARY

Although many people may think of health as simply the absence of disease, the World Health Organization defines **health** as a state of complete physical, mental, and social well-being. **Health care** is any activity intended to improve health. Two indicators of health and well-being of people in a society are the **infant mortality rate**, which is the number of deaths of infants under one year of age per 1,000 live births in a given year, and **life expectancy**, the estimate of the average lifetime of people born in a specific year. A vital part of health care is **medicine**, an institutionalized system for the scientific diagnosis, treatment, and prevention of illness. In analyzing health from a global perspective, we find that health and health care expenditures vary widely among nations. **Social epidemiology** is the study of the causes and distribution of health, disease, and impairment throughout a population. Social factors such as age, sex, race/ethnicity, social class, lifestyle choices, and drug use and abuse affect the health and longevity of people in a given area. Sexual activity, usually associated with good physical and mental health, may involve health-related hazards, such as transmission of sexually transmitted diseases. The fee-for-service method continues in the United States, which is expensive; few restrictions are placed on medical fees. The United States and the Union of South Africa are the only high-income nations without some form of **universal health care** coverage for all citizens. Private health insurance, public health insurance (consisting of Medicare and Medicaid), health maintenance organizations, and managed care compromise the major approaches for paying for U.S. health care. Advances in high-tech medicine is becoming a major part of overall health care; however, many people are turning to **holistic medicine**—an approach to health care that focuses on prevention of illness and disease and is aimed at treating the whole person—body and mind—rather than just the part or parts in which symptoms occur. According to a functionalist perspective, if a society is to function as a stable system, people must be healthy so they can contribute to that society. People who assume the **sick role** are unable to fulfill their necessary social roles and functions. Unequal access to good health and health care is a major issue of concern to conflict theorists. The **medical-industrial complex**—which encompasses both local physicians and hospitals, as well as global health-related industries—produces and sells medicine as a commodity; people below the poverty level and those just above it have great difficulty gaining access to medical care. Symbolic Interactionists, in explaining the social construction of illness, focus on the meanings social actors give their illness or disease and how that meaning affects their self-concept and their relationships with others. Interactionists also emphasize the subjective component of defining medical conditions, including **medicalization**—the process whereby nonmedical problems become defined and treated as illnesses or disorders—and **demedicalization**, which occurs when something is no longer identified as an illness or disorder. According to a postmodernist approach, doctors have gained power through the *clinical gaze*, and gained prestige through the classification of disease. **Mental illness**, a condition in which a person has a severe mental disorder requiring extensive treatment, differs from a **mental disorder**, a condition that makes it difficult or impossible for a person to cope with everyday life. **Disability**, a physical or health condition that stigmatizes or causes discrimination, is often defined in terms of work from a business perspective, or in terms of organically based impairments from a medical professional perspective. Health care in the future will be critically affected by health care technologies; however, health in the future will somewhat be up to each of us—how we safeguard ourselves against illness and how we help others who are victims of diseases and disabilities.

LEARNING OBJECTIVES

After reading chapter 18, you should be able to:

1. Define health, health care, and medicine and explain their importance for individuals and society.

2. Define the concepts infant mortality rate and life expectancy and explain their role in measuring health worldwide.

3. Compare health in a global perspective to health in the U.S.

4. Define social epidemiology and its role in explaining health and disease.

5. Explain how age, sex, race/ethnicity, and social class affect health and mortality.

6. Explain how lifestyle choices affect health, disease, and impairment.

7. Describe the key events and ideas associated with the rise of scientific medicine and professionalism.

8. Classify four methods of paying for health care and controlling health care costs in the U.S.

9. Describe the characteristics of the uninsured and underinsured in the U.S. and explain why they do not have adequate insurance.

10. Compare the U.S. method of paying for medical care to the methods of paying for medical care in other nations.

11. Summarize three major social implications of advanced medical technology.

12. Compare and contrast holistic medicine and alternative medicine with traditional, or orthodox, medical treatment.

13. Describe and analyze functionalist, conflict, symbolic interactionist, and postmodernist perspectives on health and medicine.

14. Define and explain mental illness and mental disorder and explain the treatment of each in the U.S.

15. Describe the relationships of race, class, gender, and mental disorders.

16. Discuss the impact of inequalities, stereotypes, prejudice, and discrimination related to disability.

KEY TERMS

(defined at page number shown and in glossary)

acute diseases, p. 587
chronic diseases, p. 587
deinstitutionalization, p. 609
demedicalization, p. 605
disability, p. 610
drug, p. 590
health, p. 584
health care, p. 584
health maintenance organization
(HMO), p. 597
holistic medicine, p. 602

infant mortality rate, p. 584
life expectancy, p. 584
managed care, p. 599
medical-industrial complex, p. 604
medicalization, p. 605
medicine, p. 584
sick role, p. 603
social epidemiology, p. 587
socialized medicine, p. 600
universal health care, p. 600

KEY PEOPLE

(identified at page number shown)

Abraham Flexner, p. 594
Michel Foucault, p. 607
Erving Goffman, p. 609
Talcott Parsons, p. 603

Thomas Pettigrew, p. 609
Paul Starr, p. 596
Thomas Szasz, p. 607
Ingrid Waldron, p. 590

CHAPTER OUTLINE

I. HEALTH IN GLOBAL PERSPECTIVE
    A. The World Health Organization defines **health** as a state of complete physical, mental, and social well-being.
        1. Illness is an interference to health.
        2. Health and illness are socially defined and may change over time and between cultures.
    B. **Health care** is any activity intended to improve health.
        1. **Medicine**—an institutionalized system for the scientific diagnosis, treatment, and prevention of illness—is a vital part of health care.
    C. Some measures of health and well-being are:
        1. **Life expectancy** is an estimate of the average lifetime of people born in a specific year.
        2. The **infant mortality rate** is the number of deaths of infants under one year of age per 1,000 live births in a given year.
    D. Health care and health care expenditures vary widely among nations. However, health care expenditures do not always coincide with better health outcomes.

The U.S. spends far more on health care than Switzerland, Germany, Canada, and France.

II. HEALTH IN THE UNITED STATES

A. **Social epidemiology**, the study of the causes and distribution of health, disease, and impairment throughout a population, provides some explanations of health and disease.

B. Social epidemiologists investigate disease agents, the environment, and the human host (including demographic factors) as sources of illness.

C. Younger people are more likely to have **acute diseases**, illnesses that strike suddenly and cause dramatic incapacitation and sometimes death, while older people are more likely to have **chronic diseases**, illnesses that are long-term and develop gradually or are present from birth.

D. Today, women on average live longer than men; females have a slight biological advantage over men during the prenatal stage and first month of life, and female gender roles promote a longer life span than male gender roles.

E. African Americans have lower life expectancy than whites.

1. Some research shows that social class is a greater determinant of health than race/ethnicity, as illness is related to factors such as income, neighborhood, education, and occupation.

F. Three (of many) lifestyle factors that relate to health:

1. **Drugs**—substances other than food and water that, when taken into the body, alter its functioning in some way—affect health in many ways.

a. Long-term heavy use of alcohol can damage the brain and other parts of the body, and cause nutritional deficiencies, cardiovascular problems, and alcoholic cirrhosis.

b. Nicotine (tobacco) is a toxic, dependency-producing psychoactive drug that is more addictive than heroin. Smoking tobacco is linked to cancer and other serious diseases. Environmental tobacco smoke—the smoke in the air that nonsmokers inhale from other people's tobacco smoking—is hazardous for nonsmokers.

c. The use of illegal drugs, like cocaine and marijuana, affects the lifestyle and the health of individuals.

2. Sexual activity may involve health-related hazards such as transmission of certain *sexually transmitted diseases*.

a. The most prevalent STDs are HIV/AIDS, gonorrhea, syphilis, and genital herpes. All of these affect the health of individuals; however, the most serious of these is AIDS.

b. On a global basis, the number of people infected with HIV or AIDS is increasing, but progress has been made in some countries; two-thirds of all people with HIV/AIDS live in sub-Saharan Africa.

c. HIV is transmitted through unprotected sexual intercourse with an infected partner; sharing a hypodermic needle with someone who is infected; exposure to contaminated blood or blood products; or by an infected woman passing the virus on to a child during pregnancy, childbirth, or breast feeding; it is not transmitted by casual contact such as shaking hands.

356

3. Staying Healthy: Diet and Exercise
    a. Lifestyle choices can include positive actions such as maintaining a healthy diet and good exercise program.
    b. Both of these have contributed to a significant decrease in heart disease and some cancers.

III. HEALTH CARE IN THE UNITED STATES
A. The rise of scientific medicine resulted from several significant discoveries during the nineteenth century in areas such as bacteriology and anesthesiology; at the same time, advocates of reform in medical education were promoting scientific medicine.
    1. The Flexner Report, complied by **Abraham Flexner**, included a model of medical education; medical schools that did not fit the "model" were closed. Only two of the African American medical schools survived, and only one of the medical schools for women survived.
    2. With professionalization resulting from the Flexner Report, licensed medical doctors gained control over the entire medical establishment; this continues today and may continue into the future.
B. Medicine Today
    1. Medical care in the U.S. is granted on a *fee-for-service* basis, where patients are billed individually for each service they receive.
        a. This includes treatment by doctors, laboratory work, hospital visits, other health-related expenses, and prescriptions.
        b. This payment is expensive; few restrictions are placed on the fees that doctors, hospitals, and medical providers can charge.
        c. This method, representing the "true spirit" of capitalism, has led to advances in medicine.
        d. However, this fee-for-service medicine results in its inequality of distribution.
C. Paying for Medical Care in the United States
    1. Private health insurance expansion in the United States led to escalated costs.
        a. Third party providers pay large portions of doctor and hospital bills for insured patients.
        b. Private health insurance plans are organized like other large-scale for-profit corporations.
    2. Public health insurance is subsidized by federal and state taxes; the U.S. has two nationwide public health insurance programs: Medicare, for those 65 or older who are eligible; and Medicaid, a jointly funded federal-state program to make health care more available to the poor. Both programs are in financial difficulty today.
    3. **Health maintenance organizations (HMOs)** were created to provide workers with health coverage by keeping costs down.
        a. **HMOs** provide, for a set monthly fee, total care with an emphasis on prevention to avoid costly treatment later.
        b. Supporters applaud the emphasis on preventative care.

c. Critics charge that some HMOs require their physicians to withhold vital information and medical treatment from their patients in order to reduce potential large sums of money for medical procedures or hospitalization.

4. **Managed care**, any system of cost containment that closely monitors and controls health care providers' decisions about medical procedures, diagnostic tests, and other services that should be provided to patients.

a. Patients choose a primary-care physician from a list of participating doctors.

b. Doctors must get approval before they perform any procedures or admit a patient to a hospital; if they fail to obtain advance approval, the insurance company has the right to refuse to pay for the treatment or the hospital stay.

5. The uninsured and the underinsured comprise one-third of all U.S. citizens.

a. More than 15 percent of the population had no insurance coverage in 2007, 8.7 million of whom are children.

b. The working poor constitutes a large portion; they earn too little to afford health insurance and too much to qualify for Medicaid, and their employers do not provide health insurance.

D. Paying for Medical Care in Other Nations

1. Canada has a **universal health care** system, a health care system in which all citizens receive medical services paid for by tax revenues. In Canada these revenues are supplemented by insurance premiums paid by all taxpaying citizens. It does not constitute **socialized medicine**, a health care system in which the government owns the medical care facilities and employs the physicians; the physicians are not government employees.

2. Great Britain, in 1946, passed the National Health Service Act, which provided for all health care services to be available at no charge to the entire population. The government sets health care policies, raises funds, controls the medical care budget, owns health care facilities, and directly employs physicians and other health personnel. Great Britain's health care system is socialized medicine.

3. China adopted innovative strategies after the Civil War in order to improve the health of its populace. Physician extenders, known as *street doctors* in urban areas and *barefoot doctors* in rural areas, had little formal training and worked under the supervision of trained physicians. Today, medical training is more rigorous; all doctors receive training in both Western and traditional Chinese medicine. Doctors who work in hospitals receive a salary; all other doctors are on a fee-for-service basis. The cost of health care generally remains low, but the cost of hospital care has risen. Chinese who can afford it purchase health care insurance to cover the rising cost of hospitalization. The health of its citizens is only slightly below that of most industrialized nations.

E. Social implications of advanced medical technology include:

1. New technologies create options for people and society, but alter human relationships.

2. New technologies increase the cost of medical care (e.g., CT or CAT scanner).

358

3. New technologies raise questions about the nature of life, such as duplicating mammals from adult DNA.
  F. Holistic and alternative medicine provide variations from conventional (or mainstream) medical treatment.
    1. **Holistic medicine** focuses on prevention of illness and disease, and aims at treating the whole person—body and mind—rather than just the part or parts in which symptoms occur.
    2. *Alternative medicine* involves healing practices inconsistent with dominant medical practice, which often takes a holistic approach; many people use this in addition to or in lieu of traditional medicine.

IV. SOCIOLOGICAL PERSPECTIVES ON HEALTH AND MEDICINE
  A. A Functionalist Perspective: The Sick Role
    1. According to Talcott Parsons, the **sick role** is the set of patterned expectations that define the norms and values appropriate for individuals who are sick and for those who interact with them.
    2. The sick role has four characteristics:
      a. People who are sick are not responsible for their condition.
      b. A sick person is exempted from responsibilities.
      c. A sick person must want to get well.
      d. A sick person must seek competent help from a medical professional.
    3. Illness is dysfunctional for individuals and society; sick people are unable to fulfill their social roles. It is important for society to maintain social control over people who enter the sick role.
  B. A Conflict Perspective: Inequalities in Health and Health Care
    1. Conflict theorists emphasize the political, economic, and social forces that affect health and the health care delivery system.
    2. Medicine is a commodity that is produced and sold by the **medical-industrial complex**, which encompasses both local physicians and hospitals, as well as global health-related industries such as insurance companies and pharmaceutical and medical supply companies.
    3. The U.S. is one of the few industrialized nations that relies almost exclusively on the medical-industrial complex for health care delivery and does not have universal health coverage.
    4. Medical care is linked to people's ability to pay and their position within the class structure. The affluent or those who have good medical insurance are most likely receive care; the *medically indigent*, those who don't qualify for Medicaid and don't earn enough to afford insurance, cannot afford private medical care.
    5. Physicians hold a legal monopoly over medicine and can charge inflated fees; clinics, pharmacies, laboratories, hospitals, insurance companies, and many other corporations, including large drug companies, derive excessive profits from the existing system of payment in medicine.
  C. A Symbolic Interactionist Perspective: The Social Construction of Illness
    1. Symbolic interactionists focus on the meaning that actors give their illness or disease and how this will affect their self-concept and their relationship with others.

359

2. In addition to the objective criteria for determining medical conditions, the subjective component is very important; the term **medicalization** refers to the process whereby nonmedical problems become defined and treated as illness or disorders.
3. Medicalization may occur on three levels: (a) the conceptual level; (b) the institutional level; and (c) the interactional level. Medicalization is typically the result of a lengthy promotional campaign, often culminating in legislation or other social policy changes that institutionalize a medical treatment of a new "disease."
4. **Demedicalization** refers to a problem that no longer retains its medical definition; an example is the work of women's health advocates seeking to redefine menopause as a natural process rather than as an illness or psychological disorder.
5. Symbolic interactionists provide new insights on how illness may be socially constructed, and not strictly determined by medical criteria.

D. A Postmodernist Perspective: The Clinical Gaze
1. According to Michel Foucault, the formation of clinical medicine led to the power that doctors gained over other medical personnel and everyday people.
2. Doctors gain power through the *clinical gaze*, which they use to gather information.
    a. As doctors "diagnose", they become experts.
    b. As a definition network of disease classification, new rules and new tests developed, the dominance of doctor's wisdom became enhanced.
3. Foucault's analysis is not limited to doctors who treat bodily illness, but also psychiatrists and the treatment of insanities.

V. MENTAL ILLNESS
A. **Thomas Szasz** argues mental illness is a myth, and reflects individual traits or behaviors defined as socially undesirable.
B. *Mental illness*, a condition in which a person has a severe mental disorder requiring extensive treatment, differs from a *mental disorder*, a condition that makes it difficult or impossible for a person to cope with everyday life.
C. According to the National Comorbidity Survey, nearly 50 percent of respondents between the ages of 15 and 54 years had been diagnosed with a mental disorder sometime in their lives.
D. Treatment of Mental Illness
1. The introduction of new psychoactive drugs to treat mental disorders and the **deinstitutionalization** movement have created dramatic changes in how people with mental disorders are treated; deinstitutionalization refers to the practice of rapidly discharging patients from mental hospitals into the community.
2. Deinstitutionalization is now viewed as a problem by many social scientists; in the 1960s, it was believed that many state mental patients had been deprived of their civil rights, housed in what Goffman called a *total institution*; they were released, many having no place to go, except to reside on the streets and become part of the homeless population.

3. *Involuntary commitment*, used as a social control mechanism to keep mentally ill people off the streets, does little to treat underlying medical and social conditions that contribute to mental disorders.

E. Race, Class, Gender, and Mental Disorders

1. Most studies examining race/ethnicity and mental disorders have compared African Americans and white Americans and have uncovered no significant differences in diagnosable mental illness.

2. In a qualitative study about the effects of racism, Feagin and Sikes concluded that repeated personal encounters with racial hostility deeply affect the psychological well-being of most African Americans, regardless of their level of education or social class.

3. According to Pettigrew, racism in all it forms constitutes a mentally unhealthy situation, wherein people do not achieve their full potential.

4. Researchers agree that social class has a significant impact on mental disorders; as social class increases, rates of mental disorders decrease.

5. The rate of diagnosable depression is about twice as high for women as for men; this gender difference typically emerges in puberty and increases in adulthood as women and men enter and live out their unequal adult statuses.

6. *Learned helplessness theory* posits that people become depressed when they believe they cannot control their lives.

VI. DISABILITY

A. **Disability** refers to a reduced ability to perform tasks one would normally do at a given stage of life, which may result in stigmatization or discrimination against the person with disabilities.

B. Disability is often defined in terms of work from a business perspective, or in terms of organically based impairments from a medical professional perspective.

C. An estimated 49.7 million persons in the U.S. have one or more physical or mental disabilities, and the number is increasing as medical advances make it possible for those who would have died from an accident or illness to survive, but with an impairment, and as life expectancy increases. Environment, lifestyle, and working conditions may contribute to disability.

D. Many disability rights advocates argue that persons with a disability are kept out of the mainstream of society, being denied equal opportunities in education by being consigned to special education classes or schools.

E. Sociological Perspectives on Disability

1. Functionalist Talcott Parsons focused on how people who are disabled fill the sick role. This is a *medical model* of disability; people with disabilities become chronic patients.

2. Symbolic interactionists examine how people are labeled as a result of disability; Eliot Freidson determined that the particular label results from (1) the person's degree of responsibility for the impairment, (2) the apparent seriousness of the condition, and (3) the perceived legitimacy of the condition.

3. From a conflict perspective, persons with disabilities are a subordinate group in conflict with persons in positions of power in the government, the health care industry, and the rehabilitation business. When people with disabilities are defined as a social problem and public funds are spent to purchase goods and services for them, rehabilitation becomes big business.

F. Social inequalities based on disability include prejudice and discrimination.
    1. Stereotypes of persons with a disability fall into two categories: (1) deformed individuals who also may be horrible deviants; or (2) persons who are to be pitied. Even positive stereotypes become harmful to people with a disability, such as that individuals can excel despite impairment. Disability rights advocates note that such stereotypes do not reflect the daily struggle of most people with disabilities.
    2. Employment, poverty, and disability are related; people may become economically disadvantaged as a result of chronic illness or disability; on the other hand, poor people are less likely to be educated and more likely to be malnourished and have inadequate access to health care.
    3. Disability has a stronger negative effect on women's labor participation than it does on men's.
    4. The cost of "mainstreaming" persons with disabilities closely relates to current expenditures that provide greater access to education and jobs.
VII. HEALTH CARE IN THE FUTURE
    A. U.S. Congress is currently debating passing a health care bill that would require all citizens to carry insurance, require more coverage by businesses, expand public insurance, and subsidize private insurance for those with middle incomes.
    B. In the future, advanced health care technologies for high-income countries will provide more accurate and quicker diagnosis, effective treatment techniques, and increased life expectancy; however, the inability of the poor of the world (including the U.S.) to have access to health care, the growth of managed-care companies in the U.S. making decisions formerly reserved for physicians and hospital personnel, and ethical concerns of technological advances all indicate a change in the future of health and medicine.
    C. Ultimately, health care in the future will to some degree be up to each of us—what measures we take to safeguard ourselves against illness and disorders, and how we help others who are victims of diseases and disabilities.

CRITICAL THINKING QUESTIONS

1. What is the relationship between health and health care and society? Why is health care an important social institution?
2. How does the U.S. compare to other countries in terms of health and health care?
3. Which theoretical perspective do you think provides the best understanding of health care in the U.S.? What are the strengths and weaknesses of each theory?
4. Thomas Szasz argued that mental illness is a myth. Describe and respond to his argument. Do you agree or disagree with him? Why?
5. Do you think the U.S. government should reform health care to provide universal coverage? Why or why not?

PRACTICE TESTS

MULTIPLE CHOICE QUESTIONS

Select the response that best answers the question or completes the statement:

1. The World Health Organization (WHO) defines health as the
   a. absence of disease.
   b. complete physical, mental, and social well-being of a person.
   c. body in a state of equilibrium, with all parts in balance.
   d. absence of sickness, viruses, and pains.

2. _____ refers to the positive sense of complete well-being; while _____ refers to an interference with health.
   a. Healing; disease
   b. Health; healing
   c. Health; illness
   d. Health care; disease

3. Health care in the United States
   a. is positively linked to money spent on health care and people's physical, mental, and social well-being.
   b. equates positively with longevity.
   c. is more expensive, but more productive for individuals.
   d. is defined as any activity intended to improve health.

4. _____ is an institutionalized system for the scientific diagnosis, treatment, and prevention of illness.
   a. The medical-industrial complex
   b. Universal health care
   c. Medicine
   d. Disease

5. The World Health Organization reports that almost 14 percent of all babies born in low-income countries die before they reach the
   a. first year of life.
   b. first month of life.
   c. first six weeks of life.
   d. first nine months of life.

6. Social epidemiologists are people who study the causes and distribution of
   a. health, disease, and impairment in a population.
   b. epidemics that occur throughout a population.
   c. mental illness and behavior disorders in a population.
   d. the role of degenerative diseases in a population.

7. _____ are illnesses that strike suddenly, cause dramatic incapacitation, and sometimes lead to death.
    a. Chronic diseases
    b. Disabilities
    c. Epidemics
    d. Acute diseases

8. On average, _____ live longer than _____, but have higher rates of _____ illness.
    a. men; women; chronic
    b. women; men; acute
    c. women; men; chronic
    d. men; women; acute

9. _____ is a toxic, dependency-producing drug that is more addictive than _____.
    a. Heroin; nicotine
    b. Nicotine; heroin
    c. Nicotine; marijuana
    d. Marijuana; alcohol

10. Until the 1960s, _____ and _____ were the principal STDs in this country.
    a. genital herpes; syphilis
    b. gonorrhea; genital herpes
    c. gonorrhea; syphilis
    d. genital herpes; HIV/AIDS

11. Two-thirds of all people with HIV/AIDS live in:
    a. sub-Saharan Africa
    b. South Asia
    c. Southeast Asia
    d. the United States

12. The Flexner Report stated in 1910 that
    a. new medical schools should be developed for women.
    b. more medical schools should be developed for African Americans.
    c. new medical schools should be developed in rural areas.
    d. most existing medical schools were inadequate for teaching.

13. Throughout most of the twentieth century, medical care in the United States was paid for
    a. by HMOs.
    b. on a fee-for-service basis.
    c. by third party providers.
    d. through managed care.

14. _____ is a jointly funded federal-state local health care program for the poor.
    a. Medicare
    b. Medical
    c. Medicaid
    d. Managed care

364

15. The only high-income nations without some form of universal health coverage for all its citizens are:
    a. the United States and the Union of South Africa
    b. Canada and the United States
    c. Japan and the United States
    d. Great Britain and the United States

16. Which of the following countries has socialized medicine?
    a. the United States
    b. Great Britain
    c. Canada
    d. all of these

17. An approach to health care that treats the whole person, rather than just symptoms that occur, is known as _____.
    a. mainstream medicine
    b. managed care
    c. holistic medicine
    d. medicalization

18. People in the sick role are expected to
    a. be responsible for their condition.
    b. desire to get well.
    c. continue their normal roles and obligations.
    d. ignore competent medical assistance.

19. The practice of rapidly discharging patients from mental hospitals into the community is referred to as:
    a. reinstitutionalization
    b. demedicalization
    c. deconstruction
    d. deinstitutionalization

20. In contemporary industrial societies, disability often can be attributed to
    a. epidemics related to poor sanitation and overcrowding.
    b. urban density and poverty.
    c. environment, lifestyle, and working conditions.
    d. employment in high stress jobs in the primary tier of the labor market.

21. Life expectancy is the
    a. average age people surveyed expect that they will live.
    b. average age at death in a particular year.
    c. percent of infants expected to live longer than one year.
    d. average life span of people born in a particular year.

22. The _____ is the number of deaths of infants under 1 year of age per 1,000 live births in a given year.
   a. child mortality rate
   b. infant mortality rate
   c. first year of life rate
   d. baby mortality rate

23. Sweden spends _____ per person on health care, but has a _____ infant mortality rate compared to the U. S.
   a. more; lower
   b. less; lower
   c. more; higher
   d. less; higher

24. Sociologist Ingrid Waldron argued that
   a. women are more likely to engage in the sick role than men.
   b. men are more likely to engage in risky behaviors and work in hazardous occupations that increase the likelihood of illness and injury.
   c. women are more susceptible to the clinical gaze than men.
   d. women are charged higher fees for health care services than men.

25. In relation to social epidemiology, _____ include(s) biological agents such as insects, bacteria, and viruses that carry or cause disease.
   a. the human host
   b. the environment
   c. disease agents
   d. health coefficients

26. _____ examine how social inequalities affect health and health care, power relationships between doctors and other health care workers, the dominance of the medical model, and the role of profit in the health care system.
   a. Symbolic interactionists
   b. Functionalists
   c. Postmodernists
   d. Conflict theorists

27. _____ study the processes through which particular behaviors come to be defined as illnesses.
   a. Symbolic interactionists
   b. Functionalists
   c. Postmodernists
   d. Conflict theorists

28. _____ view illness as deviant behavior that must be controlled in society.
   a. Symbolic interactionists
   b. Functionalists
   c. Postmodernists
   d. Conflict theorists

29. Recent research suggests that _____ may be a more significant factor in health and mortality than is _____.
    a. social class; race/ethnicity
    b. gender; social class
    c. race/ethnicity; gender
    d. race/ethnicity; social class

30. _____ use occurs when a person takes a drug for no purpose other than for achieving a pleasurable feeling or an altered psychological state.
    a. Recreational
    b. Chronic
    c. Therapeutic
    d. Acute

31. _____ argued that doctors gain power through the "clinical gaze" or observations of patients.
    a. Michel Foucault
    b. Erving Goffman
    c. Talcott Parsons
    d. Thomas Szasz

32. _____ argued that mental illness is a myth that actually reflects individual traits and behaviors that are considered unacceptable or deviant in society.
    a. Erving Goffman
    b. Thomas Szasz
    c. Talcott Parsons
    d. Michel Foucault

33. When attempting to treat gonorrhea and syphilis, penicillin
    a. can cure most cases of both STDs, if the disease has not spread.
    b. cannot cure most cases of either STD.
    c. is ineffective in treating gonorrhea, but effective in treating syphilis.
    d. is ineffective in treating syphilis, but effective in treating gonorrhea.

34. All of the following regarding AIDS are true, **except**:
    a. AIDS reduces the body's ability to fight diseases.
    b. No one actually dies of AIDS.
    c. Worldwide, the number of AIDS cases has dropped.
    d. AIDS almost inevitably ends in death.

35. Which of the following groups has the lowest life expectancy in the U.S.?
    a. black women
    b. white women
    c. black men
    d. white men

36. Despite public and private insurance programs, about _____ of all U.S. citizens are without health insurance or had difficulty getting or paying for medical care at some time in the last year.
    a. one-fourth
    b. one-third
    c. one-fifth
    d. one-half

37. It is estimated that slightly more than ____ percent of the U. S. population had no health insurance in 2007.
    a. 5
    b. 10
    c. 15
    d. 20

38. A _____ is a document stating a person's wishes regarding the medical circumstances under which his or her life should be terminated.
    a. living will
    b. living license
    c. living term
    d. living file

39. Examples of _____ include the removal of certain behaviors from the list of mental disorders compiled by the American Psychiatric Association and the deinstitutionalization of mental health patients.
    a. demedicalization
    b. remedicalization
    c. medicalization
    d. nonmedicalization

40. _____ is a condition in which a person has a severe mental ailment requiring extensive treatment with medication, psychotherapy, and sometimes hospitalization.
    a. Mental disorder
    b. Comorbidity
    c. Mental illness
    d. Mental dysfunctioning

TRUE/FALSE QUESTIONS

1. Studies show that racial prejudice and discrimination have a negative impact on the psychological well-being of both black and white Americans.
   T   F

2. Class is generally more of a factor for health problems than race or ethnicity.
   T   F

3. Therapeutic use of drugs occurs when a person takes a drug for a specific purpose, such as reducing fever or a cough.
   T   F

4. Cocaine is the most extensively used illegal drug in the United States.
   T   F

5. HIV can be transmitted through casual contact, such as shaking hands.
   T   F

6. As a result of the Flexner Report, medical schools for African Americans and women were expanded.
   T   F

7. Sexually transmitted diseases (STDs) did not begin to spread rapidly in the U.S. until the 1980s and 1990s.
   T   F

8. Managed care allows doctors more individual freedom in choosing the medical treatment for their patients.
   T   F

9. The Canadian health care system is an example of socialized medicine.
   T   F

10. According to conflict theorists, physicians hold a legal monopoly over medicine.
    T   F

11. The subjective component of medicalization and demedicalization reflects the major concern of the interactionist perspective of health.
    T   F

12. According to Foucault, doctors gain power through the "clinical gaze."
    T   F

13. Most people in the U.S. who do not have health insurance are unemployed.
    T   F

14. Less than 15 percent of persons with a disability today were born with it.
    T   F

15. On average, male workers with a severe disability make 50 percent of what their co-workers without disabilities earn.
    T   F

16. According to Box 18.1, the medical-industrial complex has operated in the U.S. with virtually no regulation.
    T   F

17. The field of epidemiology, according to Box 18.1, focuses primarily on how individuals acquire disease and bodily injury.
    T   F

18. According to Box 18.3, Medicare is considered to be an entitlement program, not a welfare program.
T   F

19. According to Box 18.2, it is likely that advertisements for prescription drugs will cease in the near future since this advertising is not shown to be effective at increasing sales; it is doctors rather than patients who choose which medications are appropriate.
T   F

20. According to Box 18.4, Beverly Barnes did not initially intend to start the program now known as Patient Pride.
T ,   F

FILL-IN-THE-BLANK QUESTIONS

1. The state of complete physical, mental, and social well-being is defined as
_____.

2. The average lifetime of people born in a specific year is known as the
_____.

3. The _____ Report changed the practice of medical education in the U.S.

4. A(n) _____ care system is one wherein all citizens receive medical services paid for by tax revenues.

5. The _____ is the set of patterned expectations that defines the norms and values appropriate for individuals who are sick and for those who interact with them.

6. _____ is an approach to health care that focuses on prevention of illness and disease and is aimed at treating the whole person.

7. The _____ perspective focuses on the meaning that social actors give their disease or illness.

8. The _____ perspective on health and illness states that the myth of the wise doctor was supported by the development of disease classification systems and new tests.

9. _____ is the process whereby a problem ceases to be defined as an illness or disorder.

10. A(n) _____ is a physical or mental health condition that stigmatizes or causes discrimination.

11. A(n) _____ is any substance, other than food or water, that, when taken into the body, alters its functioning in some way.

12. _____ provide, for a set monthly fee, total health care with an emphasis on prevention to avoid costly treatment later.

13. _____ is a health care program for persons age 65 or older who are covered by Social Security or who are eligible and "buy into" the program by paying a monthly premium.

14. In a(n) _____ health care program, doctors must get approval before they perform certain procedures or admit a patient to a hospital.

15. A health care system in which all citizens receive medical services paid for by tax revenues is referred to as a(n) _____.

SHORT ANSWER/ESSAY QUESTIONS

1. What is meant by health, health care, medicine, and disability, and why are these important concerns for both individuals and entire societies?
2. How is health care paid for in the United States? In some other countries? Which system do you think is better for people in a society? Explain your response.
3. What is holistic medicine and how does it differ from traditional health care? What is meant by alternative health care?
4. Explain the concepts mental illness and mental disorder. Discuss controversies regarding mental illness, including the claim that mental illness does not exist, and the deinstitutionalization movement.
5. What are some of the major issues and concerns of health care in the future?

STUDENT CLASS PROJECTS AND ACTIVITIES

1. Collect at least 10 articles dealing with current issues of health and health care found in the text, such as alcoholism, concerns about diet and exercise, AIDS, current causes of death, the American health care system, the profession of medicine (including doctors, nurses, and hospitals), health insurance, prepaid health care, managed health care, and health care in other countries. The articles can come from newspapers, news sites, professional journals, magazines, etc. Write a paper that includes: (1) a summary explaining the message of each article; (2) an evaluation of the message of each article; (3) a conclusion that compares and contrasts the messages of the different articles; and (4) a bibliographic reference for each article selected.
2. Investigate a health care system found in another modern industrial country that has a reputation of having a fairly good, reliable health care delivery system. Some suggestions are: Great Britain, Canada, Germany, Sweden, Norway, and Japan. Write a paper that describes: (1) the specific type(s) of health care system(s); (2) who receives the health care; (3) how the program is funded; (4) if personal choice is allowed—even for the higher income; (5) who the professionals are in the system; (6) their training; (7) how serious illness or disease is treated; and (8) any other information pertinent to this research. In the writing of your paper, include a bibliography, a summary, a conclusion, and a personal evaluation of this project.
3. The October 12, 2007 issue of *AARP* listed the 50 top hospitals in the United States, as cited by Consumer's Checkbook, a nonprofit consumer education organization.

This article summarizes the major findings of the publication *Consumer's Guide to Hospitals*, which rates more than 4,500 hospitals nationwide. Read this article and summarize the major criteria and the major findings of the report. Are there any surprises? What constitutes the accreditation score, the physician's rating, and special offerings/programs/research of the top-ranked hospitals? List the top 25 hospitals in the nation. Next, discuss the top-ranked hospital, noting the explanations for its ranking.

4. A current topic of controversy is whether or not the U.S. should find a way to provide some type of health insurance to all citizens. Select ten people of varying ages, genders, social classes and races and interview them on their opinions toward universal health care in the U.S. Include at least the following questions in your survey: (1) Do you think the U.S. should find a way to provide some type of health insurance to all citizens? Why or why not? (2) What do you think would be the best kind of health care system for the U.S.? Why? (3) How do you think health and health care in the U.S. compare to that in Western industrialized countries that provide health care to all citizens, such as Sweden or Great Britain? (4) How old are you? (5) Are you male or female? (6) What is your social class? (7) What is your race or ethnicity? (8) Do you currently have health insurance? If so what kind? How is it paid for? Write a paper that compares and contrasts the responses of people who support and oppose universal health care in the U.S. Include a discussion of any general trends that you notice based on the demographic factors of age, gender, social class, race, and whether the individual has health insurance or not.

INTERNET ACTIVITIES

1. The **National Institutes of Health**, **http://www.nih.gov/**, is one of eight agencies that are part of the Department of Health and Human Services. In turn, it consists of over twenty-five agencies, such as the National Institute of Allergy and Infectious Diseases, the National Institute on Alcohol Abuse and Alcoholism, and the National Institute of Drug Abuse. Investigate the wealth of information on this site.

2. To explore one of the major medical and health research institutions in the U.S., visit the web site for the **Centers for Disease Control**, **http://www.cdc.gov/**, and read about the CDC, its facilities, mission, people, budget, data, and statistics. The CDC collects data on a wide range of topics, including cancer rates, AIDS rates, violence, and rare illnesses (e.g., Ebola and Hanta viruses).

3. Health problems in other countries are often linked to poor sanitation and hygiene. To learn more about these kinds of diseases, including the guinea worm, which causes internal injuries and permanent, disfiguring scarring, visit the web site for the **World Health Organization**, **http://www.who.int/**. Be sure to note some of the differences between the kinds of diseases the CDC focuses on compared to the World Health Organization.

4. The **Department of Health and Human Services**, **http://www.os.dhhs.gov/**, is a federal agency that provides information on numerous government programs, including Medicare and Medicaid, and the State Children's Health Insurance Program, as well as medical fraud.

5. The **National Institute of Mental Health**, **http://www.nimh.nih.gov/** (also belongs to the NIH), provides information on mental illness, child and adolescent violence, and rural mental health.

INFOTRAC COLLEGE EDITION EXERCISES

Visit the **InfoTrac College Edition** web site at: **http://www.wadsworthmedia.com/webtutor/infotrac.htm**. You will arrive at a screen that enables you to search topics.

1. Search for articles (both professional and news) related to **health maintenance organizations**. How effective are **HMOs**? Are they more interested in cutting costs than in providing quality health care?

2. How pervasive is **Medicare fraud** or **medical fraud** in general? How does this impact our society? Search for news articles that expose these kinds of problems. What are the social issues involved?

3. Search the topic **disability** to get a better sense of who is affected by this condition, the legal protections provided by the government, and current issues that impact individuals who are disabled.

4. After searching for periodical references for **holistic medicine**, examine the range of journals in which this term appears. Talk about the different "backdoors" that these nontraditional methods often have to take.

SOLUTIONS

MULTIPLE CHOICE QUESTIONS

1. B, p. 584	15. A, p. 596	29. A, p. 590
2. C, p. 584	16. B, p. 600	30. A, p. 590
3. D, p. 584	17. C, p. 602	31. A, p. 607
4. C, p. 584	18. B, p. 603	32. B, p. 607
5. A, p. 584	19. D, p. 609	33. A, p. 592
6. A, p. 587	20. C, p. 610	34. C, p. 593
7. D, p. 587	21. D, p. 590	35. C, p. 590
8. C, p. 587	22. B, p. 584	36. B, p. 599
9. B, p. 584	23. B, p. 587	37. C, p. 599
10. C, p. 592	24. B, p. 604	38. A, p. 601
11. A, p. 593	25. C, p. 587	39. A, p. 605
12. D, p. 594	26. D, p. 604	40. C, p. 607
13. B, p. 595	27. A, p. 603	
14. C, p. 597	28. B, p. 607	

TRUE/FALSE QUESTIONS

1. T, p. 609	8. F, p. 599	15. T, p. 611
2. T, p. 590	9. F, p. 600	16. F, p. 586
3. T, p. 590	10. T, p. 604	17. F, p. 586
4. F, p. 592	11. T, p. 605	18. T, p. 598
5. F, p. 593	12. F, p. 607	19. F, p. 598
6. F, p. 594	13. F, p. 599	20. T, p. 615
7. F, p. 592	14. T, p. 610	

FILL-IN-THE-BLANK QUESTIONS

1. health, p. 584
2. life expectancy, p. 584
3. Flexner, p. 594
4. universal health, p. 600
5. sick role, p. 603
6. Holistic medicine, p. 602
7. symbolic interaction, p. 604
8. postmodern, p. 607
9. Demedicalization, p. 605
10. disability, p. 610
11. drug, p. 586
12. Health maintenance organizations, p. 597
13. Medicare, p. 597
14. managed care, p. 599
15. universal health care system, p. 600

19

POPULATION AND URBANIZATION

BRIEF CHAPTER OUTLINE

DEMOGRAPHY: THE STUDY OF POPULATION
 Fertility
 Mortality
 Migration
 Population Composition
POPULATION GROWTH IN GLOBAL CONTEXT
 The Malthusian Perspective
 The Marxist Perspective
 The Neo-Malthusian Perspective
 Demographic Transition Theory
 Other Perspectives on Population Change
A BRIEF GLIMPSE AT INTERNATIONAL MIGRATION THEORIES
URBANIZATION IN GLOBAL PERSPECTIVE
 Emergence and Evolution of the City
 Preindustrial Cities
 Industrial Cities
 Postindustrial Cities
PERSPECTIVES ON URBANIZATION AND THE GROWTH OF CITIES
 Functionalist Perspectives: Ecological Models
 Conflict Perspectives: Political Economy Models
 Symbolic Interactionist Perspectives: The Experience of City Life
PROBLEMS IN GLOBAL CITIES
URBAN PROBLEMS IN THE UNITED STATES
 Divided Interests: Cities, Suburbs, and Beyond
 The Continual Fiscal Crisis of the Cities
RURAL COMMUNITY ISSUES IN THE UNITED STATES
POPULATION AND URBANIZATION IN THE FUTURE

CHAPTER SUMMARY

Demography is the study of the size, composition, and distribution of the population. Population growth is the result of **fertility** (births), **mortality** (deaths), and **migration**. The **population composition**—the biological and social characteristics of a population—is affected by changes in fertility, mortality, and migration. One measure of population composition is **sex ratio**—the number of males for every hundred females in a given population. The distribution of a population by sex and age can be depicted in a **population pyramid**. Over two hundred years ago, **Thomas Malthus** warned that overpopulation would result in major global problems, such as poverty and starvation. **Karl Marx** argued that overpopulation occurs because of capitalists' demands for a surplus of workers to suppress wages and heighten workers' productivity. More recently, neo-Malthusians have re-emphasized the dangers of overpopulation, including lack of food and environmental degradation; they recommend **zero population growth**. **Demographic transition** is the process by which some societies have moved from high birth and death rates to relatively low birth and death rates as a result of technological development. Other perspectives on population change include *rational choice theory*, the epidemiological transition, economic development, and the process of "westernization." In explaining international migration theories, the *neoclassical economics approach*, the *new households economics of migration approach*, and conflict and world systems theory all add to our knowledge of the way people migrate. *Urban sociology* is the study of social relationships and political and economic structures in the city. Cities are a relatively recent innovation compared to the length of human existence. Because of their limited size, preindustrial cities tend to provide a sense of community and a feeling of belonging. The Industrial Revolution changed the size and nature of the city; people began to live close to the factories and to one another, which led to overcrowding and poor sanitation. In postindustrial cities, some people live and work in suburbs or outlying edge cities. Functionalist perspectives (ecological models) of urban growth include the *concentric zone model*, the *sector model*, and the *multiple nuclei model*. According to the political economy models of conflict theorists, urban growth is influenced by capital investment decisions, power and resource inequality, class and class conflict, and government subsidy programs. Feminist theorists suggest that cities have *gender regimes*; women's lives are affected by both public and private patriarchy. Symbolic interactionists focus on the positive and negative aspects of peoples' experiences in urban settings. Rapid population growth in many global cities is producing a wide variety of urban problems, including overcrowding, environmental pollution, and disappearance of farmland, as well as creating a limit on the availability of basic public services. Urbanization, suburbanization, **gentrification**, and the growth of *edge cities* have had a dramatic impact on the U.S. population. Many central cities continue to experience fiscal crises that have resulted in cuts in services, lack of maintenance of the infrastructure, as well as loss of resources because of terrorist attacks. Many cities and large urban areas have created a "disabling" environment; access is critical in order for persons with disabilities to be productive members of the community. Some traditional issues of the rural community include financial and emotional issues, proliferation of superstores, and increases in tourism. Rapid global population growth is inevitable in the future. The urban population will triple as increasing numbers of people in lesser developed and developing nations migrate from rural areas to mega cities that contain a high percentage of a region's population.

LEARNING OBJECTIVES

After reading Chapter 19, you should be able to:

1. Describe the study of demography and define the basic demographic concepts.

2. Explain the Malthusian perspective on population growth.

3. Discuss the Marxist perspective on population growth and compare it with the Malthusian perspective.

4. Explain the neo-Malthusian perspective on population growth, define the concept of zero population growth, and describe the relationship between the two.

5. Summarize demographic transition theory and explain why it may not apply to population growth in all societies.

6. Construct a comprehensive outlook on global migration using the new households economics of migration approach, the neoclassical economic approach, network theory, and institutional theory.

7. Trace the historical development of cities.

8. Identify the major characteristics of preindustrial, industrial, and postindustrial cities.

9. Summarize the key ideas of the major urban theorists.

10. Discuss functionalist perspectives on urbanization and outline the major ecological models of urban growth.

11. Compare and contrast conflict and functionalist perspectives on urban growth.

12. Explain the symbolic interactionist perspective on urban life.

13. Describe global patterns of urbanization in core, peripheral, and semiperipheral nations.

14. Discuss the impact of race, class, gender, and disability on city life.

15. Describe issues facing rural communities in the U.S. today.

16. Discuss the major problems facing urban areas in the United States today and in the future.

KEY TERMS

(defined at page number shown and in glossary)

crude birth rate, p. 622
crude death rate, p. 623
demographic transition, p. 632
demography, p. 621
fertility, p. 621
gentrification, p. 638
invasion, p. 638

migration, p. 624
mortality, p. 623
population composition, p. 625
population pyramid, p. 629
sex ratio, p. 629
succession, p. 638
zero population growth, p. 632

KEY PEOPLE

(identified at page number shown)

Lynn M. Appleton, p. 641
Ernest Burgess, p. 637
Joe R. Feagin and Robert Parker,
p. 639
Herbert Gans, p. 642
Harlan Hahn, p. 644
Chauncey Harris and Edward Ullman,
p. 639
Amos Hawley, p. 639

Homer Hoyt, p. 638
Thomas Robert Malthus, p. 630
Karl Marx and Frederick Engels, p. 631
Robert Park, p. 637
Georg Simmel, p. 642
Gideon Sjoberg, p. 645
Ferdinand Tönnies, p. 636
Elizabeth Wilson, p. 643
Louis Wirth, p. 642

CHAPTER OUTLINE

I. DEMOGRAPHY: THE STUDY OF POPULATION
   A. **Demography** is a subfield of sociology that examines population size, composition, and distribution.
   B. **Fertility** is the actual level of childbearing for an individual or a population.
      1. The **crude birth rate** is the number of live births per 1,000 people in a population in a given year.
      2. In most areas of the world, women are having fewer children; women who have six or more children tend to live in agricultural regions where children's labor is essential to the family's economic survival and child mortality rates are very high.
   C. A decline in **mortality**—the incidence of death in a population—has been the primary cause of world population growth in recent years.
      1. The **crude death rate** is the number of deaths per 1,000 people in a population in a given year.

2. The *infant mortality rate* is the number of deaths of infants under 1 year of age per 1,000 live births in a given year.
D. **Migration** is the movement of people from one geographic area to another for the purpose of changing residency.
   1. While *immigration* is the movement of people into a geographic area to take up residency, *emigration* is the movement of people out of a geographic area to take up residency elsewhere.
   2. The ***crude net migration rate*** is the net number of migrants (total in migrants minus total out migrants) per 1,000 people in a population in a given year.
E. **Population composition** is the biological and social characteristics of a population, including age, sex, race, marital status, education, occupation, and income.
   1. The **sex ratio** is the number of males for every hundred females in a given population; a sex ratio of 100 indicates an equal number of males and females.
   2. A **population pyramid** is a graphic representation of the distribution of a population by sex and age.
II. POPULATION GROWTH IN GLOBAL CONTEXT
A. The Malthusian Perspective
   1. According to **Thomas Robert Malthus**, the population (if left unchecked) would exceed the available food supply; population would increase in a geometric progression (2, 4, 8, 16 . . .) while the food supply would increase only by an arithmetic progression (1, 2, 3, 4 . . .).
   2. This situation could end population growth and perhaps the entire population unless positive checks (such as famines, diseases, and wars) or preventive checks (such as sexual abstinence and postponement of marriage) intervened.
B. The Marxist Perspective
   1. According to **Karl Marx** and **Frederick Engels**, food supply does not have to be threatened by overpopulation; through technology, food for a growing population can be produced.
   2. Overpopulation occurs because capitalists want a surplus of workers (an industrial reserve army) to suppress wages and force employees to be more productive.
C. The Neo-Malthusian Perspective
   1. *Neo-Malthusians* (or "New Malthusians") reemphasized the dangers of overpopulation and suggested that population growth is resulting in environmental problems.
   2. Neo-Malthusians advocate **zero population growth**—the point at which no population increase occurs from year to year because the number of births plus immigrants is equal to the number of deaths plus emigrants.
D. Demographic Transition Theory
   1. **Demographic transition** is the process by which some societies have moved from high birth and death rates to relatively low birth and death rates as a result of technological development.
   2. Demographic transition is linked to four stages of economic development:

a. Stage 1: Preindustrial Societies—little population growth occurs, high birth rates are offset by high death rates.
b. Stage 2: Early Industrialization—significant population growth occurs, birth rates are relatively high, and death rates decline.
c. Stage 3: Advanced Industrialization and Urbanization—very little population growth occurs, both birth rates and death rates are low.
d. Stage 4: Postindustrialization—birth rates continue to decline as more women are employed full time and raising children becomes more costly; population growth occurs slowly, if at all, due to a decrease in the birth rate and a stable death rate.

E. Other Perspectives on Population Change
1. Other perspectives on population change include rational choice theory, the epidemiological transition, economic development, and the process of "westernization."

III. A BRIEF GLIMPSE AT INTERNATIONAL MIGRATION THEORIES
A. In explaining international migration, the *neoclassical economics approach*, the *new households economics of migration approach*, and conflict and world systems theory all add to our knowledge of the way people migrate.
B. *Network theory*, *push or pull factors*, and *institutional theory* provide further explanation of global migrations.

IV. URBANIZATION IN GLOBAL PERSPECTIVE
A. *Urban sociology* is a subfield of sociology that examines social relationships and political and economic structures in the city.
B. Emergence and Evolution of the City
1. Cities are a relatively recent innovation compared with the length of human existence. According to **Gideon Sjoberg**, three preconditions must be present in order for a city to develop:
a. A favorable physical environment
b. An advanced technology that could produce a social surplus
c. A well-developed political system to provide social stability to the economic system.
2. Sjoberg places the first cities in the Mesopotamian region or areas immediately adjacent to it at about 3500 B.C.E.; however, not all scholars agree on this point.
C. Preindustrial Cities
1. The largest preindustrial city was Rome.
2. Preindustrial cities were limited in size because of crowded housing conditions, lack of adequate sewage facilities, limited food supplies, and lack of transportation to reach the city.
3. Many preindustrial cities had a sense of *community*—a set of social relationships operating within given spatial boundaries that provide people with a sense of identity and a feeling of belonging.
D. Industrial Cities
1. The Industrial Revolution changed the nature of the city as factories arose and new forms of transportation and agricultural production made moving to the city easier.

2. Between 1700 and 1900, the population of many European cities mushroomed; London increased to about 6.5 million.
3. New York City became the first U.S. *metropolis*—one or more central cities and their surrounding suburbs that dominate the economic and cultural life of a region.

E. Postindustrial Cities
1. Since the 1950s, postindustrial cities have emerged as the U.S. economy has gradually shifted from secondary (manufacturing) to tertiary (service and information processing) production.
2. Postindustrial cities are dominated by "light" industry, such as computer software manufacturing, information processing services, educational complexes, medical centers, retail trade centers, and shopping malls.

V. PERSPECTIVES ON URBANIZATION AND THE GROWTH OF CITIES
A. Functionalist Perspectives: Ecological Models
1. **Robert Park** based his analysis of the city on *human ecology*—the study of the relationship between people and their physical environment—and found that economic competition produces certain regularities in land use patterns and population distributions.
2. Concentric Zone Model
    a. Based on Park's ideas, **Ernest W. Burgess** developed a model that views the city as a series of circular zones, each characterized by a different type of land use, that developed from a central core: (1) the central business district and cultural center; (2) the zone of transition—houses where wealthy families previously lived that have now been subdivided and rented to persons with low incomes; (3) working-class residences and shops, and ethnic enclaves; (4) homes for affluent families, single-family residences of white-collar workers, and shopping centers; and (5) a ring of small cities and towns comprised of estates owned by the wealthy and houses of commuters who work in the city.
    b. Two important ecological processes occur: **invasion** is the process by which a new category of people or type of land use arrives in an area previously occupied by another group or land use; and **succession** is the process by which a new category of people or type of land use gradually predominates in an area formerly dominated by another group or activity.
    c. **Gentrification** is the process by which members of the middle and upper-middle classes, especially whites, move into the central city area and renovate existing properties.
3. Sector Model
    a. **Homer Hoyt**'s sector model emphasizes the significance of terrain and the importance of transportation routes in the layout of cities.
    b. Residences of a particular type and value tend to grow outward from the center of the city in wedge-shaped sectors, with the more expensive residential neighborhoods located along the higher ground near lakes and rivers, or along certain streets that stretch from the downtown area.
    c. Industrial areas are located along river valleys and railroad lines; middle-class residences exist on either side of wealthier neighborhoods; lower-

class residential areas border the central business area and the industrial areas.

4. Multiple Nuclei Model
   a. According to **Chauncey Harris** and **Edward Ullman**, cities have numerous centers of development; as cities grow, they annex outlying townships.
   b. In addition to the central business district, other nuclei develop around activities such as an educational institution or a medical complex; residential neighborhoods may exist close to or far away from these nuclei.

5. Contemporary Urban Ecology
   a. **Amos Hawley** viewed urban areas as complex social systems in which growth patterns are based on advances in transportation and communication.
   b. *Social area analysis* examines urban populations in terms of economic status, family status, and ethnic classification.

B. Conflict Perspectives: Political Economy Models
   1. According to Marx, cities are arenas in which the intertwined processes of class conflict and capital accumulation take place; class consciousness is more likely to occur in cities, where workers are concentrated.
   2. Three major themes are found in political economy models:
      a. Urban growth and decline are affected by economic factors (such as capitalist investments) and political factors (including government protection of private property and promotion of the interests of business elites and large corporations).
      b. Urban space has both an *exchange value*—profits made from buying, selling, and developing land and buildings and a *use value*—the utility of space, land, and buildings for family and neighborhood life.
      c. *Structure*—institutions such as state bureaucracies and capital investment circuits that are involved in the urban development process—and *agency*—human actors who participate in land use decisions (e.g., developers, business elites, and activists protesting development)—are both important in understanding how urban development takes place.
   3. According to political economy models, urban growth is influenced by capital investment decisions, power and resource inequality, class and class conflict, and government subsidy programs.
   4. Gender Regimes in Cities
      a. According to feminist perspectives, urbanization reflects the workings of the political economy and patriarchy.
      b. Different cities have different *gender regimes*—prevailing ideologies of how women and men should think, feel, and act; how access to positions and control of resources should be managed; and how women and men should relate to each other.
      c. Gender intersects with class and race as a form of oppression, especially for lower-income women of color who live in central cities.

C. Symbolic Interactionist Perspectives: The Experience of City Life

1. Simmel's View of City Life
    a. According to **Georg Simmel**, urban life is highly stimulating; it shapes people's thoughts and actions.
    b. However, many urban residents avoid emotional involvement with each other and try to ignore events taking place around them.
    c. City life is not completely negative; urban living can be liberating—people have opportunities for individualism and autonomy.
2. Urbanism as a Way of Life
    a. **Louis Wirth** suggested that urbanism is a "way of life." *Urbanism* refers to the distinctive social and psychological patterns of city life.
    b. Size, density, and heterogeneity result in an elaborate division of labor and in spatial segregation of people by race/ethnicity, class, religion, and/or lifestyle; a sense of community is replaced by the *mass society*—a large-scale, highly institutionalized society in which individuality is supplanted by mass media, faceless bureaucrats, and corporate interests.
3. Gans's Urban Villagers
    a. According to **Herbert Gans**, not everyone experiences the city in the same way; some people develop strong loyalties and a sense of community within central city areas that outsiders may view negatively.
    b. Five major categories of urban dwellers are: (1) cosmopolites—students, artists, writers, musicians, entertainers, and professionals who live in the city for its cultural facilities; (2) unmarried people and childless couples who live in the city for work and entertainment; (3) ethnic villagers who live in ethnically segregated neighborhoods; (4) the deprived—individuals who are very poor and see few future prospects; and (5) the trapped—those who cannot escape the city, including downwardly mobile persons, older persons, and persons with addictions.
4. Gender and City Life
    a. According to **Elizabeth Wilson**, some men view the city as *sexual space* in which women, based on their sexual desirability and accessibility, are categorized as prostitutes, lesbians, temptresses, or virtuous women in need of protection.
    b. Affluent, dominant group women are more likely to be viewed as virtuous women in need of protection, while others are placed in less desirable categories.
5. Cities and Persons with a Disability
    a. Many cities and urban areas create a "disabling" environment for many people.
    b. **Harlan Hahn** suggests that historical patterns in the dynamics of capitalism contributed to discrimination against persons with disabilities, and this legacy remains today.

VI. PROBLEMS IN GLOBAL CITIES
    A. Natural increases in population account for two-thirds of new urban growth, and rural-to-urban migration accounts for the remainder.

383

B. Rapid global population growth in Latin American and other regions is producing a wide variety of urban problems, including overcrowding, environmental pollution, and the disappearance of farmland.

C. Most cities in Africa, South America, and the Caribbean are in peripheral nations.

D. Cities in semiperipheral nations, such as India, Iran, and Mexico, are confronted with unprecedented population growth.

VII. URBAN PROBLEMS IN THE UNITED STATES

A. Divided Interests: Cities, Suburbs, and Beyond

1. Since World War II, the U.S. population has shifted dramatically as many people have moved to the suburbs.

2. Suburbanites rely on urban centers for employment and some services but pay property taxes to suburban governments and school districts; some affluent suburbs have state of the art school districts and infrastructure, while central city services and school districts lack funds.

3. Race, Class, and Suburbs

a. The intertwining impact of race and class is visible in the division between central cities and suburbs.

b. Most suburbs are predominantly white; many upper-middle- and upper-class suburbs remain virtually all white; people of color who live in suburbs often are resegregated.

4. Beyond the Suburbs

a. *Edge cities* initially develop as residential areas beyond central cities and suburbs; then retail establishments and office parks move into the area and create an unincorporated edge city.

b. Corporations move to edge cities because of cheaper land and lower utility rates and property taxes.

5. Likewise, Sunbelt cities grew in the 1970s, as millions moved from the north and northeastern states to southern and western states where there were more jobs and higher wages, lower taxes, pork-barrel programs funded by federal money that created jobs and encouraged industry, and the presence of high-technology industries.

B. The Continual Fiscal Crisis of the Cities

1. The largest cities in the United States have faced periodic fiscal crisis for many years, intensified by higher employee health care and pension costs, declining revenue, and increased expenditures for public safety and homeland safety.

2. Many cities have cut back on spending in areas other than public safety.

3. Even if the U.S. economy improves significantly in the near future, analysts believe that the positive effects of such a rebound will not improve the budgetary problems of our cities and towns for a number of years.

VIII. RURAL COMMUNITY ISSUES IN THE UNITED STATES

A. About 20 percent of the U.S. population resides in rural areas, identified as communities of 2,500 or less by the U.S. Census Bureau.

1. Rural communities today are more diverse; recently, more people from large urban areas and suburbs have moved into rural areas.

2. These recent immigrants to rural areas do not face some traditional problems experienced by long-term rural residents.

B. Individuals in rural areas whose livelihood is farming or other agricultural endeavors have experienced several difficulties, including crop failures and loss of small businesses and farms; problems of divorce, alcoholism, and abuse; limited economic opportunities; and limited health care.

C. Other Influencing Factors

1. Proliferation of superstores (e.g., Wal-Mart) putting local small businesses out of business.
2. Increase in tourism in rural America.

IX POPULATION AND URBANIZATION IN THE FUTURE

A. Rapid global population growth is inevitable: although death rates have declined in many low-income nations, there has not been a corresponding decrease in birth rates.

B. In the future, low-income countries will have an increasing number of poor people. The world's population will double, and the urban population will triple as people migrate from rural to urban areas.

C. At the macrolevel, we may be able to do little about population and urbanization; at the microlevel, we may be able to exercise some degree of control over our communities and our own lives.

CRITICAL THINKING QUESTIONS

1. Why is the global population increasing?
2. Thomas Malthus and neo-Malthusians argue that population growth has negative consequences and needs to be controlled; Marx and Engels argue that capitalism, not population growth, is the problem. Compare and contrast these two arguments. Which one do you agree with more? Why?
3. What are some ways zero population growth could be achieved?
4. Which theoretical perspective do you think is most useful for understanding urbanization?
5. In what ways do race, class, and gender shape city life?

PRACTICE TESTS

MULTIPLE CHOICE QUESTIONS

Select the response that best answers the question or completes the statement.

1. _____ is a subfield of sociology that examines population size, composition, and distribution.
   a. Population sociology
   b. Demography
   c. Human ecology
   d. Urban sociology

385

2. _____ is the actual level of childbearing for an individual or a population, while _____ is the potential number of children that could be born if every woman reproduced at her maximum biological capacity.
   a. Birth rate; fertility rate
   b. Fertility rate; birth rate
   c. Fertility; fecundity
   d. Fecundity; fertility

3. The primary cause of world population growth in recent years is a(n)
   a. increase in the birth rate.
   b. decline in the death rate.
   c. decline in all infectious diseases.
   d. increase in post–baby boom birth rates.

4. The leading cause of death in the U.S. in 2006 was:
   a. accidents
   b. pneumonia
   c. heart disease
   d. diabetes

5. _____ is the movement of people out of a geographic area to take up residency elsewhere.
   a. Immigration
   b. Emigration
   c. Transmigration
   d. Ex-migration

6. According to Thomas Malthus's perspective on population,
   a. the population would increase in a geometric progression, while the food supply would increase in an arithmetic progression.
   b. the population would increase in an arithmetic progression, while the population would increase in a geometric progression.
   c. the food supply is not threatened by overpopulation because technology makes it possible to produce the food and other goods needed to meet the demands of a growing population.
   d. societies move through a process of demographic transition.

7. According to Karl Marx and Frederick Engels's perspective on population,
   a. the population would increase in a geometric progression, while the food supply would increase in an arithmetic progression.
   b. the population would increase in an arithmetic progression, while the food supply would increase in a geometric progression.
   c. the food supply is not threatened by overpopulation because technology makes it possible to produce the food and other goods needed to meet the demands of a growing population.
   d. societies move through a process of demographic transition.

8. According to the demographic transition theory, significant population growth occurs because birth rates are relatively high while death rates decline in the _____ stage of economic development.
    a. preindustrial
    b. early industrial
    c. advanced industrial
    d. postindustrial

9. The population pyramid that depicts the sex and age of individuals in the U.S. is most similar to
    a. a classic pyramid.
    b. an upside-down pyramid.
    c. two classic pyramids stacked one on top of the other.
    d. a rectangle or barrel shape.

10. The *Gemeinschaft* and *Gesellschaft* typology originated with:
    a. Emile Durkheim
    b. Gideon Sjoberg
    c. Max Weber
    d. Ferdinand Tönnies

11. _____ refers to one or more central cities and their surrounding suburbs that dominate the economic and cultural life of a region.
    a. Urban sprawl
    b. Megalopolis
    c. Metropolis
    d. Urbanization

12. Postindustrial cities are characterized by
    a. "light" industry, information processing services, educational complexes, retail trade centers, and shopping malls.
    b. the growth of the factory system.
    c. agricultural production.
    d. "heavy" industry, such as automobile manufacturing.

13. Ecological models of urban growth are based on a _____ perspective.
    a. functionalist
    b. conflict
    c. neo-Marxist
    d. interactionist

14. All of the following are ecological models of urban growth, **except** the:
    a. concentric zone model
    b. sector model
    c. urban sprawl model
    d. multiple nuclei model

15. An upper-middle-class doctor who moves her family from the suburbs into the central city to renovate an older home is an example of:
    a. succession
    b. gentrification
    c. ex-suburbanization
    d. downward immigration

16. According to political economy models, urban growth is
    a. influenced by terrain and transportation.
    b. based on the clustering of people who share similar characteristics.
    c. linked with peaks and valleys in the economic cycle.
    d. influenced by capital investment decisions, power and resource inequality, and government subsidy programs.

17. _____ refers to the tendency of some neighborhoods, cities, or regions to grow and prosper while others stagnate and decline.
    a. Invasion
    b. Succession
    c. Gentrification
    d. Uneven development

18. The _____ model views cities as a series of circular zones that are each characterized by a particular land use.
    a. concentric zone
    b. sector
    c. multiple nuclei
    d. capitalist

19. Sociologist _____ has argued that urbanism is a "way of life."
    a. Herbert Gans
    b. Georg Simmel
    c. Louis Wirth
    d. Elizabeth Wilson

20. New urban areas that are created beyond central cities and suburbs, allow commuters to travel around rather than in and out of the metropolitan center, and attract businesses and industries through cheaper land and lower property taxes are known as:
    a. new ruralism
    b. edge cities
    c. suburbs
    d. gentrified spaces

21. _____ argued that people become somewhat insensitive to individuals and events around them as a result of the intensity of city life.
    a. Georg Simmel
    b. Robert Park
    c. Ferdinand Tönnies
    d. Thomas Robert Malthus

22. The _____ perspective emphasizes the dangers of overpopulation, including food shortages and global environmental problems.
    a. concentric zone
    b. Marxist
    c. neo-Malthusian
    d. demographic transition

23. The _____ perspective suggests that societies progress through four stages of economic development, and that each corresponds to different levels of population growth.
    a. concentric zone
    b. Marxist
    c. neo-Malthusian
    d. demographic transition

24. The most basic measure of fertility is the _____ rate.
    a. fecundity
    b. age-specific birth
    c. crude birth
    d. reproduction

25. The primary cause of world population growth in recent years has been a decline in _____—the incidence of death in a population.
    a. the crude death rate
    b. the total death rate
    c. the death quota
    d. mortality

26. In the _____ stage, little population growth occurs because high birth rates are offset by high death rates.
    a. advanced industrial
    b. early industrial
    c. preindustrial
    d. postindustrial

27. _____ is the process by which a new category of people or type of land use arrives in an area previously occupied by another group or type of land use.
    a. Invasion
    b. Gentrification
    c. Succession
    d. Concentrification

28. _____ is the number of people living in a specific geographic area.
    a. Distribution
    b. Migration
    c. Emigration
    d. Density

389

29. _____ is the movement of people out of a geographic area to take up residency elsewhere.
   a. Distribution
   b. Migration
   c. Immigration
   d. Emigration

30. _____ factors of migration at the international level include violence, war, famine, and political unrest.
   a. Pull
   b. Push
   c. Discharge
   d. Hinge

31. _____ was one of the first scholars to systematically study the effects of population. He argued that "the power of population is infinitely greater than the power of the earth to produce subsistence (food) for man."
   a. Thomas Malthus
   b. Max Weber
   c. Karl Marx
   d. Emile Durkheim

32. One measure of population composition is the _____, which is the number of males for every hundred females in a given population.
   a. sex ratio
   b. rate of gender composition
   c. population composition
   d. refined gender rate

33. Thomas Malthus argued that the population would increase in a geometric (exponential) progression while the food supply would increase only by an arithmetic progression; thus, a _____ occurs.
   a. doubling effect
   b. positive check
   c. preventive check
   d. demographic transition

34. The _____ model suggests that cities have numerous centers of development based on specific urban needs or activities rather than a single center.
   a. concentric zone
   b. multiple nuclei
   c. contemporary urban ecology
   d. sector

35. The concentric zone, sector, and multiple nuclei models of cities are part of the _____ perspective.
   a. conflict
   b. functionalist
   c. postmodern
   d. symbolic interactionist

36. A _____ refers to one or more central cities and their surrounding suburbs that dominate the economic and cultural life of a region.
    a. metropolis
    b. megalopolis
    c. conglomeration
    d. census district

37. Sociologist Robert Park based his analysis of the city on _____, which is the study of the relationship between people and their physical environment.
    a. demography
    b. urban sociology
    c. human ecology
    d. physical sociology

38. Urban ecologist _____ revitalized the ecological tradition by linking it more closely with functionalism.
    a. Herbert Gans
    b. Louis Wirth
    c. Amos Hawley
    d. Joe Fagin

39. According to _____, many residents develop strong loyalties and a sense of community in central city areas that outsiders may view negatively.
    a. Georg Simmel
    b. Herbert Gans
    c. Homer Hoyt
    d. Ernest W. Burgess

40. A lending practice of banks includes the _____ of certain properties so that acquiring a loan is virtually impossible.
    a. redlining
    b. stonewalling
    c. barricading
    d. edging

TRUE/FALSE QUESTIONS

1. Demographers analyze fertility, mortality, and migration rates.
   T   F

2. The world's population is increasing by more than 76 million people per year.
   T   F

3. In most areas of the world, women are having more children than in the past.
   T   F

4. The primary cause of world population growth in recent years has been a decline in mortality.
   T   F

5. Urban sociology is a subfield of sociology that examines social relationships and political and economic structures in the city.
   T   F

6. The largest cities in the U.S. have seen vast financial growth and prosperity in recent years.
   T   F

7. Race resegregation often occurs in the suburbs.
   T   F

8. According to the Marxist perspective, overpopulation occurs because capitalists desire to have a surplus of workers so as to suppress wages and increase workers' productivity.
   T   F

9. About 50 percent of the world's population lives in cities.
   T   F

10. The sector model views the city as a series of circular areas or zones, each characterized by a different type of land use that developed from a central core.
    T   F

11. According to conflict theorists, cities grow and decline by chance.
    T   F

12. According to feminist theorists, public patriarchy may be perpetuated by cities through policies that limit women's access to paid work and public transportation.
    T   F

13. Symbolic interactionists examine the experience of urban life rather than the political economy of the city.
    T   F

14. Herbert Gans has suggested that almost all city dwellers live in urban areas by choice.
    T   F

15. Nationally, most suburbs are predominantly white.
    T   F

16. The proliferation of superstores (Wal-Mart, Lowes) and increases in tourism have changed the face of some rural areas in recent years.
    T   F

17. According to Box 19.1, unauthorized immigrants from Mexico and Latin America represent slightly less than 50 percent of the unauthorized population in the United States.
    T   F

18. According to Box 19.3, Mexican immigrants working in the United States in mostly low-paying jobs send most of their monies back home, which is more than Mexico earns from tourism or foreign investment.
    T   F

19. In some global cities, according to Box 19.4, less than 10 percent of the population has regular collection of household waste.
T   F

20. According to Box 19.2, the Refugee Women's Alliance is a nonprofit, multiethnic organization that provides services for newly arrived immigrant families in the state of Washington.
T   F

s-THE-BLANK QUESTIONS

1. The biological and social characteristics of a population are known as the _____.

2. A graphic representation of a population by sex and age is a _____.

3. According to _____, limits to fertility are defined as _____.

4. The point at which no population increase occurs from year to year is defined as _____.

5. According to Lynn Appleton, different kinds of cities have different _____, which are prevailing ideologies of how men and women should think.

6. According to the sociologist Herbert Gans, _____ are students, artists, entertainers, and professionals who choose to live in the city.

7. The _____ model of urban development cities has numerous centers of development.

8. According to _____, urban living could have a liberating effect on people because they have opportunities for individualism and autonomy.

9. _____ is the movement of people into a geographic area to take up residency.

10. Things that attract people to move into another country are known as _____ factors.

11. The incidence of death in a population is _____.

12. The first metropolis in the U. S. was _____.

13. For Thomas Malthus, the only acceptable preventative check was _____.

14. The study of the relationship between people and their physical environment is defined as _____.

15. _____ is the process by which members of the middle and upper-middle classes (especially whites) move into the central city area and renovate existing properties.

SHORT ANSWER/ESSAY QUESTIONS

1. What is demography, and how are people affected by demographic changes?
2. What is the Malthusian perspective? Do the views of Karl Marx support this? Explain.
3. Explain the three functionalist models of urban growth.
4. What are the stages in the demographic transition theory?
5. What issues currently affect U.S. cities?

STUDENT CLASS PROJECTS AND ACTIVITIES

1. Utilizing the most current U.S. Census Bureau data, construct a population pyramid of the United States, your home state, and the city in which you live or reside (or of any city in the United States). Provide the pyramids, interpretations of the statistics, a summary of the data, a conclusion from each pyramid constructed, and bibliographical references.
2. Construct a demographic analysis of the state in which your college or university is located and compare this with an analysis of the entire United States. Present data on the size of the population, the numbers and percentage of increase or decrease from the past twenty years, the birth rate, death rate, sex ratio, and infant mortality rate, the leading cause of death, and life expectancy. Provide a summary of the data, a conclusion, and an evaluation of this project.
3. Look up current information on organizations that advocate zero population growth and investigate the kinds of activities these organizations are engaging in to try to reduce population growth. Write a paper that includes (1) a summary of the activities aimed at reducing population growth, (2) a summary of reports on the effectiveness of these activities, and (3) your analysis of the effectiveness and value of these efforts.
4. Construct a plan of a model city. In the plan, provide for all necessary public facilities, such as: an efficient transportation system, an efficient governmental system, the most efficient use of land, a provision for industrial and commercial bases, a provision for educational, medical, recreational, and religious facilities, a provision for optimum housing, promotions for attracting an equitable balance of ethnic, age, and income groups, and a provision for long-range planning for optimum population and city size. Include any other amenities you want in the planning of this optimum model city.

INTERNET ACTIVITIES

1. Explore the HUD site, **http://www.hud.gov/**, and learn about the history of the government's involvement in the life of cities. The mission of the **Department of Housing and Urban Development** is to provide "a decent, safe, and sanitary home and suitable living environment for every American."
2. Explore the **United Nations Population Information Network** site, **http://www.un.org/popin**, to learn about world population trends. It is possible from

this site to obtain statistical reports, United Nations documents, wall charts, and records of population policies of governments and world organizations.

3. **Population.com**, **http://www.population.com/**, is a huge resource for studies on population, migration, and urbanization. This site has an extensive report, as well as charts, on population trends around the world. You can apply some of the concepts you have learned in this chapter to a specific region of the globe with a distinct history and culture.

4. Visit the **Population Reference Bureau** web site, **http://www.prb.org/**, to obtain the most recent World Population Data Sheet, which provides lots of demographic information for every country in the world. You can also browse the site to find out more about the organization and read publications on other population issues, including the impact of war, aging, and family planning.

5. Explore **Population Connection**, **http://www.populationconnection.org/site/PageServer**, which is an organization that promotes zero population growth. Look at the kinds of activities this organization is engaged in and the ways it attempts to minimize population growth. Critically evaluate these practices.

INFOTRAC COLLEGE EDITION EXERCISES

Visit the **InfoTrac College Edition** web site at: **http://www.wadsworthmedia.com/webtutor/infotrac.htm**. You will arrive at a screen that enables you to search topics.

1. Investigate **global population** from a variety of perspectives. Determine the key issues based on these articles.

2. A key component in understanding **demographic transitions** is the **sex ratio** of a society. You will need to weed out articles about fig wasps and turtles, but there are a number of very informative reports and research articles on this phenomenon. Interesting comparisons can be made between biological studies of animals and those about human social behavior.

3. Conduct a keyword search for **zero population growth**. Find the article entitled: "Allowing fertility decline: 200 years after Malthus's essay on population." The economic opportunity model, in contrast to the demographic transition model, is presented as a way to reduce population growth. Analyze this new alternative in light of what you have learned from reading Chapter 19.

4. Search InfoTrac for articles about **urbanization**. Determine what the current issues are regarding urbanization and evaluate the articles.

SOLUTIONS

MULTIPLE CHOICE QUESTIONS

1. B, p. 621	15. B, p. 638	29. D, p. 624
2. C, p. 621	16. D, p. 639	30. B, p. 624
3. B, p. 623	17. D, p. 641	31. A, p. 630
4. C, p. 623	18. A, p. 637	32. A, p. 629
5. B, p. 624	19. C, p. 642	33. A, p. 631
6. A, p. 630	20. B, p. 648	34. B, p. 631
7. C, p. 631	21. A, p. 642	35. B, p. 637
8. B, p. 632	22. C, p. 631	36. A, p. 636
9. D, p. 629	23. D, p. 632	37. C, p. 637
10. D, p. 636	24. C, p. 622	38. C, p. 639
11. C, p. 636	25. D, p. 623	39. B, p. 642
12. A, p. 636	26. A, p. 632	40. A, p. 648
13. A, p. 637	27. A, p. 638	
14. C, p. 637	28. D, p. 624	

TRUE/FALSE QUESTIONS

1. T, p. 621	8. T, p. 631	15. T, p. 647
2. T, p. 621	9. T, p. 635	16. T, p. 649
3. F, p. 622	10. F, p. 639	17. T, p. 622
4. T, p. 623	11. F, p. 639	18. T, p. 634
5. T, p. 635	12. T, p. 641	19. T, p. 647
6. F, p. 649	13. T, p. 641	20. T, p. 625
7. T, p. 648	14. F, p. 642	

FILL-IN-THE-BLANK QUESTIONS

1. population composition, p. 625
2. population pyramid, p. 629
3. Thomas Malthus, p. 630; preventive checks, p. 631
4. zero population growth, p. 632
5. gender regimes, p.641
6. cosmopolites, p. 642
7. multiple nuclei, p. 639
8. Georg Simmel, p. 642
9. Immigration, p. 624
10. pull, p. 624
11. mortality, p. 623
12. New York City, p. 636
13. moral restraint, p. 631
14. human ecology, p. 637
15. Gentrification, p. 638

20

COLLECTIVE BEHAVIOR, SOCIAL MOVEMENTS, AND SOCIAL CHANGE

BRIEF CHAPTER OUTLINE

CHAPTER SUMMARY

Social change is the alteration, modification, or transformation of public policy, culture, or social institutions over time. Such change usually is brought about by **collective behavior**—voluntary, often spontaneous activity that is engaged in by a large number of people and typically violates dominant group norms and values. A **crowd** is a relatively large number of people who are in one another's immediate vicinity. Five categories of crowds have been identified: (1) *casual crowds* are relatively large gatherings of people who happen to be in the same place at the same time; (2) *conventional crowds* are comprised of people who specifically come together for a scheduled event and thus share a common focus; (3) *expressive crowds* provide opportunities for the expression of some strong emotion; (4) *acting crowds* are collectivities so intensely focused on a specific purpose or object that they may erupt into violent or destructive behavior; and (5) *protest crowds* are gatherings of people who engage in activities intended to achieve specific political goals. Acting crowds can take the form of a **mob**—a highly emotional crowd whose members are prepared to engage in violence; a **riot**—violent crowd behavior fueled by deep-seated emotions not directed at a target; or a **panic**—which occurs when a large number of people react with strong emotions and self-destructive behavior to a real or perceived threat. Protest crowds sometimes participate in **civil disobedience**—nonviolent action that seeks to change a policy or law by refusing to comply with it. Explanations of crowd behavior include contagion theory, social unrest and circular reaction, convergence theory, and emergent norm theory. Examples of **mass behavior**—collective behavior that takes place when people respond to the same event in much the same way—include **rumors**, **gossip**, mass hysteria, fads, fashions, and public opinion. The major types of **social movements**—organized groups that act consciously to promote or resist change through collective action—are reform movements, revolutionary movements, religious movements, alternative movements, and resistance movements. Sociological theories explaining social movements include relative deprivation theory, value-added theory, resource mobilization theory, social constructionist theory, political opportunity theory, and new social movement theory. Social change produces many challenges that remain to be resolved: environmental problems, changes in the demographics of the population, and new technology that benefits some—but not all—people. As we head into the future, we must use our sociological imaginations to help resolve the issues of the twenty-first century.

LEARNING OBJECTIVES

After reading Chapter 20, you should be able to:

1. Define collective behavior and describe the conditions necessary for such behavior to occur.

2. Distinguish between crowds and masses, and identify casual, conventional, expressive, acting, and protest crowds.

3. Distinguish among mobs, riots, and panics.

4. Distinguish the key elements of these four explanations of collective behavior: contagion theory, social unrest and circular reaction, convergence theory, and emergent norm theory.

5. Define mass behavior and describe the most frequent types of this behavior.

6. Describe the difference between rumors and gossip, fads and fashions, and public opinion and propaganda.

7. Integrate Simmel's, Veblen's and Bourdieu's perspectives on fashion.

8. Describe social movements and note when and where they are most likely to develop.

9. Differentiate among the five major types of social movements based on their goals and the amount of change they seek to produce.

10. Evaluate the effectiveness of different types of social movements.

11. Determine why social movements may be an important source of social change.

12. Identify the stages in social movements.

13. Compare relative deprivation theory and value-added theory as explanations of why people join social movements.

14. State the key assumptions of resource mobilization theory.

15. State the key assumptions of the social constructionist approach to understanding social movements and explain how it incorporates frame analysis.

16. Explain the main ideas of political opportunity theory.

17. Describe new social movement theory and the types of social movements on which it focuses.

18. Show how the current concern with global climate change can develop into a social movement.

19. Assess the predictions about social change presented in the conclusion of this chapter.

KEY TERMS

(defined at page number shown and in glossary)

civil disobedience, p. 660
collective behavior, p. 656
crowd, p. 659
environmental racism, p. 674
gossip, p. 664
mass, p. 659
mass behavior, p. 663
mob, p. 660

panic, p. 660
propaganda, p. 665
public opinion, p. 665
riot, p. 660
rumors, p. 663
social change, p. 656
social movement, p. 666

KEY PEOPLE

(identified at page number shown)

Herbert Blumer, p. 659
Pierre Bourdieu, p. 665
Rachel Carson, p. 656
Lois Gibbs, p. 657
Erving Goffman, p. 665
Gustave Le Bon, p. 657

John Lofland, p. 659
Robert E. Park, p. 661
Georg Simmel, p. 665
Neil Smelser, p. 669
Thorstein Veblen, p. 665

CHAPTER OUTLINE

I. COLLECTIVE BEHAVIOR
   A. **Social change** is the alteration, modification, or transformation of public policy, culture, or social institutions over time; such change usually is brought about by **collective behavior**—relatively spontaneous, unstructured activity that typically violates established social norms.
   B. Conditions for Collective Behavior
      1. Collective behavior occurs as a result of some common influence or stimulus that produces a response from a *collectivity*—a relatively large number of people who mutually transcend, bypass, or subvert established institutional patterns and structures.
      2. Major factors that contribute to the likelihood that collective behavior will occur are:
         a. Structural factors that increase the chances of people responding in a particular way
         b. Timing
         c. A breakdown in social control mechanisms and a corresponding feeling of normlessness
         d. A common stimulus, for example, the publication of the book *Silent Spring*

400

C. Dynamics of Collective Behavior
1. People may engage in collective behavior when they find that their problems are not being solved through official channels; as the problem appears to grow worse, organizational responses become more defensive and obscure.
2. People's attitudes are not always reflected in their political and social behavior.
3. People act collectively in ways they would not act singly due to:
    a. The noise and activity around them
    b. A belief that it is the only way to fight those with greater power and resources
D. Distinctions Regarding Collective Behavior
1. People engaging in collective behavior may be a:
    a. **Crowd**—a relatively large number of people who are in one another's immediate, face-to-face presence.
    b. **Mass**—a number of people who share an interest in a specific idea or issue but who are not in one another's immediate physical vicinity.
2. Collective behavior also may be distinguished by the *dominant emotion* expressed (e.g., fear, hostility, joy, grief, disgust, surprise, or shame).
E. Types of Crowd Behavior
1. **Herbert Blumer** divided crowds into four categories:
    a. *Casual crowds*—relatively large gatherings of people who happen to be in the same place at the same time; if they interact at all, it is only briefly.
    b. *Conventional crowds*—people who specifically come together for a scheduled event and thus share a common focus.
    c. *Expressive crowds*—people releasing their pent up emotions in conjunction with others who experience similar emotions.
    d. *Acting crowds*—collectivities so intensely focused on a specific purpose or object that they may erupt into violent or destructive behavior. Examples:
        i. A **mob**—a highly emotional crowd whose members engage in, or are ready to engage in, violence against a specific target, which may be a person, a category of people, or physical property.
        ii. A **riot**—violent crowd behavior fueled by deep-seated emotions, but not directed at a specific target.
        iii. A **panic**—a form of crowd behavior that occurs when a large number of people react with strong emotions and self-destructive behavior to a real or perceived threat.
2. To these four types of crowds, Clark McPhail and Ronald T. Wohlstein added *protest crowds*—crowds that engage in activities intended to achieve specific political goals.
    a. Protest crowds sometimes take the form of **civil disobedience**—nonviolent action that seeks to change a policy or law by refusing to comply with it.
    b. At the grassroots level, protests often are seen as the only way to call attention to problems or demand social change.

F. Explanations of Crowd Behavior
   1. According to *contagion theory*, people are more likely to engage in antisocial behavior in a crowd because they are anonymous and feel invulnerable; **Gustave Le Bon** argued that feelings of fear and hate are contagious in crowds because people experience a decline in personal responsibility.
   2. According to **Robert Park**, social unrest is transmitted by a process of *circular reaction*—the interactive communication between persons in such a way that the discontent of one person is communicated to another whom, in turn, reflects the discontent back to the first person.
   3. *Convergence theory* focuses on the shared emotions, goals, and beliefs many people bring to crowd behavior.
      a. From this perspective, people with similar attributes find a collectivity of like-minded persons with whom they can release their underlying personal tendencies.
      b. Although people may reveal their "true selves" in crowds, their behavior is not irrational; it is highly predictable to those who share similar emotions or beliefs.
   4. According to *emergent norm theory*, crowds develop their own definition of the situation and establish norms for behavior that fits the occasion.
      a. Emergent norms occur when people define a new situation as highly unusual or see a long-standing situation in a new light.
      b. Emergent norm theory points out that crowds are not irrational; new norms are developed in a rational way to fit the needs of the immediate situation.
G. Mass Behavior
   1. **Mass behavior** is collective behavior that takes place when people (who often are geographically separated from one anther) respond to the same event in much the same way.
   2. The most frequent types of mass behavior are **rumors**—unsubstantiated reports on an issue or subject—and **gossip**—rumors about the personal lives of individuals.
   3. *Mass hysteria* is a form of dispersed collective behavior that occurs when a large number of people react with strong emotions and self-destructive behavior to a real or perceived threat; many sociologists believe this behavior is best described as a panic with a dispersed audience.
   4. Fads and Fashions
      a. A *fad* is a temporary but widely copied activity enthusiastically followed by large numbers of people.
      b. *Fashion* is a currently valued style of behavior, thinking, or appearance. Fashion also applies to art, music, drama, literature, architecture, interior design, and automobiles, among other things.
   5. **Public opinion** consists of the attitudes and beliefs communicated by ordinary citizens to decision makers (as measured through polls and surveys based on interviews and questionnaires).
      a. Even on a single topic, public opinion will vary widely based on characteristics such as race, ethnicity, religion, region of the country, urban or rural residence, social class, education level, gender, and age.

b. As the masses attempt to influence elites and vice versa, a two-way process occurs with the dissemination of **propaganda**—information provided by individuals or groups that have a vested interest in furthering their own cause or damaging an opposing one.

II. SOCIAL MOVEMENTS

A. A **social movement** is an organized group that acts consciously to promote or resist change through collective action. Social movements are categorized based on their goals and the amount of change they seek to produce.

B. Types of Social Movements

1. *Reform movements* seek to improve society by changing some specific aspect of the social structure.

2. *Revolutionary movements* seek to bring about a total change in society.

3. *Religious movements* seek to produce radical change in individuals and typically are based on spiritual or supernatural belief systems.

4. *Alternative movements* seek limited change in some aspect of people's behavior (e.g., a movement that attempts to get people to abstain from drinking alcoholic beverages).

5. *Resistance movements* seek to prevent change or undo change that already has occurred.

C. Stages in Social Movements

1. In the *preliminary stage*, widespread unrest is present as people begin to become aware of a threatening problem. Leaders emerge to agitate others into taking action.

2. In the *coalescence stage*, people begin to organize and start making the threat known to the public. Some movements become formally organized at local and regional levels.

3. In the *institutionalization stage*, an organizational structure develops, and a paid staff (rather than volunteers) begins to lead the group.

III. SOCIAL MOVEMENT THEORIES

A. *Relative deprivation theory* asserts that people who suffer relative deprivation are likely to feel that a change is necessary and join a social movement in order to bring about that change.

B. According to Neil Smelser's *value-added theory*, six conditions are necessary and sufficient to produce social movements when they combine or interact in a particular situation:

1. structural conduciveness

2. structural strain

3. spread of a generalized belief

4. precipitating factors

5. mobilization for action

6. social control factors

C. *Resource mobilization theory* focuses on the ability of a social movement to acquire resources (money, time and skills, access to the media, etc.) and mobilize people to advance the cause.

D. *Social constructionist theory* is based on the assumption that social movements are interactive, symbolically defined, and a negotiated process.

403

1. This process involves participants, opponents, and bystanders.
2. Research based on this perspective often investigates how problems are framed and what names they are given.
3. Distinct frame alignment processes occur in social movements:
   a. Frame bridging
   b. Frame amplification
   c. Frame extension
   d. Frame transformation

E. *Political opportunity theory* states that potential protestors and movement organizers will engage in collective action when such opportunities are most readily available and will produce the most favorable outcome.
   1. *Opportunity* refers to options for collective action, with chances and risks attached to them that depend on factors outside the mobilizing group.
   2. This approach highlights the interplay of opportunity, mobilization, and influence in determining when certain types of behavior may occur.

F. *New social movement theory* looks at a diverse array of collective actions and the manner in which those actions are based in politics, ideology, and culture.
   1. It incorporates sources of identity, including race, class, gender, and sexuality, as sources of collective action and social movements.
   2. Examples of already existing "new social movements" include ecofeminism and environmental justice movements.
      a. According to *ecofeminism*, patriarchy is a root cause of environmental problems because it contributes to a belief that nature is to be possessed and dominated, rather than treated as a partner.
      b. Environmental justice movements focus on the issue of **environmental racism**—the belief that a disproportionate number of hazardous facilities (including industries such as waste disposal/treatment and chemical plants) are placed in low-income areas populated primarily by people of color.

IV. SOCIAL CHANGE IN THE FUTURE
   A. The Physical Environment and Change: Changes in the physical environment often produce changes in the lives of people; in turn, people can make dramatic changes in the physical environment, over which we have only limited control.
   B. Population and Change: Changes in population size, distribution, and composition affect the culture and social structure of a society and change relationships among nations.
   C. Technology and Change: Advances in communication, transportation, science, and medicine have made significant changes in people's lives, especially in developed nations; however, these changes also have created the potential for new disasters, ranging from global warfare to localized technological disasters at toxic waste sites.
   D. Social Institutions and Change: During the past twentieth century, many changes occurred in the family, religion, education, the economy, and the political system; at the beginning of the new century, the U.S. government seemed less able to respond to the needs and problems of the country.

E. Changes in physical environment, population, technology, and social institutions operate together in a complex relationship, sometimes producing consequences we must examine by using our sociological imagination.

F. A Final Thought: One purpose of this text is to facilitate understanding of different viewpoints in order to deal with the issues of the twenty-first century, thereby producing a better way of life in this country and worldwide.

CRITICAL THINKING QUESTIONS

1. Which forms of collective behavior have you engaged in? What were the circumstances?
2. Many of the examples of collective behavior and social movements throughout the chapter deal with environmental issues. What do you think are the most effective methods of mobilizing a population to engage in more activities to protect the environment?
3. Compare and contrast the different types of social movements and think of an example of each that does not appear in your textbook.
4. Which theory do you think best explains social movements? What are the strengths and weaknesses of each?
5. In what ways do you think U.S. society will change throughout the twenty-first century?

PRACTICE TESTS

MULTIPLE CHOICE QUESTIONS

Select the response that best answers the question or completes the statement.

1. _____ typically relies on structural factors, timing, breakdown in social control mechanisms, and a common stimulus.
   a. Social change
   b. Institutional behavior
   c. Collective behavior
   d. Organizational behavior

2. A _____ is a number of people who share an interest in a specific issue but are not in close proximity to each other.
   a. crowd
   b. mass
   c. riot
   d. mob

3. Casual crowds are
   a. comprised of people who specifically come together for a scheduled event and thus share a common focus.
   b. situations that provide an opportunity for the expression of some strong emotion.
   c. comprised of people who happen to be in the same place at the same time.
   d. comprised of people who are so intensely focused on a specific purpose or object that they may erupt into violent or destructive behavior.

4. Revelers assembled at Mardi Gras or on New Year's Eve at Times Square in New York are an example of:
   a. protest crowds
   b. expressive and acting crowds
   c. conventional crowds
   d. panics

5. When the residents of Love Canal burned both the governor and the health commission in effigy, they were taking part in a(n):
   a. acting crowd
   b. casual crowd
   c. conventional crowd
   d. panic

6. Convergence theory focuses on
   a. the social-psychological aspects of collective behavior, including how moods, attitudes, and behavior are communicated.
   b. how social unrest is transmitted by a process of circular reaction.
   c. the importance of social norms in shaping crowd behavior.
   d. the shared emotions, goals, and beliefs many people bring to crowd behavior.

7. Emergent norm theory is based on the _____ perspective.
   a. symbolic interactionist
   b. functionalist
   c. conflict
   d. postmodernist

8. Rumors, gossip, fashions, and fads are examples of _____ behavior.
   a. mob
   b. mass
   c. irrational
   d. casual

9. A form of dispersed collective behavior that occurs when a large number of people react with strong emotions and self-destructive behavior to a real or perceived threat is known as:
   a. mob behavior
   b. mass behavior
   c. mass hysteria
   d. contagious behavior

10. "Streaking"—students taking off their clothes and running naked in public—in the 1970s is an example of a:
    a. panic
    b. trend
    c. fashion
    d. fad

11. According to the text, _____ is information provided by individuals or groups that have a vested interest in furthering their own cause or damaging an opposing one.
    a. propaganda
    b. public opinion
    c. political rhetoric
    d. a press release

12. Which of the following statements regarding social movements is true?
    a. Social movements are more likely to develop in preindustrial societies, where there is an acceptance of traditional beliefs and practices.
    b. Social movements have become institutionalized and are a part of the political mainstream.
    c. Social movements make democracy less accessible to excluded groups.
    d. Social movements offer "outsiders" an opportunity to have their voices heard.

13. _____ crowds are likely to engage in civil disobedience.
    a. Conventional
    b. Protest
    c. Expressive
    d. Casual

14. According to relative deprivation theory,
    a. people who are satisfied with their present condition are more likely to seek social change.
    b. certain conditions are necessary for the development of a social movement.
    c. people who feel they have been deprived of their "fair share" are more likely to believe change is necessary and join a social movement.
    d. some people bring more resources to a social movement than others.

15. _____ theory is based on the assumption that six conditions, including structural conduciveness and structural strain, must be present for the development of a social movement.
    a. Value-added
    b. Relative deprivation
    c. Resource mobilization
    d. Emergent norm

16. _____ theory focuses on the ability of members of a social movement to acquire resources and mobilize people in order to advance their cause.
    a. Value-added
    b. Relative deprivation
    c. Resource mobilization
    d. Emergent norm

17. _____ is a social movement based on the belief that patriarchy is a root cause of environmental problems.
    a. Ecology Today
    b. Conflict Ecologists
    c. Environmental Justice
    d. Ecofeminism

18. In the _____ stage of a social movement, people begin to organize and publicize the problem.
    a. preliminary
    b. coalescence
    c. institutionalization
    d. deinstitutionalization

19. The belief that a disproportionate number of hazardous facilities are placed in low-income areas populated by people of color is known as:
    a. environmental racism
    b. environmental justice
    c. reverse environmentalism
    d. racial pollution

20. All of the following statements regarding natural disasters are true, **except**:
    a. Major natural disasters can dramatically change the lives of people.
    b. Trauma that people experience from disasters may outweigh the actual loss of physical property.
    c. Natural disasters are not affected by human decisions.
    d. Disasters may become divisive elements that tear communities apart.

21. Unlike _____ behavior (for example, in education, religion, or politics), collective behavior lacks established norms to govern behavior.
    a. organizational
    b. group
    c. government
    d. institutional

22. _____ refer(s) to rumors about the personal lives of individuals.

23. Mass hysteria
    a. Blog
    b. Slander
    c. Gossip

24. In discussing rumors, the text points out that
    a. while rumors may spread through an assembled collectivity, they also may be transmitted among people who are dispersed geographically.
    b. although they initially may contain a kernel of truth, as they spread, rumors may be modified to serve the interests of those repeating them.
    c. rumors thrive when tensions are high and little authentic information is available on an issue of great concern.
    d. all of the above are characteristics of rumors.

25. Sociologist _____ asserted that fashion serves mainly to institutionalize conspicuous consumption among the wealthy.
    a. Thorstein Veblen
    b. Pierre Bourdieu
    c. William A. Gamson
    d. Robert E. Park

26. _____ consist(s) of the attitudes and beliefs communicated by ordinary citizens to decision makers.
    a. Polls
    b. Public opinion
    c. Masses
    d. Gossip

27. According to sociologist John Lofland, _____ refers to the "publicly expressed feeling perceived by participants and observers as the most prominent in an episode of collective behavior."
    a. casual crowds
    b. dominant emotion
    c. aggregate opinion
    d. expressive crowds

28. _____ crowds are collectivities so intensely focused on a specific purpose or object that they may erupt into violent or destructive behavior.
    a. Acting
    b. Casual
    c. Expressive
    d. Conventional

29. A(n) _____ is violent crowd behavior that is fueled by deep-seated emotions but not directed at one specific target.
    a. riot
    b. mob
    c. fight
    d. aggregate

30. The _____ theory views social movements as interactive, symbolically defined, and negotiated processes that involve participants, opponents, and bystanders.
    a. social constructionist
    b. political opportunity
    c. value-added
    d. relative deprivation

31. Social movements that aim to produce radical change in individuals through "inner change" are _____ movements.
    a. revolutionary
    b. alternative
    c. reform
    d. religious

32. Examples of _____ movements are groups organized since the 1950s to oppose school integration, civil rights and affirmation action legislation, and domestic partnership initiatives.
    a. resistance
    b. revolutionary
    c. reform
    d. religious

33. The text identifies _____ as the most widely known resistance (regressive) movement.
    a. the Christian Temperance Union
    b. the National Rifle Association (NRA)
    c. MADD
    d. pro-life advocates (Operation Rescue)

34. In the _____ stage of a social movement, widespread unrest is present as people become aware of a problem. At this stage, leaders emerge to agitate others into taking action.
    a. coalescence
    b. preliminary
    c. secondary
    d. institutionalization

35. In the _____ stage of a social movement, people begin to organize and to publicize the problem. At this stage, some movements become formally organized at local and regional levels.
    a. secondary
    b. institutionalization
    c. preliminary
    d. coalescence

36. Some religious movements are _____—that is, they forecast that "the end is near" and assert that an immediate change in behavior is imperative
    a. foreseers
    b. Scientologists
    c. millenarian
    d. prophetic

37. People must become aware of a significant problem and have the opportunity to engage in collective action. This illustrates sociologist Neil Smelser's value-added theory condition of:
    a. spread of generalized beliefs
    b. mobilization for action
    c. structural strain
    d. structural conduciveness

38. The _____ reflects the influence of sociologist Erving Goffman's "Frame Analysis", in which he suggests that our interpretation of the particulars of events and activities is dependent on the framework from which we perceive them.
    a. social constructionist theory
    b. value-added theory
    c. relative deprivation theory
    d. political opportunity theory

39. Sociologists have identified ways in which grievances are framed. _____ framing provides a vocabulary of motives that compel people to take action.
    a. Motivational
    b. Frame Alignment
    c. Bridge
    d. Transformation

40. According to _____ theory, people will choose options for collective action that are most readily available to them, producing the most favorable outcome.
    a. social constructionist
    b. relative deprivation
    c. political opportunity
    d. value-added

41. In a study of a contaminated landfill in the Carver Terrace neighborhood of Texarkana, Texas, sociologist Stella Capek found that residents
    a. saw no relationship between their plight and that of residents in areas such as the Love Canal.
    b. were able to mobilize for change and win a federal buyout and relocation.
    c. were powerless when confronted with the federal bureaucracy.
    d. believed that they had been included in previous environmental cleanups.

TRUE/FALSE QUESTIONS

1. Collective behavior lacks an official division of labor, hierarchy of authority, and established rules and procedures.
   T   F

2. People are more likely to act as a collectivity when they believe it is the only way to fight those with greater power and resources.
   T   F

3. People gathered for religious services and graduation ceremonies are examples of casual crowds.
   T   F

4. Panics often arise when people believe that they are in control of a situation.
   T   F

5. Protest crowds engage in activities intended to achieve specific political goals.
   T   F

6. Widespread unrest is present and people begin to become aware of a problem in the coalescence stage of a social movement.
   T   F

7. New social movements incorporate factors of identity, such as race, class, and gender, as sources of collective action and social movements.
   T   F

8. For mass behavior to occur, people must be in close proximity geographically.
   T   F

9. Public opinion does not always translate into action by decision makers in government and industry or by individuals.
   T   F

10. Grassroots environmental movements are an example of reform movements.
    T   F

11. Political opportunity theory investigates how problems are framed and what names they are given.
    T   F

12. Revolutionary movements are also referred to as expressive movements.
    T   F

13. Movements based on relative deprivation are most likely to occur when people have unfulfilled rising expectations.
    T   F

14. Changes in population size, distribution, and composition have little impact on the culture and social structure of a society.
    T   F

15. Natural disasters can be made worse by human decisions, as in the case of Hurricane Katrina.
    T   F

16. According to Box 20.1, scientists are forecasting a global warming of between 2 and 11 degrees Fahrenheit over the next century.
    T   F

17. According to Box 20.1, most social movements are reform movements that focus on improving society by changing some specific aspect of the social structure.
    T   F

18. In Box 20.2, sociologists suggest that flash mobs have become a fashion and will continue to remain as long-term activities.
    T   F

19. According to Box 20.3, riparian rights refers to the rights of a person or group to use water by virtue of owning or occupying the bank of a river or a lake.
    T   F

20. According to Box 20.4, there is nothing we can do individually to reduce the amount of greenhouse gases.
    T   F

FILL-IN-THE-BLANK QUESTIONS

2. A highly emotional crowd whose members are ready to engage in violence against a specific target is a _____.

3. The person most connected to the grassroots campaign of exposing sites of the chemical dump at Love Canal is _____.

4. Nonviolent action that seeks to change a policy or law by refusing to comply with it is known as _____.

5. _____ are unsubstantiated reports on an issue or subject.

6. The radio dramatization *The War of the Worlds* produced a type of _____.

7. A(n) _____ is a temporary but widely copied activity enthusiastically followed by large numbers of people.

8. The attitudes and beliefs communicated by ordinary citizens to decision makers are defined as _____.

9. Information provided by individuals or groups that have a vested interest in furthering their own course or damaging an opposing one is defined as _____.

10. The most common type of crowd behavior wherein people seek to escape from a perceived danger is known as _____.

11. Rumors about the personal lives of individuals are defined as _____.

413

12. Behavior that is voluntary, often spontaneous, and engaged in by a large number of people is known as _____.

13. The alteration, modification, or transformation of a public policy, culture, or social institution over time is identified as _____.

14. A(n) _____, which is short for web log, is an online journal maintained by an individual who frequently records entries that are maintained in a chronological order.

15. Sociologist _____ divided crowds into four categories: casual, conventional, expressive, and acting.

16. A(n) _____ is a currently valued style of behavior, thinking, or appearance.

SHORT ANSWER/ESSAY QUESTIONS

1. What is collective behavior, and what causes people to engage in collective behavior?
2. What are some types of crowd behavior? What causes crowd behavior? How could crowd behavior influence individual behavior?
3. What are the five major theories that explain social movements? Which do you think is most applicable to solving future social problems? Why?
4. What is a social movement? What are the major types of social movements?
5. Why do people engage in social movements? Are social movements effective avenues for social change? Explain.

STUDENT CLASS PROJECTS AND ACTIVITIES

1. The uprising against socialism in the Soviet Union in 1991 is an example of a revolutionary movement; it sought, and apparently produced, an entire restructuring of the Soviet political and economic system by overthrowing the existing social order and creating a new one. This could be considered the second of the revolutions that have occurred in this country since the Russian Revolution of 1917. Choose one of these two revolutions and trace the development of the movement, including the conditions necessary for the development of a social movement according to Smelser, as mentioned in the text. In your paper, provide examples of the six conditions of revolution and critique the success of the movement.
2. Collect at least ten articles dealing with some form of collective behavior discussed in the text. The articles could focus on fads, fashions, mobs, public opinion, or specific types of social movements. Write a paper that provides: (1) a summary of each article; (2) an explanation of the significance of each article; (3) a personal reaction to each article; (4) a conclusion that compares and contrasts the information in the different articles or in some way relates the articles to each other; and (5) a bibliographic reference for each article selected. Provide a copy of each article with your paper.

3. Research the impact of the U.S. war in Iraq upon the political, economic, religious, educational, and medical institutions in both the U.S. and Iraq. What are the present national and international reactions to the war? What U.S. and international social movements have been created in response to the war? What are the levels of support for the war both nationwide and in international communities? What has been the economic cost in both the U.S. and Iraq? What is the political cost? How have the major social institutions in both countries been affected? What is the level of support on your college campus? In addition to providing the information above in your paper or project, conduct a survey of thirty or more people on your campus to gain some insight of the level of support. Write a summary of your research findings.

4. Research a social movement that has taken place in the United States, such as the Civil Rights Movement, the feminist movement, the gay pride movement, the United Farm Workers movement, etc. Write a paper that summarizes the history of the movement, the social and political climate in which the movement took place, the kinds of activities the movement engaged in, the goals of the movement, the success or failure of the goals of the movement, and the current state of the movement.

INTERNET ACTIVITIES

1. **Civil disobedience** is a form of collective behavior. This web site contains a history of civil disobedience in America, as well as a training manual for those wishing to organize their own social movement: **http://www.actupny.org/documents/CDdocuments/CDindex.html**

2. **PETA** (People for the Ethical Treatment of Animals), **http://www.peta.org/**, represents an animal rights activist movement. Members of PETA often engage in tactics that border on violent behavior as they seek to eliminate cruelty to animals. Discern the "cruelty-free" companies that do not use animals in testing their products.

3. Look at what different organizations are doing to **protect the environment** at the following web sites: **http://www.greenpeace.org/**, **http://www.sierraclub.org/**

4. These are just a few of the sites on the Internet dealing with **riots**. Conduct a more extensive search for historical accounts and news stories of riots here in the U.S. and around the world: **http://www.67riots.rutgers.edu/d_index.htm/**, **http://brainz.org/riots/**, **http://www.cato.org/pubs/journal/cj14n1-13.html**

INFOTRAC COLLEGE EDITION EXERCISES

Visit the **InfoTrac College Edition** web site at: **http://www.wadsworthmedia.com/webtutor/infotrac.htm**. You will arrive at a screen that enables you to search topics.

1. Search for references on InfoTrac about **environmental movements**. Bring to class an article or brief report on a specific movement. Focus attention on the social aspects of these movements.

2. There are periodical references on the **"social aspects" of collective behavior**. Read one and then answer this question: *How can this article add to my knowledge about collective behavior?*
3. Search for the word **propaganda**. There are a number of categories, such as propaganda in television, communist propaganda, and war propaganda. Explore the different categories of propaganda.
4. Look up the concept **urban legend**. InfoTrac has a number of periodical references to this phenomenon. Bring an example to class. How do urban legends fit into the study of collective behavior?

SOLUTIONS

MULTIPLE CHOICE QUESTIONS

1. C, p. 656
2. B, p. 659
3. C, p. 660
4. B, p. 660
5. A, p. 660
6. D, p. 661
7. A, p. 662
8. B, p. 659
9. C, p. 664
10. D, p. 664
11. A, p. 665
12. D, p. 666
13. B, p. 660
14. C, p. 669

15. A, p. 669
16. C, p. 670
17. D, p. 673
18. B, p. 668
19. A, p. 674
20. C, p. 676
21. D, p. 656
22. D, p. 664
23. D, p. 663
24. A, p. 665
25. B, p. 665
26. B, p. 659
27. A, p. 660
28. A, p. 660

29. A, p. 670
30. D, p. 668
31. A, p. 668
32. D, p. 668
33. B, p. 668
34. D, p. 668
35. C, p. 668
36. D, p. 669
37. A, p. 670
38. A, p. 670
39. C, p. 671
40. B, p. 673

TRUE/FALSE QUESTIONS

1. T, p. 656
2. T, p. 659
3. F, p. 660
4. F, p. 660
5. T, p. 660
6. F, p. 668
7. T, p. 673

8. F, p. 663
9. T, p. 665
10. T, p. 667
11. F, p. 671
12. F, p. 668
13. T, p. 669
14. F, p. 678

15. T, p. 676
16. T, p. 658
17. T, p. 658
18. F, p. 672
19. T, p. 678
20. F, p. 681

FILL-IN-THE-BLANK QUESTIONS

1. mob, p. 660
2. Lois Gibbs, p. 657
3. civil disobedience, p. 660
4. Rumors, p. 663
5. mass hysteria, p. 663
6. fad, p. 664
7. public opinion, p. 665
8. propaganda, p. 665
9. panic, p. 660
10. gossip, p. 664
11. collective behavior, p. 656
12. social change, p. 656
13. blog, p. 659
14. Herbert Blumer, p. 659
15. fashion, p. 664